New challenges to health: the threat of virus infection

Viruses continually evolve and adapt, posing new threats to health. This book discusses the ecology of viruses including the emergence of devastating haemorrhagic disease, and reviews the molecular and cell biological basis of the pathogenesis of several virus diseases. An introduction is given to the mathematical analysis of recurrent epidemic virus disease, such as measles. Neurological and psychological disease is discussed in relation to the pathological mechanisms that may underlie prion disease (such as new variant CJD) and to the possible virus involvement in human psychiatric illness. Virus infections that have come to prominence recently (HIV, bunyaviruses, morbilliviruses and caliciviruses) or that remain a threat (influenza and hepatitis viruses) are discussed. There are also chapters on new and potential niches for virus infections in the immunocompromised, and the problem of the emergence of antiviral drug resistance in viruses for which therapies exist.

Geoffrey Smith is Professor of Virology and Wellcome Trust Principal Research Fellow at the Wright-Fleming Institute, Imperial College School of Medicine, St. Mary's Campus, London, UK.
Will Irving is a Reader in Virology in the Department of Microbiology at the University of Nottingham, UK.
John McCauley is a researcher in the Division of Molecular Biology at the Institute for Animal Health, Compton, UK.
Dave Rowlands is Professor of Molecular Virology in the Division of Microbiology at the University of Leeds, UK.

Symposia of the Society for General Microbiology

Managing Editor: Dr Melanie Scourfield, SGM, Reading, UK
Volumes currently available:

28	Structure and function of prokaryotic membranes
29	Microbial technology: current state, future prospects
30	The eukaryotic microbial cell
32	Molecular and cellular aspects of microbial evolution
33	Virus persistence
35	Interferons from molecular biology to clinical applications
36	The microbe 1984 – Part I: Viruses. Part II: Prokaryotes and eukaryotes
37	Viruses and cancer
38	The scientific basis of antimicrobial chemotherapy
39	Regulation of gene expression – 25 years on
41	Ecology of microbial communities
42	The nitrogen and sulphur cycles
43	Transposition
44	Microbial products: new approaches
45	Control of virus diseases
46	The biology of the chemotactic response
47	Prokaryotic structure and function: a new perspective
49	Molecular biology of bacterial infection: current status and future perspectives
50	The eukaryotic genome: organisation and regulation
51	Viruses and cancer
52	Population genetics of bacteria
53	Fifty years of antimicrobials: past perspectives and future trends
54	Evolution of microbial life
55	Molecular aspects of host–pathogen interactions
56	Microbial responses to light and time
57	Microbial signalling and communication
58	Transport of molecules across microbial membranes
59	Community structure and co-operation in biofilms

SIXTIETH SYMPOSIUM OF THE
SOCIETY FOR GENERAL MICROBIOLOGY
HELD AT HERIOT-WATT UNIVERSITY MARCH 2001

Edited by
G. L. Smith, W. L. Irving, J. W. McCauley, D. J. Rowlands

New challenges to health: the threat of virus infection

Published for the Society for General Microbiology

CAMBRIDGE
UNIVERSITY PRESS

PUBLISHED BY THE PRESS SYNDICATE OF THE UNIVERSITY OF CAMBRIDGE
The Pitt Building, Trumpington Street, Cambridge, United Kingdom

CAMBRIDGE UNIVERSITY PRESS
The Edinburgh Building, Cambridge CB2 2RU, UK
40 West 20th Street, New York, NY 10011-4211, USA
10 Stamford Road, Oakleigh, VIC 3166, Australia
Ruiz de Alarcón 13, 28014 Madrid, Spain
Dock House, The Waterfront, Cape Town 8001, South Africa

http://www.cambridge.org

First published 2001

Printed in Great Britain at the University Press, Cambridge

Typeface Sabon (Adobe) 10/13.5pt *System* QuarkXPress™ [SE]

A catalogue record for this book is available from the British Library

ISBN 0 521 80614 3 hardback

Front cover illustration: False-colour transmission electron micrograph of the Ebola virus.
CREDIT: A. B. DOWSETT/SCIENCE PHOTO LIBRARY.

CONTENTS

CONTRIBUTORS

Barrett, T.
Institute for Animal Health, Pirbright Laboratory, Ash Road, Pirbright, Surrey GU24 0NF, UK

Bode, L.
Project 23: Bornavirus infections, Robert Koch-Institut, Nordufer 20, D-13353 Berlin, Germany

Boshoff, C.
Departments of Oncology and Molecular Pathology, The CRC Viral Oncology Group, Wolfson Institute for Biomedical Research, Cruciform Building, Gower Street, University College London, London WC1E 6BT, UK

Cozzio, A.
Dept of Pathology, Stanford University School of Medicine, Stanford, CA 94304, USA

Darby, G.
Glaxo Wellcome Research and Development, Gunnels Wood Road, Stevenage SG1 2NY, UK

Elliott, R. M.
Division of Virology, Institute of Biomedical and Life Sciences, University of Glasgow, Church Street, Glasgow G11 5JR, UK

Feldmann, H.
Institut für Virologie, Philipps-Universität Marburg, Postfach 2360, 35011 Marburg, Germany

Fischer, M.
Dept für Innere Medizin, Abt. für Infektionskrankheiten, Universitätsspital Zürich, Zürich 8091, Switzerland

Flechsig, E.
MRC Prion Unit/Neurogenetics, Imperial College School of Medicine at St Mary's, London W2 1PG, UK

Grenfell, B. T.
Zoology Department, University of Cambridge, Downing Street, Cambridge CB2 3EJ, UK

Gubler, D. J.
Division of Vector-Borne Infectious Diseases, National Center for Infectious Diseases, Centers for Disease Control and Prevention, Public Health Service, US Department of Health and Human Services, PO Box 2087, Fort Collins, CO 80522, USA

Hay, A. J.
Division of Virology, National Institute for Medical Research, The Ridgeway, Mill Hill, London NW7 1AA, UK

Hegyi, I.
Institut für Neuropathologie, Universitätsspital Zürich, 8091 Zürich, Switzerland

Klenk, H.-D.
Institut für Virologie, Philipps-Universität Marburg, Postfach 2360, 35011 Marburg, Germany

Leimeroth, R.
Dept of Cell Biology, ETH, Hönggerberg, 8093 Zürich, Switzerland

Lemon, S. M.
Departments of Microbiology & Immunology and Internal Medicine, The University of Texas Medical Branch, Galveston, TX 77555-1019, USA

Ludwig, H.
Institute of Virology, Free University of Berlin, Königin-Luise-Strasse 49, D-14195 Berlin, Germany

Peters, C. J.
Departments of Microbiology and Immunology and of Pathology, Member, Center for Tropical Diseases, University of Texas Medical Branch, Galveston, TX 77555-0609, USA

Richardson, B. J.
Centre for Biostructural and Biomolecular Research, University of Western Sydney Hawkesbury, Richmond, NSW 2783, Australia

Rossi, D.
MRC Prion Unit/Neurogenetics, Imperial College School of Medicine at St Mary's, London W2 1PG, UK

Shmerling, D.
Core Technologies Dept, Novartis Pharma AG, 4002 Basel, Switzerland

Stoye, J. P.
Division of Virology, National Institute for Medical Research, The Ridgeway, Mill Hill, London NW7 1AA, UK

Volchkov, V. E.
Institut für Virologie, Philipps-Universität Marburg, Postfach 2360, 35011 Marburg, Germany

Volchkova, V. A.
Institut für Virologie, Philipps-Universität Marburg, Postfach 2360, 35011 Marburg, Germany

Weiss, H. A.
Infectious Disease Epidemiology Unit, London School of Hygiene and Tropical Medicine, Keppel Street, London WC1E 7HT, UK

Weiss, R. A.
Windeyer Institute of Medical Sciences, University College London, 46 Cleveland Street, London W1P 6DB, UK

Weissenhorn, W.
European Molecular Biology Laboratory (EMBL) Grenoble Outstation, 6 rue Jules Horowitz, 38000, Grenoble, France

Weissmann, C.
MRC Prion Unit/Neurogenetics, Imperial College School of Medicine at St Mary's, London W2 1PG, UK

EDITORS' PREFACE

This book presents the proceedings of a meeting '*New Challenges to Health: The Threat of Virus Infection*' organized by the Society for General Microbiology at Heriot-Watt University, Edinburgh, in March 2001. The purpose of the meeting was to review the continuing threat of viruses (and prions) to human and animal health. Although several virus diseases have been controlled by vaccination (such as polio, measles, mumps, rubella and yellow fever) and one (smallpox) has been eradicated, viruses remain a potent threat to human and animal health due to their ability to evolve and adapt rapidly. For viruses such as influenza and HIV, the ability to undergo rapid antigenic variation enables them to evade existing immunity and cause disease. Viruses may adapt to new situations, such as changes in the population density of human, animal or insect hosts, or the presence of substantial numbers of immunosuppressed individuals, and cause disease where hitherto they were unable to do so. Rapid virus evolution also enables virus strains to arise that are resistant to existing antiviral drugs. The book starts with a consideration of the mathematical modelling of epidemic virus infections, and the surveillance and emergence of virus infections. There follow chapters that review the molecular and cell biological mechanisms by which viruses and prions may induce disease. These include influenza and HIV, the devastating haemorrhagic diseases caused by Ebola and Marburg viruses, prion diseases such as BSE and variant CJD, certain psychiatric illness, and how the emergence of drug-resistant strains of virus poses a major problem for antiviral chemotherapy. The editors are most grateful to all the authors for their contributions to the meeting and to this book, and thank Melanie Scourfield and Josiane Dunn at the Society for General Microbiology for invaluable assistance in the production of the book and in the arrangements for the meeting. The book will be of interest to all those interested in virus and prion disease and is an important reminder to all that viruses are and will remain a continual threat to human and animal health.

Geoffrey Smith
Will Irving
John McCauley
Dave Rowlands

The viruses in our past, the viruses in our future

C. J. Peters

Departments of Microbiology and Immunology and of Pathology, Member, Center for Tropical Diseases, University of Texas Medical Branch, Galveston, TX 77555-0609, USA

THE PREMISE

One often hears that money cannot buy love, good health or happiness. Like most aphorisms, these sayings bear a kernel of truth, but unfortunately as far as health is concerned data from most of the world demonstrate the opposite (Fig. 1). Public health officials would say that the average lifespan decreases precipitously as the per capita gross natural product falls below $5000. If this were a laboratory experiment, the scientist would say that the mean time to death on the vertical axis increases rapidly as the 'treatment' on the horizontal increases. The role for public health in the traditional sense is to convince uninterested rulers, populations with well-established cultural attitudes, and international donors to alter public health practices and allocation of money to obtain a shift in position of their country to an improved longevity.

The relationship between wealth and lifespan has consequences for our ideas about emerging infections and how we may best deal with them. The countries at the left side of the graph obviously do not have an adequate health infrastructure and it will be difficult to establish surveillance for monitoring and recognizing emerging diseases because of the lack of diagnostic capability and communications. Their efforts will best be directed to implementing simple, inexpensive, high-yield procedures such as reliable stool cultures or malaria smears. Rich countries requesting cooperation in surveillance should be cognizant of the life and death financial situation these countries face. Nevertheless, we also recognize that the poorer countries, particularly in tropical climes, may be a disproportionate source of 'new' emerging threats.

SGM symposium 60: New challenges to health: the threat of virus infection.
Editors G. L. Smith, W. L. Irving, J. W. McCauley, D. J. Rowlands. Cambridge University Press. ISBN 0 521 80614 3 ©SGM 2001.

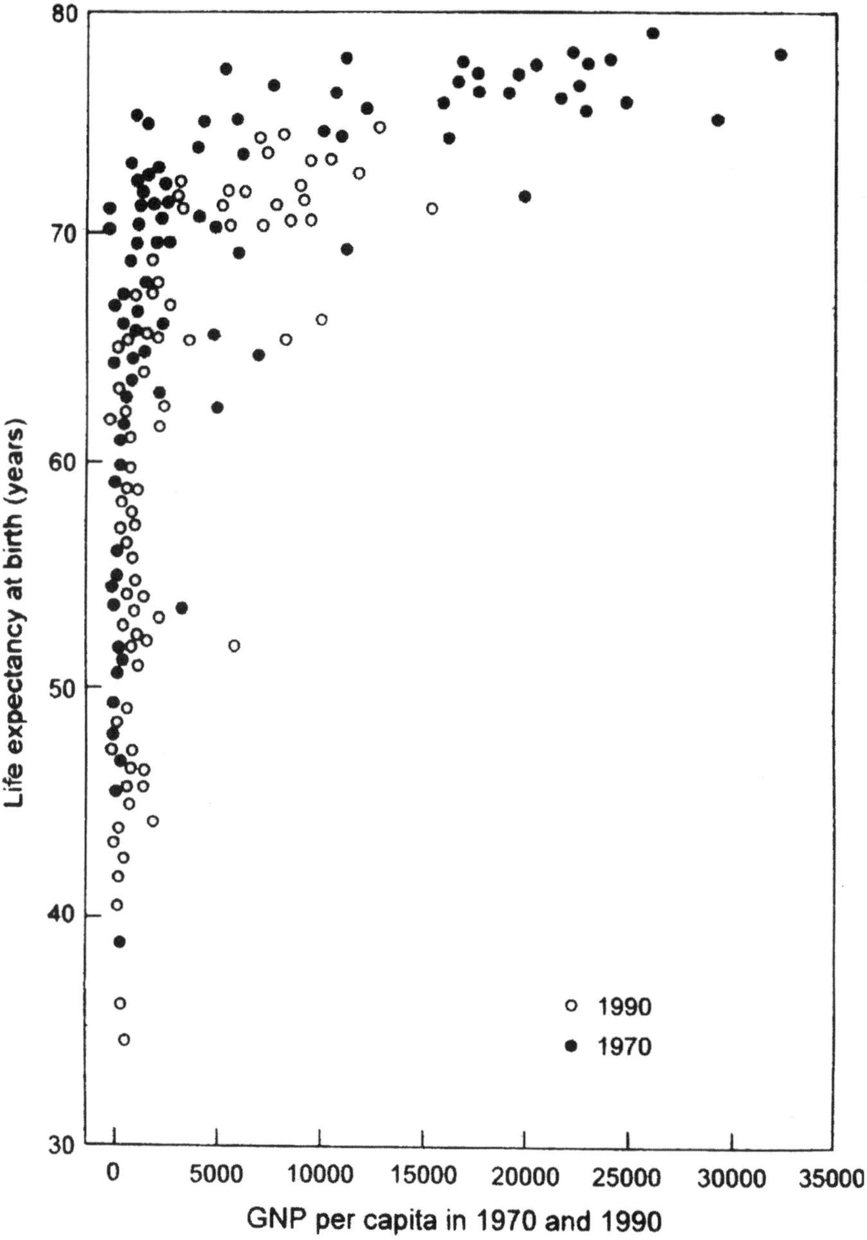

Fig. 1. Length of your life depends on how rich your country is [from Holland, H. D. & Petersen, U; *Living Dangerously: the Earth, its Resources, and the Environment.* Copyright © 1995 by Princeton University Press. Reprinted by permission of Princeton University Press.

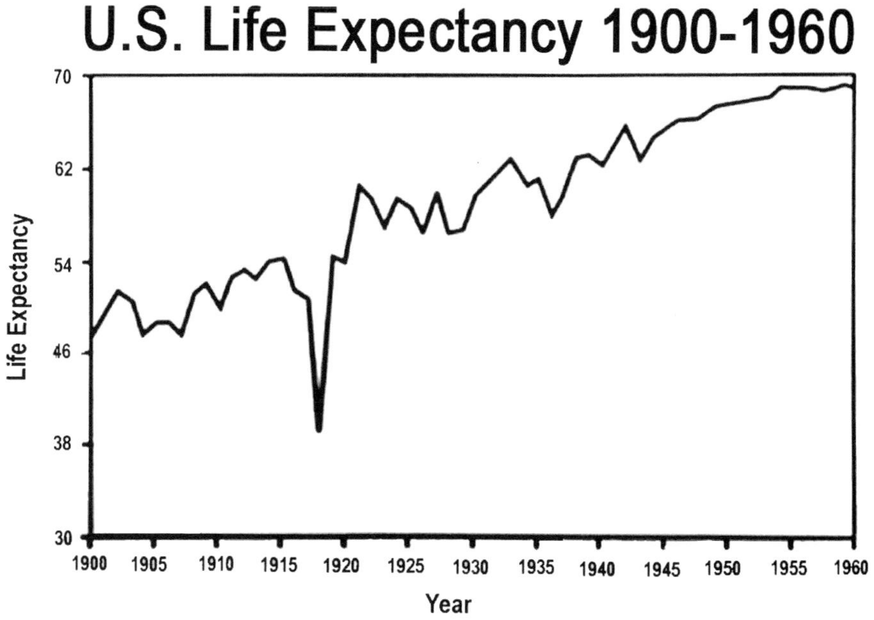

Fig. 2. US life expectancy was seriously decreased by the 1918 influenza epidemic. Courtesy of Nancy Cox, CDC, Atlanta, GA, USA.

Poverty is not the only issue or we could immediately stop worrying about emerging infections and turn to economics or social policy. There are many examples of how infectious diseases have skewed the life expectancy curve. One of the most important is the 1918 influenza epidemic (Crosby, 1990). The trend in life expectancy in the US over the years was gradually increasing until the precipitous dip of more than 12 years as a consequence of this epidemic (Fig. 2). We will witness a similar impact from the human immunodeficiency virus type 1 (HIV-1) epidemic in Africa and later in Asia. Recognizing and dealing with such dramatic and less dramatic newly emerging or re-emerging infectious diseases has only recently and belatedly become a focus of the biomedical community (Lederberg *et al.*, 1992) and I plan to discuss here some ideas that may help in our response to these very real threats.

Two additional well-established ideas need to be examined as well: the history of the continents and the social ecology of humans. The evolution of our earth's biota and ecological zones was strongly determined by the geological history of the continents and the subsequent rapid alterations wrought by human changes in the biosphere. The impact of humans on the earth's ecology was enhanced by their social organization and this organization also had direct effects on patterns of disease emergence.

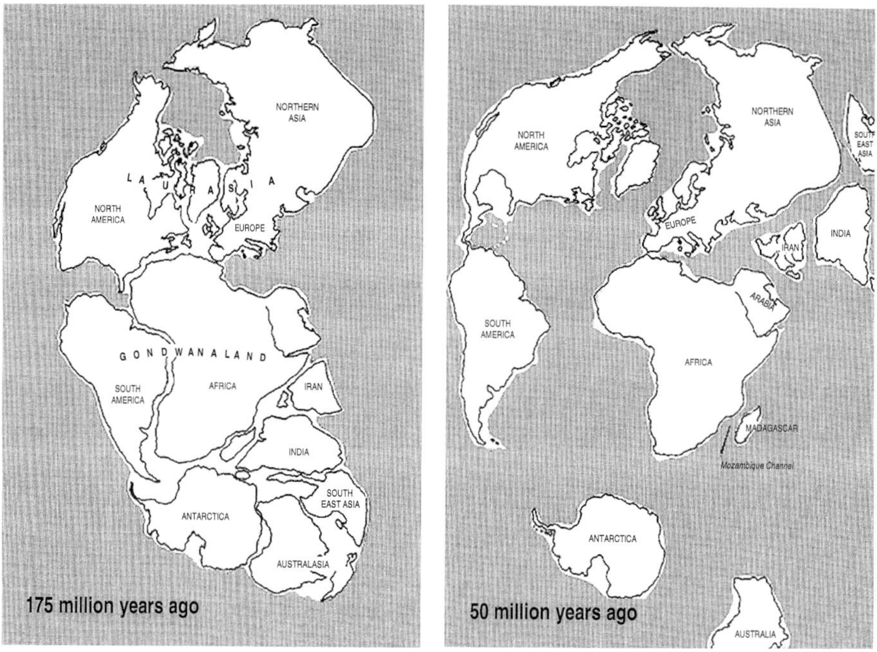

Fig. 3. History of the continents [from *The Penguin Atlas of African History* by Colin McEvedy (Penguin Books 1980, revised edition 1995) copyright © Colin McEvedy, 1980, 1995, pp. 9, 11]. Reproduced by permission of Penguin Books Ltd.

HISTORY OF THE CONTINENTS

We know that more than 200 million years ago the earth's surface was composed of a single continent and that its separation into the continents we recognize today began around 150 million years ago when a northern land mass (Laurasia) and a southern land mass (Gondwanaland) began separating at an infinitesimal rate (Fig. 3) (McEvedy, 1995; Osborne & Tarling, 1996). In Laurasia, the evolutionary stock of mammals evolved into the presumably more fit placentals. When the fused continents of Australia, South America and Antarctica drifted away from Gondwanaland 30–50 million years ago, marsupials were the predominant mammals there. Subsequently the modern continents formed as tectonic plates continued their movements. Later still, South America rejoined with North America 3–7 million years ago concomitant with an extensive placental mammal southward migration (the so-called 'Great Faunal Interchange') (Gould, 1993). Today, the isolated Australian continent remains the only land mass with a predominantly marsupial mammal fauna. This is a very well-known difference among the continents with the lay public, but biologists are aware of many differences in plant and animal life among the continents and their ecological regions (Udvardy, 1969; Bailey, 1998). Some of the evolutionary radiations in the biota were facilitated by continental separations and movements and others by intracontinental barriers such as mountains, deserts or river drainages. Although microbial species and infectious diseases are routinely disregarded in ecological studies, significant associations between microbial species and the local composition of other plant and animal species exist. For example, a virologist can tell the viruses of Australia from those of other areas as surely as a school child can tell you that a kangaroo has a pouch.

The characteristic flora and fauna of a given region evolved in the presence of their companion species. Each area is comprised of a characteristic array of species, which in some cases appear to form an integrated system with mutually supportive interactions (Elton, 1958; Gould, 1993; Vitousek *et al.*, 1996). One example of this principle is Wallace's line, separating the Asian and the Australopacific biogeographic regions. Although there is considerable blurring of the interface, species predominantly of Asian origin are found to the west and Australopacific to the east. An island on one side of Wallace's line may be visible to the naked eye from an island on the other side, but have a different set of trees, flowers or birds which are more closely related to their spatially distant cousins in the same biogeographic region than to their nearby neighbours which evolved on the other side of the line. Indeed, two virus families have a significant division point at Wallace's line (Tesh *et al.*, 1975). On the Asian side, the main togavirus transmitted is Chikungunya virus, while on the Australopacific side the indigenous species is Ross River virus. Among flaviviruses, Japanese encephalitis is the Asian virus and the related Murray Valley encephalitis virus occupies a similar niche in the riverine areas of Australia.

These basic ecological patterns laid down over millions of years and changed only gradually over the past few thousand years are being altered rapidly by anthropogenic movement of species and alteration of land use patterns, which accelerated rapidly with the age of exploration (Crosby, 1994). These historically recent changes have had and are having profound consequences for the distribution of viruses.

HISTORIC CHANGES IN HUMAN ECOLOGY AND PATTERNS OF INFECTION

Humans have their own social ecology that also contributes to patterns of virus transmission and evolution. We began our existence as relatively isolated groups of hunter-gatherers. Viral infections were mainly of two types: persistent viruses or zoonotic viruses. Herpesviruses and papovaviruses such as BK or JC are important examples of persistent human virus infections which survive in their host species by latent infection. Subsequent reactivation occurs to extend the chain of infection to the next generation within their small group. The ability of certain viruses to establish close relations to their host is often reflected by phylogenetic relationships that parallel those of their hosts (Agostini *et al.*, 1997; Sugimoto *et al.*, 1997; Gao *et al.*, 1999; Monroe *et al.*, 1999; Bowen *et al.*, 1997). Such infections and their evolutionary relationships are best documented in chronic productive infections of rodents or lentivirus infections of non-human primates but have not been as carefully analysed in humans.

Infections transmitted from animals to humans were an important facet of the viral diseases of early human life on our planet. Such viruses were transmitted from rodents and other mammals as well as from biting arthropods. Humans were not part of the natural survival strategies of these viruses and that largely continues to be the case. A few agents such as the dengue viruses have largely abandoned their suspected primordial monkey–mosquito cycle for dominant human–mosquito transmission and other viruses may have crossed species in the past to become human viruses exclusively (Fenner, 1970).

The coalescence of emerging agricultural villages into large towns made possible by irrigated agriculture provided the conditions for the appearance of viruses such as measles, smallpox, poliomyelitis and others (Fenner, 1970). Epidemiological studies of island populations, Amazonian tribes and persons living in other isolated circumstances have established the need for substantial numbers of susceptible individuals for their maintenance. The larger, interconnected populations appearing in the Fertile Crescent ~5000 years ago could produce the numbers needed for the human-to-human hopping of these non-latent, non-zoonotic viruses. The 'new' viruses were presumably derived from viruses of animals. For example, rinderpest virus of cattle is the closest known relative of measles virus and cattle were domesticated in the Fertile Crescent about 5–10 000 years ago when measles was thought to have emerged (see the chapter by Barrett, this volume). These viruses are highly infectious among humans and have been formidable foes over the centuries, but their peripatetic lifestyle provides a weakness that is being exploited to eradicate or attempt to eradicate some of them. Other human ecology related changes occurred, as well. The coalescence of large towns into modern cities resulted in the intensification of transmission of many viruses, particularly those relying on exposure to human waste; and later this situation was ameliorated by improvements of sanitation.

I include the age of exploration as another human behaviour that has had profound effects on disease ecology. Many diseases that evolved in Africa or Eurasia were brought to the 'New World' in the 1500s. Highly infectious viruses such as measles and smallpox were unleashed on the native populations of the Americas. For years the prevailing modern historical opinion was that the technology of the Spanish conquistadores and the social disruption wrought among native Americans were the major factors in the conquest of the Americas. However, careful demographic studies coupled with information on the carrying capacity of the land show that the major problem of the Amerinds was not cruel foreign soldiers, the horse in warfare, savage dogs or the 'holy' fires of the priests (Dobyns, 1983). Introduced viral and other diseases swept through populations without any pre-existing acquired immunity, without prior genetic selection, and with no knowledge of nursing or treatment (Cook, 1998). Rampant sickness and death left them unable to resist the invaders; the great military loss of the Aztecs in their battle to retain Mexico City was really lost to smallpox and not the Spaniards. Similar dramatic occurrences are well documented in the Pacific Islands (Moorehead, 1966).

Another facet of the age of exploration was the purposeful and the accidental introduction of many plants and animals around the world (Crosby, 1994; Elton, 1958). The extent and historical impact of these movements has been well established by the careful scholarship of Crosby (Crosby, 1994, 1995). These ecological intruders often made themselves at home in their new environment, established their own altered ecosystems, and could not be dislodged by native species, effectively changing the face of the area indefinitely (Elton, 1958).

THE YELLOW FEVER EXAMPLE

To appreciate some of the important variables contributing to changes in disease patterns it is useful to look at the history of yellow fever using the tools currently available to reconstruct past events, admittedly with an element of speculation and simplification. Phylogenetic studies identify yellow fever as the ancestral species for the mosquito-borne flaviviruses (Gould *et al.*, 1997). We can say with some modest certainty that it arose in the African tropical forest and evolved into a virus transmitted by *Aedes* (*Stegomyia*) mosquitoes among non-human primates living in the canopy. Later the virus apparently found a significant new arthropod vector with relevance to humans, *Aedes* (*Stegomyia*) *aegypti*. This mosquito had been a forest dweller but development of villages and cities gave it a niche, which it was evidently quite capable of occupying. As the mosquito adapted to breeding in the water containers in human villages, it became paler in colour and increasingly anthropophilic. Thus humans became an important new vertebrate host for the virus thanks to the vector competence and changing habits of *Ae. aegypti*.

Severe epidemics thought to be yellow fever were repeatedly observed in Africa. Probably with the onset of the slave trade the virus was brought to the New World and phylogeny supports the West African origin (Deubel & Drouet, 1997). One can imagine that the source of the imported virus was a mosquito stowed away on a ship, perhaps even from Goree Island off the coast of Senegal and the shipping point for so many slaves. Alternatively, there could have been continuous transmission on shipboard with *Ae. aegypti* breeding in the fresh water supply on board (Hargis, 1880). The first historical epidemics in the Americas were identified in the Yucatan from 1648, based on the characteristic 'black vomit' and epidemiological setting (Carter, 1931).

We assume that *Ae. aegypti* colonization of American ports must have preceded the successes of the virus. And the virus was successful, indeed. It raged throughout the Americas and the Caribbean, reaching as far north as Philadelphia and as far south as Buenos Aires. Its Caribbean transmission sites provided an excellent seed for passing boats moving along the coast of the Gulf of Mexico and returning to points as far away as Spain, England, France or Gibraltar. Of course the inability of the *Ae. aegypti* mosquito to persist beyond the northern limits of 35–45° N, depending on the region of the world, and the southern limits of 32–34° S, limited the endemic areas and this caused great mystification among early theorists. After 1900, the work of several key scientists led by Walter Reed elucidated yellow fever as the first human viral disease and the first mosquito-transmitted viral disease. The intensive attacks on *Ae. aegypti* led by Gorgas and carried forward by the Rockefeller Foundation (Strode, 1951) began to rid the great cities of the Americas of the virus; however, their success in increasingly rural areas led to the discovery that there was a jungle yellow fever cycle involving American tree-top mosquitoes and American monkeys.

Some of the most interesting features of the several century career of yellow fever as an emerging virus centre around its vector–vertebrate host relationships. The virus was not indiscriminant and the several *Aedes* vectors it utilizes all belong to the subfamily *Stegomyia*. It could, however, adapt to arboreal mosquitoes in the New World and that are quite distantly related to its Old World vectors. The invasive, flexible, anthropophilic *Ae. aegypti* mosquito vector made yellow fever a fearsome viral disease: witness our problems today with dengue and Chikungunya viruses transmitted among humans by the same mosquito species. Nor has yellow fever ceased to be a threat. Travellers from rich, highly developed, medically sophisticated countries travel to endemic areas and return with fatal infections; hundreds of jungle yellow fever cases are reported annually from Africa and South America; and urban yellow fever smoulders in African cities killing thousands every year. We still do not understand why yellow fever has never invaded Asia and we are awaiting the verdict of time on whether jungle yellow fever in South America can reinvade large population centres where extensive *Ae. aegypti* populations actively transmit the four dengue viruses.

SOME LESSONS FROM TODAY

Although many different examples of emerging infections have been documented in the last decade and the detailed circumstances are different for each one, I feel confident in saying that none of the determinants of emergence are fully understood. This is not surprising because of the different agent and host genomic structures, modes of

Table 1. Viral haemorrhagic fevers and their virus families

Arenaviridae	Lassa fever
	South American HF (Argentine, Bolivian, Venezuelan)
Bunyaviridae	
Phlebovirus	Rift Valley fever (RVF)
Nairovirus	Crimean–Congo haemorrhagic fever (CCHF)
Hantavirus	Haemorrhagic fever with renal syndrome (HFRS)
	Hantavirus pulmonary syndrome (HPS)
Filoviridae	Marburg HF
	Ebola HF
Flaviviridae	Yellow fever
	Dengue HF*
	Kyasanur Forest, Omsk, Al Fakrah

*Differs from others in that haemorrhagic fever and shock result from immune manifestations related to infections with more than one dengue virus. Also dengue virus is not aerosol infectious.

transmission, and the complexities of the problems. I would like to discuss some of the recent experiences of Special Pathogens Branch at the National Center for Infectious Diseases in Atlanta, Georgia, with outbreaks of emerging viral diseases, particularly the viral haemorrhagic fevers (VHF). There are other agents which deserve study and may differ in important respects: influenza A virus with its dependence on a mixed animal–human strategy and its strong interaction with the human immune system; enterovirus 71, a person-to-person virus; canine parvovirus, an important DNA virus that apparently mutated to change its host range and produce a pandemic; lentiviruses with their complexities; and the spongiform encephalopathies (see the chapter by Weissmann and others, this volume), which of course are not viruses at all, but are nevertheless subtle, dangerous and poorly understood.

Viruses from four different RNA virus families cause a syndrome referred to as viral haemorrhagic fever (VHF) (Table 1) (Peters & Zaki, 1999). The viruses differ in all the ways one might expect from finding they belong to four different families: replication, genomic structure, morphology, morphogenesis, human pathogenesis, reservoir host, survival strategy in nature. Nevertheless, they all cause the VHF syndrome in humans and, interestingly, all are infectious by small particle aerosols (Peters *et al.*, 1996). Over the last several years they have repeatedly surprised us and new viruses have been added to the list. Special Pathogens Branch at CDC has been involved peripherally or centrally in a number of these (Table 2), although many outbreaks are unnoticed or undiagnosed

Table 2. Recent novel findings with VHF and other BSL-4 virus diseases

1993	Rift Valley	Epidemic in Egypt
1993	Bolivian HF	First case since 1975
1993	HPS	Discovery of the disease and the virus
1994	HPS	New viruses found
1994	Bolivian HF	Family epidemic – interpersonal transmission
1994	Sabia virus	Laboratory infection, US
1994	Lassa fever	Case imported into city of Lagos
1994	CCHF	UAE epidemic in stockyards
1994–96	Al Fakrah	Tick-borne flavivirus discovered in butchers in Saudi Arabia
1994	Ebola virus	New subtype in Ivory Coast
1994–96	Ebola virus	Zaire subtype epidemics in Democratic Republic of Congo (DRC), Gabon
1996	Ebola virus	'Exported' to South Africa
1995	Hendra virus	Newly discovered paramyxovirus in Australia
1995	Lassa fever	War-related epidemics in Sierra Leone
1996	Andes virus	Person-to-person transmission hantavirus
1997	HPS	South American totals: 229 cases, 6 countries
1997	Rift Valley	East African epidemic; El Niño/southern oscillation (ENSO)
1997–8	Nipah virus	Discovered in Malaysia: 289 cases, 37 % fatal
1998	Smallpox	Potential role in bioterrorism
1998–99	HPS	ENSO in SW US leads to increased human cases
1999	Marburg	Active in DRC; multiple genotypes
1999	CCHF	Reports of activity in Central Asia, Russia
2000	HPS	Panama cases
2000	Rift Valley	Epidemic in Yemen, Saudi Arabia
2000	Ebola virus	Sudan subtype causes Uganda epidemic

outside their local area, worked up by other organizations, or are found to be caused by typhus, relapsing fever or other non-viral agents. I have also listed Hendra and Nipah viruses, newly discovered paramyxoviruses causing respiratory disease and encephalitis (Murray *et al.*, 1998; Chua *et al.*, 2000), and smallpox virus, eradicated from nature but still a bioterrorist threat (Henderson *et al.*, 1999).

I can summarize my feelings about the essentials for recognition and control of these very dangerous diseases simply: there is no substitute for competent and alert clinicians who have local routine diagnostic laboratories available. The clinicians should be integrated with a national public health service that performs surveillance on the local diseases of importance and is capable of alerting national authorities to unusual change in incidence or disease type. There must be access to a reference capability, which includes clinical, laboratory and other expertise to apply to unusual or complex problems. The surveillance of established diseases and the investigation of presumed 'new' diseases should be accomplished by multidisciplinary teams composed of clinicians, microbiologists, veterinarians, epidemiologists, ecologists, and others with a coordinated goal to bring together the means to control outbreaks and elucidate all feasible aspects of the disease. Other more generic modalities such as supporting patient care, logistics and press information are readily achieved by other international partners but should be done in close coordination with the subject matter experts involved. Leadership is essential to achieve disease control and to obtain the information that will prepare to fight the 'next' epidemic. Leadership must be based, as far as possible, on scientific knowledge and goals and not on narrower short-term considerations that may apply to local and international organizations.

Rather than make a matrix or a laundry list from Table 2 to support these assertions, I will comment briefly on some virus outbreaks and how they illuminate topics in the preceding paragraph.

Hantavirus pulmonary syndrome (HPS)

In 1993, an alert clinician and an excellent state medical examiner system recognized unusual cases of a pulmonary disease in the southwestern US. State health departments investigated the situation with first-rate epidemiological and laboratory methods but failed to find the cause. CDC assistance was requested in this multi-state problem. Because of the population's fearful response, stigmatization of some population groups, the unknown nature of the disease and the unknown mode of transmission, CDC supplemented state efforts with a massive response of field epidemiologists and specialized laboratory efforts in Atlanta. As is often the case in zoonotic diseases, the laboratory proved crucial in ascertaining the aetiology of the 'mystery disease'. Within 48 hours of receiving the first materials, a hantavirus was suspected as the cause, and within 8 days the aetiological agent was 'wrapped up' with serological, viral genetic and immunohistochemical evidence (Nichol *et al.*, 1993).

Details of the events help us understand some aspects of emerging disease epidemics that are seen time and time again:

(1) Alert clinicians are key to making the critical observations that something 'new' is happening, and they must be listened to. In this case, an excellent system of autopsies and experienced pathologists was present and made the critical observations as well. The pathology findings were very important in decision making, both

Table 3. Social accompaniments of pestilence

Rumours, exaggerations
Press melee
Panic, fright and flight
Political demands, political turmoil
Economic losses
Religious revivals
Stigmatization of subgroups, sufferers, families
Medical profession: best and worst behaviour

ordinary histological examination and immunohistochemical staining which demonstrated the pathophysiology and allowed a reasonable hypothesis of the unexpected lung pathogenesis (Ksiazek *et al.*, 1995; Zaki *et al.*, 1995).

(2) Serology was key in the recognition of the hantaviral aetiology (Nichol *et al.*, 1993). Cross-reactive antibodies have consistently provided robust and sensitive clues to the causative agents of new diseases, including other hantaviruses (Peters, 1998), new arenaviruses (Fulhorst *et al.*, 1996), new subtypes of Ebola and the identification of Nipah virus (Chua *et al.*, 2000).

(3) Subsequently, reverse transcriptase and polymerase chain reaction (RT-PCR) analysis quickly solidified the serological findings and greatly extended our understanding of the genetics of the causative virus. This analysis allowed us to rapidly identify the rodent host (Childs *et al.*, 1994). Interestingly, previously published hantavirus PCR primers based on cell culture analyses were not sufficiently sensitive to reveal hantaviral RNA in infected tissues so that careful individual design of the optimally sensitive primers to detect low concentrations of viral RNA in tissues was critical. Later, after experience with several different American hantaviruses, it was possible to design highly sensitive primer sets to amplify a broad array of these viruses with high sensitivity (Johnson *et al.*, 1999).

(4) Discovery of the causative agent (Sin Nombre virus or SNV), definition of its reservoir, and knowledge of other hantaviruses allowed suggestions to be made for control, even as more specific studies of local risk factors were undertaken.

(5) In every VHF outbreak, going back at least to yellow fever in Philadelphia in 1793 (Carey, 1794), a number of public reactions have been the rule (Table 3), and use of a coordinated public affairs response through the press was essential in minimizing the negative aspects.

After a careful examination of the clinical picture of HPS, it was apparent that we had discovered not only a new virus, but a new disease (Duchin *et al.*, 1994; Moolenaar *et al.*, 1995). The dissemination of this information resulted in the recognition of clinical disease and the subsequent rapid discovery of new hantaviruses causing HPS both in the US and elsewhere (Peters, 1998). Discovery often followed a local upsurge of disease, usually in relation to climatic events that increased rodent populations. In the US, alert clinicians have extended the clinical spectrum of HPS to include milder cases as well (Kitsutani *et al.*, 1999). The less severe hantavirus infections would not have been recognized without the careful work on full-blown HPS and the availability of diagnostic tests.

The discovery of emerging virus epidemics not only provides an opportunity for immediate research but may also demand longer-term projects for understanding and control of the disease. These projects are more easily initiated on the heels of an epidemic rather than trying to make the funding argument in an inter-epidemic period, regardless of the need. In the case of SNV in the southwestern US, it has been possible to follow rodent populations, rodent infection and human cases longitudinally for more than 5 years and valuable data have been obtained (Mills *et al.*, 1999). In particular, the cascade of events following El Niño/southern oscillation (ENSO) perturbations of the world's climate have been traced sequentially through weather changes, environmental vegetation responses, rodent population changes, subsequent rodent infection increases and finally human disease (T. Yates and others, unpublished). It has also been possible to begin to examine the Four Corners area from space using higher resolution systems and develop algorithms that predict local risk of disease (Glass *et al.*, 2000).

Ebola virus

After the 1976 discovery of the Zaire and Sudan subtypes of the virus, no activity was detected in Africa until 1994–1996, when epidemics were discovered in Gabon and Democratic Republic of Congo (DRC, former Zaire) (Peters & LeDuc, 1999). This long gap in epidemic activity is characteristic of many of the zoonotic viruses and makes it difficult to evaluate the global risk of the viruses and also hard to obtain support for research activities. Interestingly, the apparent surge in activity included discovery of the 'new' Côte d'Ivoire subtype of Ebola in the Tai forest, presumably reflecting widespread changes in African forest conditions that give rise to increased Ebola transmission. Several attempts have been made to elucidate the reservoir of Ebola viruses without success, and in the case of the Kikwit, DRC epidemic in 1995 more than 3000 vertebrates and 30 000 invertebrates were collected and analysed. This was a huge effort involving 40 people with costs of hundreds of thousands of dollars and found no trace of Ebola virus or antibodies directed against it. The magnitude of the problem is complicated by the extravagant complexity of the tropical forest; the relevant areas have usually never had a complete inventory of their flora and fauna so that new and rare species comprise particular challenges. Although there are speculations and a small amount of laboratory data suggesting that bats may be important in Ebola ecology (Swanepoel *et al.*, 1996), there are really no good data that permit one to ignore the many other possibilities in the search (Murphy & Peters, 1998). I think it is likely that the key observations will come from systematic studies in a single site such as conducted in the Tai forest or from some fortuitous observation. Discovering the maintenance strategy of the virus is important in predicting virus activity and in understanding what will happen to virus distribution and activity as the forest continues to be eroded. Will the reservoirs be eliminated or will the reservoirs seek new habitat?

Each Ebola or Marburg virus outbreak usually reflects a single virus introduction into an African milieu, which is then amplified by inadequate hospital hygiene. Thus there are two aspects to emergence: (1) the initial spillover into the human population, which occurs in a natural setting (for Ebola, the tropical forest), and (2) the inadequate protection afforded the typical African hospital worker plus the pernicious effects of disposable needles and syringes. The technology that protects the developed world from blood-borne diseases is lethal in the African setting.

The 1995 Kikwit episode provided the first opportunity to examine the role of barrier nursing in protection of hospital staff, to obtain sufficient human autopsy material for modern analysis, to obtain follow-up information on human cases, and for other studies (Peters & LeDuc, 1999). The research established several important findings objectively:

(1) Barrier nursing can halt transmission, even in hospitals with ongoing disease and multiple in-patients
(2) Virus is present in the skin and can explain some of the epidemiological findings (role of contact, risk of post-mortem preparation and mourning practices, need for gloves and gowns) as well as provide a mechanism for formalin-fixed skin biopsy surveillance
(3) Virus persists in genital secretions for modest periods of time
(4) Human convalescent bone marrow contains genetic information to develop recombinant anti-Ebola antibodies with potential therapeutic value.

This small information explosion demonstrates the need for every epidemic investigation to consider that new information can be gained while control is instituted. Rapid response, coordination, and additional resources are important to take full advantage of these events.

The unfortunate exportation of an unsuspected Ebola case from Gabon to South Africa resulted in the unexpected infection of a nurse in the receiving hospital. The episode would have been missed if the alert laboratory at the National Institute of Virology had not cultured samples from the nurse for Ebola virus and made the diagnosis (R. Swanepoel, personal communication, 1997). The lack of additional secondary or of tertiary Ebola infections is an important data point for risk-assessment in hospitals in developed countries that receive these patients.

A negative lesson should also be drawn from the Kikwit data and the nurse's progressive, fatal clinical course in the face of first-rate supportive care. It is often said that better supportive care could improve the ~90 % mortality of Zaire subtype human infections, and this may be so, but the typical mortality associated with Zaire and Sudan (~50 %) subtypes seems to suggest that there are important virus-specific factors at work. Patients dying with Zaire subtype infections in Kikwit showed no serum antibodies to evidence an ongoing immune response and indeed had such extensive damage to T and B cell dependent areas of lymphoid organs that it seems unlikely their immune system could respond in the short term. In addition, the organs showed extensive necrosis and very extensive viral involvement (Zaki & Goldsmith, 1999). These findings taken together emphasize the need for an antiviral drug to limit viral replication and hopefully then allow the body to respond physiologically and immunologically.

Field diagnostics are also an issue. The 1995 Kikwit epidemic is illustrative of some of the considerations. The epidemic began in January and samples were not sent for diagnosis until May. Diagnosis of acute Ebola infection was made by antigen detection ELISA within 8 hours of receipt of samples at CDC and the genotype was available within 48 hours. The most reasonable approach to speeding the detection of Ebola outbreaks is to improve clinical acuity. An Ebola test would be much less valuable to clinicians than reliable tests for shigellosis, typhoid, and other local diseases that would allow them to accurately diagnose common diseases of great public health importance and then focus on the unusual. Furthermore, if the infrastructure for basic laboratory tests such as bacterial blood and stool cultures is not readily available, it seems unlikely that Ebola virus tests would be readily translated to or sustained in such a medical setting.

Once the diagnosis was made, control of the outbreak was not impeded by the lack of a diagnostic test because in this epidemic setting when so many of the severely ill patients in the town were infected with Ebola virus, the clinical case definition was both sensitive and specific. However, there is an important need for diagnosis in sporadic cases. This can be fulfilled by skin biopsy of fatal cases (~90 % of Zaire subtype cases are fatal). This material is readily obtained and safely transported at room temperature because of the formalin fixation. In the African setting, this is far more practical than attempting to maintain a cold chain for virological tests. The newer regulations for

Table 4. Selected significant Rift Valley fever outbreaks

1918	Kenya	Earliest recognized outbreak; clinically diagnosed
1936	Kenya	Virus isolated: sheep, human disease described, reproduced
1950	Republic of South Africa	First in southern Africa
1974	Republic of South Africa	Haemorrhagic fever and encephalitis first recognized
1977	Egypt	First exportation outside sub-Saharan Africa
1987	Mauritania	First West African epidemic; dam and flooding
1990–91	Madagascar	Ecological and social change
1997–98	East Africa	ENSO-related; large number of cases, large area involved
2000	Arabia	Yemen and Saudi Arabia; first epidemic outside Africa

transporting frozen infectious samples are confusing to shipper and local airline representatives alike and require expensive packing material not usually locally available.

Rift Valley fever (RVF)

This disease first became an emerging pathogen in 1931 when studies of the virology, immunology and ecology of RVF during an epidemic in sheep and humans in Kenya laid the groundwork for our current understanding of the virus (Daubney *et al.*, 1931). Epidemics, presumably related to the importation and extensive farming of European breeds of sheep and cattle, occurred in sub-Saharan Africa and beyond (Table 4). The virus has continued to surprise us by appearing in new areas with large epidemics occurring in domestic sheep and cattle, usually related to ecological or climate change (Peters & Linthicum, 1994; Peters, 1997). The most alarming is the exportation to the Arabian peninsula in the year 2000 (CDC, 2000b, c). The epidemic may well have been seeded following the very extensive epidemic of 1997–98 in East Africa (C. W. Woods and others, unpublished; R. Swanepoel & S. Nichol, personal communication, 2000). The relatively moist area along the western coast of Yemen and Saudi Arabia is apparently fertile ground for transmission.

It has been recognized since RVF spread from sub-Saharan Africa to Egypt in 1977 that the virus poses a threat for introduction to areas outside sub-Saharan Africa. Mosquitoes from North American and other areas tested are often efficient vectors for the virus in the laboratory. Once an epidemic begins, it seems likely that the explosive spread of the virus will only be stopped by the quirks of ecology and weather or by a highly efficient vaccine for sheep and cattle, by analogy with Venezuelan equine encephalitis virus. There are two vaccines available for veterinary use. One is a live, attenuated vaccine that is highly teratogenic in sheep, may be insufficiently immunogenic in cattle, and which carries an unacceptably high likelihood of reversion. The other is an inactivated vaccine that was designed for use in cattle but has had little field use in sheep

Laboratory workers, veterinarians, vaccine production staff and others require direct protection. The US Department of Defense has developed a candidate live attenuated human vaccine for RVF but after successful inoculation of more than 60 human volunteers the project has been abandoned because of its lack of priority within the military system (Caplen *et al.*, 1985). No other agency has proceeded with testing of this vaccine. The only alternative is an inactivated vaccine developed by the Department of Defense that is in short supply, requires three injections for solid immunity, and is not currently accessible (Pittman *et al.*, 1999). Lack of human vaccine, of course, is a severe impediment to surveillance or additional work on the virus. It is also worth mentioning that RVF is a disease with considerable implications for either bioterrorism or state-sponsored biowarfare attack and lack of vaccine as well as other restrictions severely hamper research efforts (Peters, 2000).

Diseases such as RVF, Crimean–Congo HF and Nipah encephalitis emphasize the need for coordination between the medical and veterinary sectors. Surveillance is usually much more sensitive if it is based on the target domestic animal species, and in the case of epidemic RVF we believe that control is more readily achieved by vaccination of domestic animals. In spite of the established facts, veterinary and medical coordination is not usually adequate and veterinary laboratories often do not have the agent-specific laboratory tests to definitively diagnose micro-organisms that are pathogenic for humans and of economic significance. Parenthetically, wildlife diseases are also potentially useful sentinels and control points for human disease but are even less-often included in the biological and ecological loop.

It is likely that RVF epidemics in sub-Saharan Africa can be predicted from climatological conditions and/or remote sensing data (Linthicum *et al.*, 1999), although there is still much work to be done establishing this link. Exportations of the virus are, however, harder to predict. Many different mosquitoes are capable of transmitting the virus in nature and concentrations of livestock to amplify the virus are commonly found in many of the same areas. Lack of distant exportations to date may well be a consequence of the rural distribution of active RVF virus transmission limiting the opportunities for viraemic humans or infected mosquitoes to travel in airplanes.

Nipah

The emergence of Nipah virus in Malaysia illustrates once again the need for strong local institutions and immediate laboratory diagnosis of epidemic disease. The encephalitis outbreak was initially thought to be caused by the locally endemic flavivirus, Japanese encephalitis virus, but the epidemiology of the disease belied this explanation. Fortunately, a Malaysian scientist isolated an agent that caused cytopathology in cell culture suggestive of a paramyxovirus and not a flavivirus. Working with a reference laboratory, the virus was shown to be a 'new' paramyxovirus related to the previously known Hendra virus (Chua *et al.*, 2000). The CDC reagents for Hendra virus, prepared in anticipation of future emerging problems, proved useful, not only for initial diagnosis, but for the subsequent studies to diagnose infection of human patients and to study the spread and control of the epidemic.

This virus has extremely dangerous potential. It is believed to be a parasite of fruit bats that moved into swine populations and spread among pigs, causing disease but a low mortality in pigs. Humans in contact with infected pigs were at risk for serious encephalitis. The virus also spread to dogs and cats. This broad host range suggests that the virus should be treated with extreme caution and closely monitored. The spread of the virus within pig populations and spillover into humans was only controlled by killing and burying more than 1 million pigs within the major infected areas.

Of course the question remains 'why did Nipah virus emerge?'. This is under study but the answer will not be easy to obtain. It is possible that the drought and fires in Southeast Asia influenced by ENSO in 1997 disturbed the ecology or physiology of the local bats, with behavioural or virological consequences. One can also speculate that increasingly intensive pig-raising practices may have made swine populations more likely to support continuous virus transmission.

One of the most important but under-appreciated aspects of investigations of large or unusual epidemics is the size and diversity of the teams needed for a satisfactory outcome. For example, in the 1997–98 Nipah investigation, CDC initially sent eight persons to assist the national ministry of health. This large effort was undertaken because of the need for a broad-based response to a clearly recognized but poorly understood problem that threatened to spread to swine-raising areas throughout Asia. Because the pig was already epidemiologically identified as being involved, there was a need for veterinary as well as medical expertise. Nothing was known of the disease or its pathogenesis so that coordinated human and veterinary medical observations were essential. Immunohistochemical studies of humans and swine were particularly critical to understand the pathogenesis and form hypotheses on the route(s) of transmission. Because of the rapidly expanding areas of infection, geographic information systems were desirable for following and analysing the epidemic; although we were unable to get the system functioning in time for monitoring, it served as a tool for retrospective analysis of the patterns of spread. In addition to the CDC consultants, Australian colleagues came with their Hendra virus experience and test reagents to collect faunal samples for the reservoir search and provide additional strength in veterinary surveillance and epidemiology; they also performed important confirmatory inoculations of potential hosts in their containment laboratories in Australia.

WHAT SHOULD WE EXPECT IN THE FUTURE?

I have attempted to set the stage for understanding of the future by reviewing the past. Millions of years of development of the earth's surface and evolution of its denizens resulted in an ecological configuration that was undergoing only slow change until a few thousand years ago. About 5–10 thousand years ago, humans began to increasingly impact on the environment and accelerate the naturally occurring alterations in the flora and fauna of the planet. In the last 1500 years, human activities have stepped up markedly, particularly in the arena of environmental modification and purposeful or unintended transport of animal, plant and viral species. The loss of species from habitat destruction and modification is alarming, and the biological invasions contribute independently and additively to the loss (Vitousek *et al.*, 1996). Such species losses have occurred in geologic time, but never at the rate we are precipitating today (Leakey & Lewin, 1995). These changes are not only accelerating but are accelerating at an increasingly rapid pace. Because over half the land surface of the earth has already been modified by humans, the environment itself is becoming increasingly susceptible to invasions by exotic species (Elton, 1958) and we have fewer and fewer options to ameliorate adverse events. The driving force behind this is humankind. We are the major modifiers of the environment and our numbers are ever-increasing (Holland & Petersen, 1995).

We can only expect the global trends to worsen in many significant aspects (Table 5). One issue that was not emphasized previously will be one of the major problems in the future: homogeneous microbial target populations or 'monocultures'. Their major impact so far has been in plant, land animal, and aquatic food production. The intensified agriculture practiced in the Malaysian swine industry may have provided the substrate for spread of Nipah virus, and one wonders what would happen in the US in states such as North Carolina if such a pathogen emerged. Not only would the economic impact have been severe, but it is difficult to imagine how more than a million swine could have been eliminated to control the disease within regulatory strictures. Plant crops are highly vulnerable: the Irish potato famine is the classic historical example but the virtual destruction of the US corn crop by a fungus in 1970 provides a remarkable recent example (Ullstrup, 1972). One of the most instructive case studies of the issue comes from Bali and shows how introduction of new 'green' rice strains and locally untried intensive cultivation practices resulted in repeated epidemics in the target crop (Lansing, 1991).

Table 5. Global influences on emerging infections that will increase in coming years

Increasing environmental modification
Increasing population pressure on resources decreases options for amelioration
Changing human behaviour, customs, traditional social structures
Travel of people, domestic animals, plants
Movement of viruses, vectors and reservoir species
Wars, disasters, refugees
Crowded 'monocultures': cattle, chickens, aquaculture, crops, people in megacities
More immunosuppressed humans
Decreases in effective public health microbiology infrastructure, particularly developing country laboratories and laboratories based in developed countries and working overseas
Climatic change?

The 'monocultures' that will be most important in the next decades may well be the megacities. These agglomerations of millions of people could provide the next quantum change in the interrelationships of humans and micro-organisms, just as the larger population centres made possible by irrigated agriculture in the Middle East led to the emergence of diseases transmitted from human to human ~5000 years ago (Fenner, 1970). The third-world megacities provide a substrate in which unprecedented numbers of humans of all ages are in close proximity for faecal–oral, fomite, droplet and aerosol transmission of microbes with almost no medical supervision or surveillance. City-dwellers are in communication with kin from the same rural areas that provide the increasing numbers of migrants and also potentially dangerous viruses. An additional consideration is the breakdown in social structure, sexual behaviour and other human variables inherent in migration.

In the US, there will be similar trends with different emphases (Table 6). Changes in our natural and social environment will continue to open new niches for microbes. It is notable that the short-term, profit-oriented focus of managed care will further weaken our first line of defence, the individual practitioner. It is not judged to be cost-effective in the limited perspective and the short term to investigate diseases in depth or seek aetiologic diagnoses if there are no immediate treatment implications. This attitude will make it even more important that we strengthen our public health system to fill the gaps. The public health system may need to pay for some diagnostic tests in order to understand the changing distribution of viruses and other human pathogens.

As shifts in the natural world influence the patterns of zoonotic and arboviral diseases, we are witnessing changes in human-to-human transmitted diseases as well.

Table 6. Factors enhancing infectious diseases that will worsen in the US in coming years

Increasing environmental modification
Social change, 'Sex, drugs, rock and roll'
Crowded 'monocultures': domestic animals, crops, aquaculture, day care centres
Increasing foreign travel, food imports, immigration
Managed care: medicine as a business
Loss of human vaccine production infrastructure
Diseases of medical technology, including risks from xenotransplantation
Anti-vaccine movement among citizens unaccustomed to seeing infectious diseases
Public health infrastructure decay?

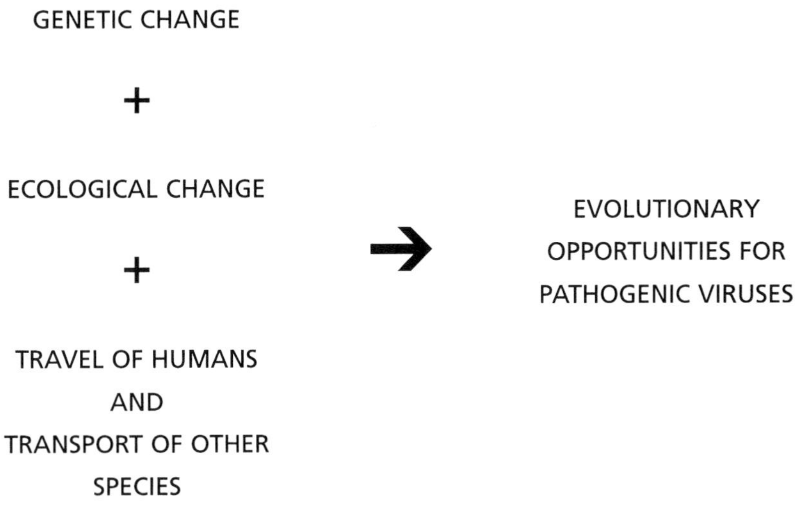

Fig. 4. A schema for emerging infections.

However, viruses transmitted in the natural world around us remain a major problem because they are potentially direct sources of human infection and the fount of past and future diseases transmitted among humans. We should understand that in the broad sense many of the changes in inter-human disease transmission are driven by changes in human ecology; after all, *oikos* (the Greek root of 'ecology') means 'house' and humans serve this function for many viruses. In fact, I divide the driving forces behind the appearance and increasing incidence of diseases into three broad categories (Fig. 4): viral genetics, ecological shift, and transposition. Of these, ecological changes are by far the most important, with travel and the movement of vectors providing opportunities for the viruses to take advantage of differing ecological settings.

Viral mutation in general has had only a permissive role in disease emergence. Influenza is, of course, the major, predictable exception for this assertion. We can expect both continued drift and sooner or later shift from influenza A. Other emergences driven by mutation may occur, as presumably was the case for canine parvovirus disease. Even HIV-1 probably emerged as a major human pathogen as it spread under the influence of changing sexual behaviour related to urbanization and travel. Well-documented phylogenetic differences compared to the chimpanzee virus progenitor are presumably a consequence of adaptation to inter-human transmission, which in turn was driven by human ecological changes (Gao *et al.*, 1999). The short-term sequential changes within the human body do not 'drive' viral change but rather occur randomly and are then selected by the immune pressures of the host.

It has also been surprising how infrequently suppression of the human immune system has played a role in emergence of disease in the general population. Immunosuppression from AIDS and other sources has been increasingly common and has made sufferers more and more susceptible to infectious agents in the environment, but has had a surprisingly small effect on the general spread and importance of diseases to normal hosts, except with a few organisms such as tuberculosis and possibly zoonotic visceral leishmaniasis. Immunosuppressed patients are far more likely to develop the disseminated disease after infection with *Leishmania infantum* or *Leishmania chagasi* and are capable of transmitting the parasite to sandflies, thus potentially bypassing the role of the usual canine intermediate host (Alvar *et al.*, 1997; Molina *et al.*, 1994). I suspect that sooner or later we will have more examples of an intersection of immunosuppressed humans and the ecology of an important pathogen that will lead to increased danger to immunologically intact hosts. Indeed, we may have found an example with immunosuppressed patients who chronically excrete poliovirus, thus endangering our poliovirus eradication programmes (Kew *et al.*, 1998; Bellmunt *et al.*, 1999).

WHAT CAN THE SCIENTIFIC COMMUNITY DO?

We have three major tools to deal with the increasing brunt of emerging infectious diseases: research, public health and vaccines.

Research

We must stop regarding emerging infections as isolated processes of modest importance. They are integral to a greater problem in planetary biology, that of destruction of environments, movement of species around the planet, and ever-shifting ecological vistas. Human and natural ecology must be integrated, as well. As a very simple example, the deer mouse is a quantitatively important rodent species and also is the reservoir of a pathogenic virus. Studies of this and other rodent species concentrate on the undisturbed environment and are distinctly silent on issues of extreme public health importance such as rodent behaviour in disturbed environments that occur around rural residences or ranches. Microbes should be included in research on biodiversity and evolution. A simple example refers to the attempts to reintroduce the black-footed ferret to the North American west. Plague, first brought to the Americas in the early 1900s, has become a common infection of the rodents the ferret preys upon and is highly lethal for the endangered predator species, threatening the ability of the ferret to reoccupy its previous natural range.

There should be a rogue's gallery of viruses that receive special attention. An improved inventory of viruses in the environment will help us select the most important potential pathogens and could give us an improved understanding of both acute and chronic infectious diseases of unknown aetiology. Of course, this approach has to be selective and might best be applied to relatives of known pathogens and important examples of poorly studied virus groups. The isolation and characterization of Whitewater Arroyo virus, an arenavirus of no known pathogenic significance when it was initially discovered, exemplifies this approach (Fulhorst *et al.*, 1996). RT-PCR studies using optimal primers suggest that the virus may be the cause of severe human disease in southern California (CDC, 2000a). This type of thinking also led to the availability of Hendra virus reagents to provide diagnostic capability for Nipah virus when it emerged 3 years later.

We should do the groundwork now and not when we are faced with a 'mystery disease' or a new, challenging pathogen. In addition to a library of viruses, we need to understand their diversity, basic microbiology, immunology, prototype experimental vaccines, diagnostic principles and basic maintenance cycles. HIV provides an excellent example of how such forethought could have possibly saved millions of lives. Some of the findings could well be exercised on select sets of patients with acute, febrile diseases of unknown aetiology.

A fundamental problem of disease-oriented microbiology is to understand why certain micro-organisms infect and cause disease in a given host. This type of research does not yield easily to current reductionist research approaches and is neither popular nor easily funded. Recent advances in identification of receptors and molecular participants in 'natural immunity' may allow advances in this field that could be very useful in predicting emerging viruses on which to focus.

The genetic properties that link a given virus with its natural reservoir and with its ability to 'jump' to another reservoir are as important as understanding the ability of a virus to cause disease in a given species. For example, hantaviruses are tightly linked to their specific reservoir hosts and indeed the phylogenetic relationships bespeak a long co-evolutionary relation (Monroe *et al.*, 1999). We do not understand the basis for these restrictions and of course they are quite important in understanding the risk of emergence.

One example of how pressing questions may be simultaneously scientifically challenging and public health relevant lies within the proposed new genus of *Paramyxoviridae* that contains Hendra and Nipah viruses as established human pathogenic members. How many members of this genus exist and what are their hosts? The natural history of the known viruses is not known, even though Nipah and Hendra viruses seem to be natural parasites of flying foxes (Murray *et al.*, 1998; Chua *et al.*, 2000). Why did the viruses 'cross' species to horses or pigs and result in disease? What are the determinants of human disease? Why can these viruses apparently infect so many animal species and cause disease? What is the pathogenesis of the diseases in humans and other animals and the infection of the natural hosts? What are the minimum determinants of vaccine protection if a seemingly highly transmissible virus should escape into a large pig-raising area such as Thailand or China?

Public health

Our first line of defence is, and will remain, surveillance. There is no substitute for knowing what's going on around the world. In addition, there is a need for culturally and financially appropriate technologies to minimize disease transmission. Finally, there should be interventions into dangerous situations. For example, continuous inter-human transmission of Ebola virus is undesirable because of the, admittedly unlikely, possibility of its adapting to inter-human transmission on a more efficient basis (Murphy & Peters, 1998). RVF epizootics and epidemics enhance the likelihood of distant spread of the virus and should be curtailed with appropriate vaccines (Shope *et al.*, 1982; Peters, 1997).

Public health is the traditional stepchild of funding and this may be one of reasons that the discipline remains largely reactive. Our public health infrastructure needs to be assured by receiving sufficient funding to execute the traditional programmes as well as new, innovative approaches that are prioritized according to broad scientific goals. One of the problems will be the lack of infrastructure in the countries perhaps at most need of surveillance and study, and some of the slack will have to be taken up by the developing countries. The risk to us from uncontrolled disease elsewhere in the world has been shown too many times to need additional discussion.

Diagnostic infrastructure would be an extremely important contribution to surveillance for emerging infectious diseases. The public health and research communities have typically provided limited assistance in the form of reference reagents or selected diagnostic tests, but they do not have the financial depth or the expertise to effect broad change and support testing around the world. Rapid and accurate diagnostics are unlikely to contribute in a sufficient way unless we can overcome some of the same hurdles that face vaccines (*vide infra*). Modern commercial diagnostics are aimed at high throughput automated systems for modern hospital markets and will often not be technologically appropriate and capable of maintenance in the African context; indeed, the spectrum of organisms identified in automated blood culture systems neglects such pathogens as tularaemia and anthrax. The development of simple antigen or IgM antibody tests for viral diseases is well within our technological capability, but even the markets of the US Department of Defense or the civilian bioterrorism community have failed to induce commercial production, which would ensure a continuous supply of standardized diagnostic kits. Alternative sources should consider the difficulties of interlot standardization, control sera, distribution, expiry dates and regulatory strictures. Simplified PCR technology is also becoming a practicality and many primer sets have been developed for the known viruses. The technology is here, but it will be some years before the available simplified machines are rugged, standardized and affordable for the developed world, if ever.

The public health and the research community both need to interface with national governments to reduce some of the obstacles to surveillance. Within the developed countries, there are several important concerns that the biomedical community has not succeeded in voicing effectively. These often involve regulations designed to achieve a general good that is not in dispute but which are not really useful in achieving these ends. One excellent example is the shipping of virus strains and samples suspected of containing infectious agents. The regulations surrounding the packing and transport of such materials are becoming increasingly more stringent, but I am unaware of any example of such shipments ever having infected a bystander. The special agent category in the US was designed to limit the distribution of potential bioterrorism or biowarfare agents, but most of the agents on the list are widely available in nature; thus the gain from the regulations as written is minimal, although the impact on research to combat these agents is quite negative. Various governmental agriculture ministry restrictions on movement of antibodies, cell cultures and other materials often seem excessively broad, capricious and poorly founded, particularly given the negligible impact from these practices in the past. The animal activists have succeeded in placing such broad requirements on animal care and use that many scientists prefer to confine themselves to *in vitro* experiments, regardless of what is needed to answer the most important questions at hand. No one would speak out against safe shipping methods, limitation of distribution of particular

Table 7. Obstacles to vaccine development

Inadequate financial incentives/market
No constituency for vaccine
Scientific interest limited by funding, infrastructure required
High risk (financial, legal)
High risk (hazardous viruses)
Difficult translation of preclinical research into human products
Immunosuppressed and other special population

strains of micro-organisms, protection of the environment and domestic animals from infectious agents, ensuring humane care of laboratory animals, or ethical treatment of humans involved in research; but the regulations seem excessive in many areas and working scientists are not given the resources to recoup the inroads on budget and time made by the mandates.

Problems such as noted above indirectly rob citizens of the protection they expect from efficient public health and basic research resources, but there is a much more obvious role in which governments may be irresponsible. Denial of the presence of a disease is common when the economic consequences are high, leading to human suffering within the country and additional risk to other countries. Bureaucratic and/or political hurdles often delay the movement of outside public health experts and may also unreasonably limit the deployment of helpful treatment or prevention modalities.

Vaccines

If public health and general improvement in hygiene do not control an infectious disease, the mainstay is vaccination. We have an excellent set of conventional live, attenuated vaccines against the most prevalent virus diseases of high impact, and the private sector has much to be proud of in this regard. Nevertheless, there are a series of disincentives for further vaccine development (Table 7). The remarkable advances in molecular biology have given us an extraordinary set of tools to analyse the pathogenesis and immunology of viral disease, but the only mainstream vaccine from biotechnology is based on the expression of the self-assembling hepatitis B particle, which was the basis of the original plasma-derived vaccine. One of the newly introduced vaccines in the US is that for Lyme disease. The vaccine is a biotechnological product, but is only modestly successful and is directed at a non-fatal disease susceptible to antibiotic therapy. It is perhaps no accident that the highest incidence of the disease is in areas of the US with a very high per capita income.

In my opinion, we must fix this problem. There has to be a national vaccine production capability that can be used to make vaccines that are 'orphan' in the sense of profits, but which are important to all of us in terms of their potential for averting global harm. This facility could also direct its attention to geographically limited disease problems that are locally stultifying even though their global significance may at any given time not be judged to be great.

We also must address the gap between the very efficient process of analysis of the immune response and development of prototype immunogens versus the very long and inefficient process that leads to development of vaccines that can be used in humans. This may require research attention devoted specifically to this area. We may also need to look again at vaccine regulations and to perform a closer examination of risk benefit considerations, including the risks or negative aspects of not making a vaccine.

RVF provides a case study. Like yellow fever and West Nile viruses, RVF virus produces high viraemia in susceptible hosts, has efficient mosquito vectors in different areas of the world, and is a threat for distant introduction. RVF virus could have much more impact on animal and human health than WN but has never garnered the attention it deserves as a potential problem for the developed world. After the 1977 introduction into Egypt, it was believed that many receptive areas, such as the Fertile Crescent and parts of the US and Europe, would clearly be at risk (Shope *et al.*, 1982). The virus did not oblige these 'expert' predictions, but in the arena of emerging viruses, patience is a necessary quality. With each large outbreak, the potential of the virus seems to be even more firmly established, but no action on vaccine development has been forthcoming (Peters, 1997a, b). The presence of active virus transmission on the Arabian Peninsula in the year 2000 once again increases the threat from this versatile virus (CDC, 2000b, c).

Although antiviral drugs for emerging diseases have received little attention in this review or in the research and development arena, they really fall within the same category as vaccines. Their development is expensive and the return is small in most cases. To emphasize how unlikely it is that a commercial enterprise would undertake this research and development, one only has to consider the evaporation of effective antiparasitic drugs from the US market and even from the developing countries (White, 2000). Yet, new approaches to mass screening, combinatorial strategies and advances in structural chemistry make the possibilities for drug discovery greater than ever. Ebola virus is only one case in point. The virus would not justify vaccination of any given population in my opinion, but the availability of an effective antiviral would offer an important safety net for medical personnel and virus researchers as well as an intervention for those already infected in African epidemics before preventive measures can be instituted.

It would seem that a social problem that we must solve collectively is how to bring the fruits of modern science (and not just biotechnology) to the third world. In the field of viral diseases, diagnostics, vaccines and drugs are clearly not going to be available unless academic institutions can develop prototypes and a way is found to produce them for actual use. The loss to the most affected countries is great, but with the movement of infectious diseases we have seen recently and anticipate to continue at an increasing pace, the self-interest of the developed countries should dictate attempts to solve these problems.

IN CONCLUSION

The shifting equation that determines the emergence of virus diseases (Fig. 4) is moving increasingly to favour the viruses. Our responses are limited by the poverty of many sites of maximum risk and the inexorable changes that are occurring worldwide in both human and natural ecology. The major options available to combat the viruses are research, public health and vaccines. We need a strong programme to investigate disease outbreaks around the world intelligently and also a well-grounded academic research programme to study the basic science of the agents and their field biology. Public health institutions require increased resources and a broader remit for surveillance and control. Finally, vaccines are the real safeguards for otherwise uncontrollable diseases and we must move to make them a practical alternative for humans and animal disease control. This may require considerable changes in our thinking toward human vaccine development and deployment.

ACKNOWLEDGEMENTS

I would like to thank my many colleagues who have educated me (at least partially) over the years and particularly Luis Ruedas, Stuart Nichol, Bob Tesh and John McCauley for their contributions to this paper.

REFERENCES

Agostini, H. T., Yanagihara, R., Davis, V., Ryschkewitsch, C. F. & Stoner, G. L. (1997). Asian genotypes of JC virus in Native Americans and in a Pacific Island population: markers of viral evolution and human migration. *Proc Natl Acad Sci USA* **94**, 14542–14546.

Alvar, J., Canavate, C., Gutierrez-Solar, B., Jimenez, M., Laguna, F., Lopez-Velez, R., Molina, R. & Moreno, J. (1997). Leishmania and human immunodeficiency virus coinfection: the first 10 years. *Clin Microbiol Rev* **10**, 298–319.

Bailey, R. G. (1998). *Ecoregions: the Ecosystem Geography of the Oceans and Continents.* New York: Springer.

Bellmunt, A., May, G., Zell, R., Pring-Akerblom, P., Verhagen, W. & Heim, A. (1999). Evolution of poliovirus type I during 5.5 years of prolonged enteral replication in an immunodeficient patient. *Virology* **265**, 178–184.

Bowen, M. D., Peters, C. J. & Nichol, S. T. (1997). Phylogenetic analysis of the Arenaviridae: patterns of virus evolution and evidence for cospeciation between arenaviruses and their rodent hosts. *Mol Phylogenet Evol* **8**, 301–316.

Caplen, H. C., Peters, C. J. & Bishop, D. H. L. (1985). Mutagen directed attenuation of Rift Valley fever virus as a method for vaccine development. *J Gen Virol* **66**, 2271–2277.

Carey, M. (1794). *A Short Account of the Malignant Fever, Lately Prevalent in Philadelphia, with a Statement of the Proceedings that Took Place on the Subject in Different Parts of the United States*. Philadelphia: Mathew Carey.

Carter, H. R. (1931). *Yellow Fever: an Epidemiological and Historical Study of its Place of Origin*. Baltimore: Waverly Press.

CDC (2000a). Fatal illnesses associated with a New World arenavirus – California, 1999–2000. *Morb Mortal Wkly Rep* **49**, 709–711.

CDC (2000b). Outbreak of Rift Valley fever – Saudi Arabia, August-October 2000. *Morb Mortal Wkly Rep* **49**, 905–908.

CDC (2000c). Update: Outbreak of Rift Valley Fever – Saudi Arabia, August-November 2000. *Morb Mortal Wkly Rep* **49**, 982–985.

Childs, J. E., Ksiazek, T. G., Spiropoulou, C. F. & 10 other authors (1994). Serologic and genetic identification of *Peromyscus maniculatus* as the primary reservoir for a new hantavirus in the southwestern United States. *J Infect Dis* **169**, 1271–1280.

Chua, K. B., Bellini, W. J., Rota, P. A. & 19 other authors (2000). Nipah virus: a recently emergent deadly paramyxovirus. *Science* **288**, 1432–1435.

Cook, N. D. (1998). *Born to Die. Disease and New World Conquest, 1492–1650*. New York: Cambridge University Press.

Crosby, A. W. (1990). *America's Forgotten Pandemic. The Influenza of 1918*, 2nd edn. New York: Cambridge University Press.

Crosby, A. W. (1994). *Germs, Seeds, & Animals: Studies in Ecological History*. Armonk, NY: M. E. Sharpe.

Crosby, A. W. (1995). *Ecological Imperialism The Biological Expansion of Europe, 900–1900*. New York: Cambridge University Press.

Daubney, R., Hudson, J. R. & Garnham, P. C. (1931). Enzootic hepatitis or Rift Valley fever: an undescribed virus disease of sheep, cattle and man from East Africa. *J Pathol Bacteriol* **34**, 545–579.

Deubel, V. & Drouet, M.-T. (1997). Biological and molecular variations of yellow fever virus strains. In *Factors in the Emergence of Arbovirus Diseases*, pp. 157–165. Edited by J. F. Saluzzo & B. Dodet. Paris: Elsevier.

Dobyns, H. F. (1983). *Their Number Became Thinned. Native American Populations in Eastern North America*. Knoxville: University of Tennessee Press.

Duchin, J. S., Koster, F. T., Peters, C. J., Simpson, G. L., Tempest, B., Zaki, S. R., Ksiazek, T. G., Rollin, P. E., Nichol, S., Umland, E. T., Moolenaar, R. L., Reef, S. E., Nolte, K. B., Gallaher, M. M., Butler, J. C., Breiman, R. F. & the Hantavirus Study Group (1994). Hantavirus pulmonary syndrome: a clinical description of 17 patients with a newly recognized disease. *N Engl J Med* **330**, 949–955.

Elton, C. S. (1958). *The Ecology of Invasions by Animals and Plants*. London: Methuen.

Fenner, F. (1970). Chapter in *The Impact of Civilisation on the Biology of Man*. Edited by S. V. Boyden. Canberra: Australian National University Press.

Fulhorst, C. F., Bowen, M. D., Ksiazek, T. G., Rollin, P. E., Nichol, S. T., Kosoy, M. Y. & Peters, C. J. (1996). Isolation and characterization of Whitewater Arroyo virus, a novel North American arenavirus. *Virology* **224**, 114–120.

Gao, F., Bailes, E., Robertson, D. L. & 9 other authors (1999). Origin of HIV-1 in the chimpanzee *Pan troglodytes troglodytes*. *Nature* **397**, 436–441.

Glass, G. E., Cheek, J. E., Patz, J. A. & 9 other authors (2000). Using remotely sensed data to identify areas at risk for hantavirus pulmonary syndrome. *Emerg Infect Dis* **6**, 238–247.

Gould, E. A., de Andrade Zanotto, P. M. & Holmes, E. C. (1997). The genetic evolution of flaviviruses. In *Factors in the Emergence of Arbovirus Diseases*, pp. 51–63. Edited by J. F. Saluzzo & B. Dodet. Paris: Elsevier.

Gould, S. J. (editor) (1993). *The Book of Life. An Illustrated History of the Evolution of Life on Earth*. New York: Norton.

Hargis, R. (1880). *Yellow Fever: its Ship Origin and Prevention*. Philadelphia: D. G. Brinton.

Henderson, D. A., Inglesby, T. V., Bartlett, J. G. & 12 other authors (1999). Smallpox as a biological weapon. Working Group on Civilian Biodefense. *JAMA* **281**, 2127–2137.

Holland, H. D. & Petersen, U. (1995). *Living Dangerously: the Earth, its Resources, and the Environment*. Princeton, NJ: Princeton University Press.

Johnson, A. M., de Souza, L. T. M., Ferreira, I. B., Pereira, L. E., Ksiazek, T. G., Rollin, P. E., Peters, C. J. & Nichol, S. T. (1999). Genetic investigation of novel hantaviruses causing fatal HPS in Brazil. *J Med Virol* **59**, 527–535.

Kew, O. M., Sutter, R. W., Nottay, B. K., McDonough, M. J., Prevots, D. R., Quick, L. & Pallansch, M. A. (1998). Prolonged replication of a type 1 vaccine-derived poliovirus in an immunodeficient patient. *J Clin Microbiol* **36**, 2893–2899.

Kitsutani, P. T., Denton, R. W., Fritz, C. L. & 8 other authors (1999). Acute Sin Nombre hantavirus infection without pulmonary syndrome, United States. *Emerg Infect Dis* **5**, 701–705.

Ksiazek, T. G., Peters, C. J., Rollin, P. E. & 10 other authors (1995). Identification of a new North American hantavirus that causes acute pulmonary insufficiency. *Am J Trop Med Hyg* **52**, 117–123.

Lansing, J. S. (1991). *Priests and Programmers. Technologies of Power in the Engineered Landscape of Bali.* Princeton, NJ: Princeton University Press.

Leakey, R. & Lewin, R. (1995). *The Sixth Extinction*. New York: Anchor/Doubleday.

Lederberg, J., Shope, R. E. & Oaks, S. C. (editors) (1992). *Emerging Infections. Microbial Threats to Health in the United States*. Washington, DC: National Academy Press.

Linthicum, K. J., Anyamba, A., Tucker, C. J., Kelley, P. W., Myers, M. F. & Peters C. J. (1999). Climate and satellite indicators to forecast Rift Valley fever epidemics in Kenya. *Science* **285**, 397–400.

McEvedy, C. (1995). *The Penguin Atlas of African History*, 2nd edn. New York: Viking Penguin.

Mills, J. N., Ksiazek, T. G., Peters, C. J. & Childs, J. E. (1999). Long-term studies of hantavirus reservoir populations in the southwestern US: a synthesis. *Emerg Infect Dis* **5**, 135–142.

Molina, R., Canavate, C., Cercenado, E., Laguna, F., Lopez-Valez, R. & Alvar, J. (1994). Indirect xenodiagnosis of visceral leishmaniasis in 10 HIV-infected patients using colonized Phlebotomus perniciosus. *AIDS* **8**, 277–279.

Monroe, M. C., Morzunov, S. P., Johnson, A. M. & 7 other authors (1999). Genetic diversity and distribution of Peromyscus-borne hantaviruses in North America. *Emerg Infect Dis* **5**, 75–85.

Moolenaar, R. L., Dalton, C., Lipman, H. B. & 11 other authors (1995). Clinical features that differentiate hantavirus pulmonary syndrome from three other acute respiratory illnesses. *Clin Infect Dis* **21**, 643–649.

Moorehead, A. (1966). *The Fatal Impact. The Invasion of the South Pacific. 1767-1840.* New York: Harper & Rowe.

Murphy, F. A. & Peters, C. J. (1998). Ebola virus: where does it come from and where is it going? In *Emerging Infections*, pp. 375–410. Edited by R. M. Krause. New York: Academic Press.

Murray, K., Eaton, B., Hooper, P., Wang, L., Williamson, M. & Young, P. (1998). Flying foxes, horses, and humans: a zoonosis caused by a new member of the Paramyxoviridae. In *Emerging Infections 1*, pp. 43–58. Edited by W. M. Scheld, D. Armstrong & J. M. Hughes. Washington, DC: American Society for Microbiology.

Nichol, S. T., Spiropoulou, C. F., Morzunov, S. & 7 other authors (1993). Genetic identification of a hantavirus associated with an outbreak of acute respiratory illness. *Science* **262**, 914–917.

Osborne, R. & Tarling, D. (1996). *The Historical Atlas of the Earth*. New York: Henry Holt.

Peters, C. J. (1997a). Emergence of viral hemorrhagic fevers: genetic sequences and social sequences. In *Vaccines 97*, pp. 81–86. Edited by F. Brown, D. Burton, P. Doherty, J. Mekalanos & E. Norrby. Plainview, NY: Cold Spring Harbor Laboratory.

Peters, C. J. (1997b). Emergence of Rift Valley fever. In *Factors in the Emergence of Arbovirus Diseases*, pp. 253–264. Edited by J. F. Saluzzo & B. Dodet. Paris: Elsevier.

Peters, C. J. (1998). Hantavirus pulmonary syndrome in the Americas. In *Emerging Infections 2*, pp. 17–64. Edited by W. M. Scheld, W. A. Craig & J. M. Hughes. Washington, DC: American Society for Microbiology.

Peters, C. J. (2000). Are hemorrhagic fever viruses practical agents for biological terrorism? In *Emerging Infections 4*, pp. 203–211. Edited by W. M. Scheld, W. A. Craig & J. M. Hughes. Washington, DC: American Society for Microbiology.

Peters, C. J. & LeDuc, J. W. (editors) (1999). An introduction to Ebola: the virus and the disease. *J Infect Dis* **179** (suppl. 1).

Peters, C. J. & Linthicum, K. J. (1994). Rift Valley fever. In *Handbook Series of Zoonoses, Section B: Viral Zoonoses*, 2nd edn, pp. 125–138. Edited by G. W. Beran. Boca Raton, FL: CRC Press.

Peters, C. J. & Zaki, S. R. (1999). Viral hemorrhagic fever: an overview. In *Tropical Infectious Diseases: Principles, Pathogens, & Practice*, pp. 1180–1188. Edited by R. L. Guerrant, D. H. Walker & P. F. Weller. New York: W. B. Saunders.

Peters, C. J., Jahrling, P. B. & Khan, A. S. (1996). Management of patients infected with high-hazard viruses: scientific basis for infection control. *Arch Virol Suppl* **11**, 1–28.

Pittman, P. R., Liu, C. T., Cannon, T. L., Makuch, R. S., Mangiafico, J. A., Gibbs, P. H. & Peters, C. J. (1999). Immunogenicity of an inactivated Rift Valley fever vaccine in humans: a 12-year experience. *Vaccine* **18**, 181–189.

Shope, R. E., Peters, C. J. & Davies, F. G. (1982). The spread of Rift Valley fever and approaches to its control. *Bull WHO* **60**, 299–304.

Strode, G. K. (editor) (1951). *Yellow Fever*. New York: McGraw-Hill.

Sugimoto, C., Kitamura, T., Guo, J. & 16 other authors (1997). Typing of urinary JC virus DNA offers a novel means of tracing human migrations. *Proc Natl Acad Sci U S A* **94**, 9191–9196.

Swanepoel, R., Leman, P. A., Burt, F. J., Zachariades, N. A., Braack, L. E. O., Ksiazek, T. G., Rollin, P. E., Zaki, S. R. & Peters, C. J. (1996). Experimental inoculation of plants and animals with Ebola virus. *Emerg Infect Dis* **2**, 321–325.

Tesh, R. B., Gajdusek, D. C., Garruto, R. M., Cross, J. H. & Rosen, L. (1975). The distribution and prevalence of group A arbovirus neutralizing antibodies among human populations in Southeast Asia and the Pacific Islands. *Am J Trop Med Hyg* **24**, 664–675.

Udvardy, M. D. F. (1969). *Dynamic Zoogeography with Special Reference to Land Animals*. New York: Van Nostrand Reinhold.

Ullstrup, A. J. (1972). Corn blight in the US. *Annu Rev Phytol* **10**, 27–56.

Vitousek, P. M., D'Antonio, C. M., Loope, L. L. & Westbrooks, R. (1996). Biological invasions as global environmental change. *Am Sci* **84**, 468–478.

White, A. C. (2000). The disappearing arsenal of antiparasitic drugs. *N Engl J Med* **343**, 1273–1274.

Zaki, S. R. & Goldsmith, C. S. (1999). Pathologic features of filovirus infections in humans. *Curr Top Microbiol Immunol* **235**, 97–116.

Zaki, S. R., Greer, P. W., Coffield, L. M. & 12 other authors (1995). Hantavirus pulmonary syndrome: pathogenesis of an emerging infectious disease. *Am J Pathol* **146**, 552–579.

Dynamics and epidemiological impact of microparasites

Bryan T. Grenfell

Zoology Department, University of Cambridge, Downing Street, Cambridge CB2 3EJ, UK

INTRODUCTION

The current and historical impact of infectious disease on human, animal and plant populations is immense. In addition, the relationship between host and pathogen has provided fertile ground for fundamental progress in population dynamics and evolutionary biology (Anderson & May, 1991). The impact of pathogens on host populations begs important questions on both population dynamic and evolutionary timescales.

The ultimate evolutionary questions here – notably, the coevolutionary trajectory of pathogen transmissibility and virulence – depend in turn on the nonlinear population dynamic interaction between host and pathogen (Van Baalen & Sabelis, 1995). In this chapter, we focus mainly on the population dynamic timescale – discussing the pattern of pathogen impact on human host populations arising from different pathogen life histories. The chapter focuses on microparasites, especially viruses and bacteria.

The impact of pathogens is a two-edged sword – pathogen population dynamics depends at least as much on the demographic and other characteristics of the host population as vice versa. To explore these issues, we therefore need a system where both host and pathogen dynamics are extremely well understood empirically and the natural history of infection and immunity is sufficiently straightforward to permit modelling.

The paradigm of these characteristics is provided by the great childhood viral and bacterial infections, notably measles. The population dynamics and natural history of

SGM symposium 60: New challenges to health: the threat of virus infection.
Editors G. L. Smith, W. L. Irving, J. W. McCauley, D. J. Rowlands. Cambridge University Press. ISBN 0 521 80614 3 ©SGM 2001.

measles are particularly well studied (Anderson & May, 1991; Black, 1984; Bolker & Grenfell, 1995; Cliff *et al.*, 1993; Grenfell & Harwood, 1997). Essentially, infection occurs in epidemics, which extinguish themselves as waves of infection immunize susceptible individuals. Epidemics tend to be episodic in small communities, though measles can be endemic in urban centres above a Critical Community Size (CCS) of 300–500 000 (Bartlett, 1960a). Before the vaccine era, measles epidemics in large, developed country communities often showed regular oscillations, both seasonally and at longer – typically biennial – major epidemic timescales (Anderson *et al.*, 1984).

The advent of mass vaccination in the late 1960s reduced both mean measles incidence and the relative amplitude and regularity of major epidemics (Anderson & May, 1991; Bolker & Grenfell, 1996; Cliff *et al.*, 1992). Measles is now at a very low level in the UK (Miller, 1996). However, the infection is still a major killer in developing countries and epidemics are still a significant public problem in developed countries (Babad *et al.*, 1995; Morse *et al.*, 1994). Recent experience in the USA with international movement of infection stresses, in particular, that we need to understand the spatial dynamics of infection and its interaction with vaccination (Bolker & Grenfell, 1996).

In the next section of the article, we give a graphical description of basic models for the population dynamics of measles. We then review recent work on the forces which drive spatio-temporal patterns of measles morbidity in developed countries and the impact of vaccination on the pattern of epidemics.

One limitation of these case studies is that they are only directly relevant to relatively acute infections, with relatively conserved and strong cross-immunity across the genetic diversity of the pathogen. The next section therefore extends the discussion to explore what the childhood disease case study can tell us about the impact of microparasites in general and in particular the dynamics of novel 'emergent' infections. We end with a synthesis of the chapter's conclusions.

POPULATION DYNAMICS OF CHILDHOOD INFECTIONS – OBSERVED PATTERNS

The data

Acute childhood infections such as measles generate characteristic recurrent epidemics in large communities. Probably the best illustration is provided by the historical pattern of measles morbidity for England and Wales. Fig. 1(a) shows the total notified weekly cases for England and Wales for the period 1944 (the start of national notification) to 1994, along with temporal changes in measles vaccine uptake; cases are square-root-transformed to clarify the shape of epidemic troughs. These data are also richly disaggregated spatially, with records for around 1400 rural and urban locations (Grenfell & Bolker, 1998). We illustrate the spatio-temporal dynamics of the system by plotting a contour map of measles notifications for 60 cities ranked by size (Fig. 1b). Inspection of this time series reveals four fairly distinct eras of spatio-temporal dynamics:

- 1944–1950: a mainly annual pattern of epidemic dynamics, with some de-synchronization of epidemics between cities.
- 1950s–1960s: strongly synchronized biennial epidemic cycles in most large cities; more irregular epidemics in smaller centres. Small towns also characteristically show 'fade-out' (local extinction) of infection in the troughs between major epidemics; by contrast, pre-vaccination measles persisted in large centres above a CCS of around 300 000 (Bartlett, 1960a).
- 1970s–1990s: the vaccine era showed declining incidence due to vaccination, with much more irregular, lower amplitude and spatially de-synchronized epidemic patterns (Babad *et al.*, 1995; Bolker & Grenfell, 1996; Fine & Clarkson, 1983). At the high level of recent vaccine uptake, cases become very sporadic and irregular and the incidence of mis-notification is much higher (Tait *et al.*, 1996).

More broadly, measles dynamics also depends on the recruitment rate of susceptible infants following the waning of maternal immunity – hence ultimately epidemic patterns depend on birth rate. Much the most important case is the situation in many high birth rate countries where measles is still a major killer. Here, the high recruitment rate of susceptibles generated characteristically annual epidemics in the pre-vaccination era (Mclean & Anderson, 1988a, b). Relatively *low* birth rates can also generate significant departures from the paradigm of biennial epidemics (Earn *et al.*, 2000).

THE SEIR MODEL FOR CHILDHOOD DISEASE DYNAMICS

Understanding the dynamical clockwork underlying these striking patterns has spawned a large body of distinguished work in both epidemiology and population biology (Kermack & McKendrick, 1927; Bartlett, 1956; Anderson & May, 1991; Black, 1984; Cliff *et al.*, 1993). The main conceptual tool has been a family of compartmental dynamical models, based on the SIR (**S**usceptible, **I**nfectious, **R**ecovered) paradigm. These successfully illustrate the basic dynamics of epidemics and the general impact of age structure and spatial heterogeneity (Anderson & May, 1991; Grenfell & Harwood, 1997; Schenzle, 1984).

The basic SEIR model embodies the following assumptions (Fig. 2a) (Black, 1984). After the waning of maternally derived passive immunity, infants move into the Susceptible class – note that, for measles, births and immigration are the major input to

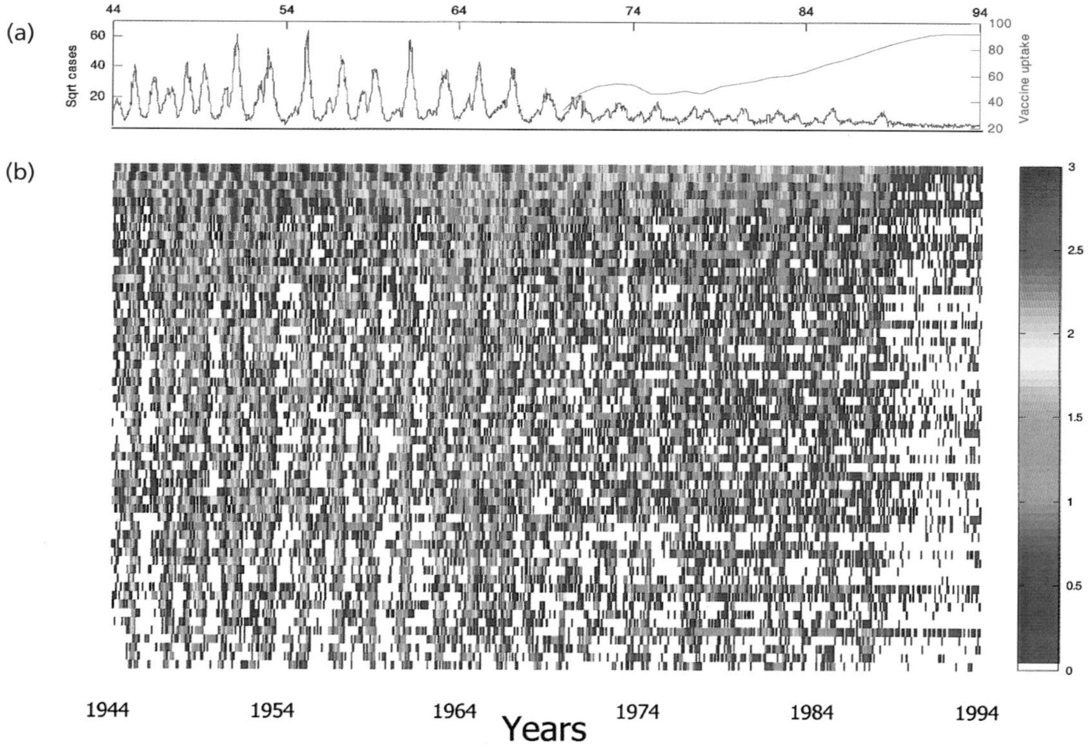

Fig. 1. Observed spatio-temporal dynamics of measles in England and Wales. (a) Observed weekly aggregate time series of notifications for London (blue) and measles vaccine uptake for England and Wales (red). (b) Measles dynamics for 60 towns and cities, in descending order, from London (top) to Teignmouth (bottom, pop. 11 000). Cases are colour-coded on a visual light scale (blue = few to red = many; scale = $\log_{10}$cases per week); white represents no reported cases. Reprinted with permission from Rohani *et al.* (1999). Copyright 1999 American Association for the Advancement of Science.

susceptibles in the population. Susceptibles then acquire infection by close contact with infectious individuals (mainly via respiratory aerosol for measles), entering the Exposed class. After incubating the infection for around a week, individuals in the model enter the Infectious class for about a week, then Recover. Recovered individuals are then assumed to be protected from further infection for life – note here that the model does not preclude *subclinical* infection which does not lead to retransmission of infection to others. In the simplest versions of the model, successful vaccination is also assumed to confer lifelong immunity.

Although these assumptions are very simplistic, the resulting family of SEIR models has proved very successful in capturing the dynamics of measles in the pre-vaccination era and qualitative changes in dynamics driven by mass vaccination. In the following section, we assess how well models capture the dynamical transitions noted above for measles in England and Wales.

MODELLING OBSERVED EPIDEMIC PATTERNS AND DYNAMICAL TRANSITIONS

The most straightforward dynamical case is the so-called 'simple epidemic', where we consider an epidemic in a closed population of susceptibles, ignoring births over the course of the epidemic. At the start of the epidemic (Fig. 2b), we assume that a single introduced infectious individual will produce a mean of R_0 secondary cases over their infectious period. R_0 is the well-known – not to say talismanic – *basic reproduction ratio* of infection: the number of secondary cases following the introduction of an infectious individual into a wholly susceptible population (Anderson & May, 1991). For measles, R_0 is around 18 (Anderson & May, 1991; Schenzle, 1984), so that our model epidemic takes off rapidly and approximately exponentially (Fig. 2). However, this also depletes the number of susceptibles, lowering the effective value of R_0. If – as for measles – the infectious period is short, the epidemic eventually declines and extinguishes itself. The simple epidemic thus illustrates a tendency for cyclical epidemic dynamics in short-duration diseases which follow the SEIR pattern. As discussed in a later section, epidemics are much less likely when these criteria are not met – in particular when the infectious period is relatively long.

Recurrent epidemics

In real populations, births will replenish the density of susceptibles following a major epidemic, generating the potential for *recurrent* epidemics, as so dramatically observed in historical measles records (Fig. 1). However, the basic SEIR model fails to capture the relatively stable pattern of violent recurrent epidemics. Instead (Fig. 3a), it shows a relatively violent epidemic following the introduction of infection into a population (corresponding to our simple epidemic), followed by a sequence of successively smaller

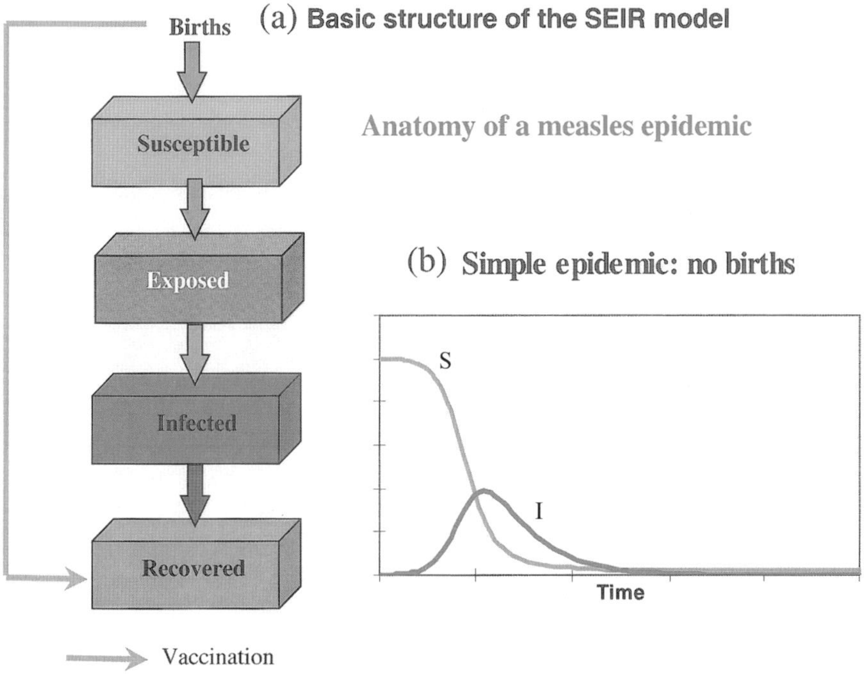

Fig. 2. (a) Basic flows of individuals captured by the SEIR model. (b) Schematic time series of a simple epidemic (see text) following the introduction of one infectious individual into a wholly susceptible population; the effect of birth rate on susceptible density is ignored. Line marked 'S', proportion of the population susceptible; line marked 'I', infected proportion.

epidemics, tending towards a steady endemic state. This well-known problem with the model (May, 1986) essentially arises because the upper limit on the density of susceptibles in a population of constant size acts as a density-dependent constraint on epidemic cycles.

This question of the maintenance of recurrent epidemics (Anderson & May, 1991; Bartlett, 1956) parallels similar controversies about population cycles in ecology. As is generally the case with models, the inadequacy of the SEIR model implies that an important biological process is missing from the formulation. The main missing ingredient is powerful seasonal 'forcing' of transmission, arising from the increased contact rate between children during school terms (Fine & Clarkson, 1982). In short, including seasonality can generate sustained cycles (Fig. 3b), essentially by resonating with the biennial epidemic tendency of major measles epidemics (Anderson & May, 1991; Bartlett, 1956; Dietz, 1976; Schenzle, 1984). High levels of seasonality can also generate complex and subtle epidemic patterns and even irregular chaotic epidemic time

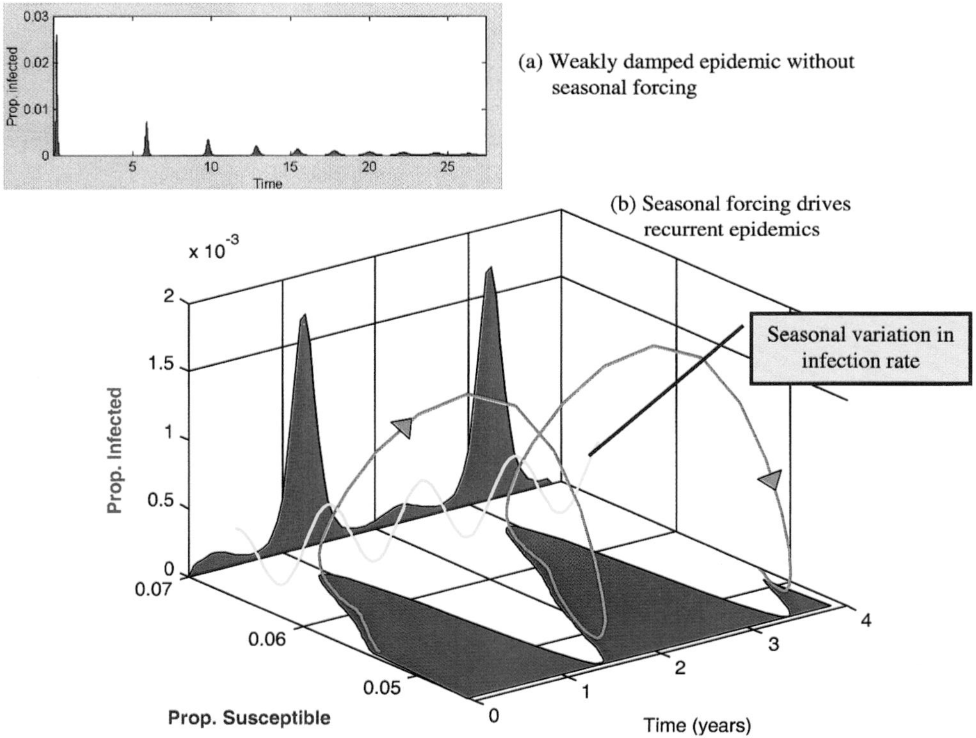

Fig. 3. (a) Simulated SEIR dynamics for measles with a birth rate, but in the absence of seasonal forcing of transmission rate; for full model specification and parameters, see Grenfell & Bolker (1998). (b) Same model, with sinusoidal seasonal forcing of the infection as shown (yellow – forcing amplitude set to 0.2; Grenfell & Bolker, 1998). The green trajectory shows the joint dynamics of susceptible and infectious densities through time; red and blue area plots show the dynamics of infectious and susceptible individuals. Reprinted from *Trends in Ecology and Evolution*, vol. 12, Grenfell, B. T. & Harwood, J., (Meta)population dynamics of infectious diseases, pp. 395–399, Copyright 1997, with permission from Elsevier Science.

series (Earn *et al.*, 2000; Ellner *et al.*, 1998; Grenfell, 1999; Grenfell *et al.*, 1994, 1995; Schaffer & Kot, 1985; Schwartz & Smith, 1983; Sugihara & May, 1990). The jury is still out on whether measles epidemics can be chaotic, though both regular cycles (see below) and more complex dynamics seem possible (Earn *et al.*, 1998).

Seasonally forced models can generate a remarkably accurate portrayal of measles dynamics (Bolker & Grenfell, 1993; Finkenstädt & Grenfell, 2000; Schenzle, 1984) (Fig. 4; see below for a further explanation). Variations across the epidemic in the age structure of epidemics are also important (Schenzle, 1984), though simple 'pulsed' seasonality captures the pattern of school terms relatively well (Earn *et al.*, 2000). In his seminal work on measles dynamics, Bartlett (1956) also observed that stochastic perturbations could also drive roughly biennial dynamics in measles (by throwing the system off its steady state and on to the biennial transient shown in Fig. 3a). Seasonality probably dominates the story here; however, Bartlett's work illustrates the potential importance of external perturbations.

Impact of birth rate variations and vaccination

This brings us to arguably the major source of external perturbations to measles dynamics – temporal variations in the recruitment rate of susceptibles due to secular variations in birth rate and/or vaccine uptake. The classical theory shows that mass vaccination acts to reduce the effective value of the basic reproduction ratio of infection – the vaccination proportion which reduces this quantity below unity provides an approximate criterion for disease elimination (Anderson & May, 1991). Recent work builds on this, to show that temporal changes in birth and vaccination rates act synonymously to affect measles dynamics (O. N. Bjørnstad and others, unpublished; Earn *et al.*, 2000; B. Finkenstädt and others, unpublished; Finkenstädt & Grenfell, 1998, 2000; Finkenstädt *et al.*, 1998). We see the major effects in Fig. 1(a). First, high birth rates (associated with the baby boom in 1944–1950) generate predominantly annual epidemics. Second, lower birth rates in the 1950s and 1960s led to biennial epidemics. This transition is analysed for London in Fig. 4, using a discrete time series version of the SEIR model – the TSIR model – which we can fit to observed epidemiological time series to estimate the seasonality in transmission and the time series of susceptible density (B. Finkenstädt and others, unpublished; Finkenstädt & Grenfell, 2000). Fig. 4 shows a projection of the model for the pre-vaccination era, starting from the observed disease incidence in 1944 and predicting the subsequent epidemic sequence, based on the observed birth rate (Fig. 4). Overall, the model captures the pattern of epidemics and the transition from annual to biennial epidemics very closely – essentially, higher birth rates decrease the inter-epidemic interval by replenishing susceptible density to the threshold for the next epidemic more quickly.

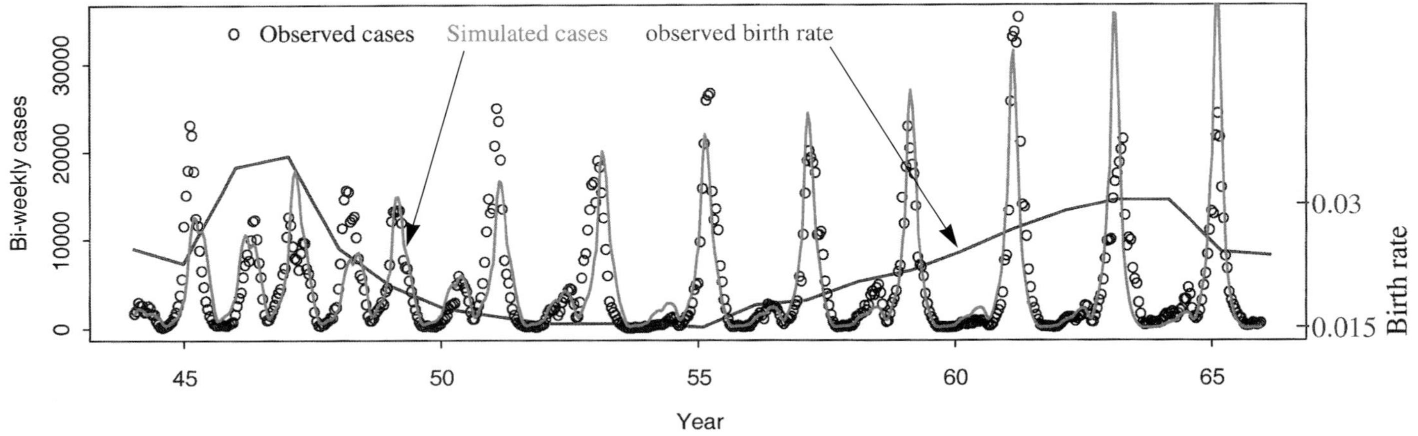

Fig. 4. Observed pre-vaccination measles dynamics for London, corrected for under-reporting, along with observed birth rate and predictions of an autoregressive time series version of the SEIR model, starting at the observed initial density of cases and susceptibles (for more details, see Finkenstädt & Grenfell, 2000).

In the vaccine era, classical theory predicts the reverse – vaccination reduces the 'birth rate' of susceptibles and should increase the epidemic interval (Anderson & May, 1991). Overall, this is borne out by the observed epidemics in England and Wales (Fig. 1; Anderson *et al.*, 1984). However, the interaction of vaccination and seasonality in transmission is complex – recent work indicates that vaccination against measles in developed countries is likely to have generated irregular and unpredictable sequences of epidemics (Earn *et al.*, 2000). This is likely to be a major cause of the irregular pattern of epidemics in most of the vaccine era in the UK (Fig. 1); corroboration is also provided by irregularities in measles epidemics in US cities in the 1930s, when birth rates were relatively low (Earn *et al.*, 2000). As discussed in the next section, these complexities have potentially profound implications for the spatio-temporal dynamics of childhood infections and the design of vaccination strategies.

Stochasticity, space and synchrony

So far, we have considered mainly the deterministic dynamics of measles, assuming that populations are large enough so that we can ignore *demographic stochasticity*: individual variations in the rate of disease transmission, infectious period, mortality rate, etc. This is a reasonable assumption for pre-vaccination epidemics in large cities (Fig. 4); however, in smaller populations, cases in the trough between epidemics may fall to very low numbers or even disappear. Such local extinctions were first analysed in detail by Bartlett (1956, 1957, 1960a, b, 1966); he demonstrated, both empirically and theoretically, an urban CCS of 300–500 000 below which measles frequently or always becomes extinct in the troughs between major epidemics. The pre-vaccination CCS for England and Wales is illustrated in Fig. 5. As population size falls below the CCS, the regular biennial pre-vaccination epidemics of measles seen in large cities such as London give way to more and more irregular epidemics with intervening periods of local disease extinction.

A recent body of work illustrates that stochastic versions of the standard measles models introduced above greatly overestimate the level of the CCS (Fig. 5; Bolker & Grenfell, 1995, 1996; Keeling & Grenfell, 1997, 1998). The problem arises because standard models assume that the variation between hosts in incubation and infectious periods follows an exponential distribution; thus for an incubation period of around a week, a substantial minority of individuals will have very long (e.g. 3 weeks) or very short (e.g. 1 day) periods. In fact, the empirical information indicates that these infection periods are generally much more tightly grouped, with a variation of only a few days around the mean (Hope Simpson 1952). Models of the exponential and relatively constant infection scenarios indicate that the constant infectious period produces a much higher level of measles persistence (lower CCS), given the same epidemic dynamics (Keeling & Grenfell, 1997). This dramatic difference arises because the exponential

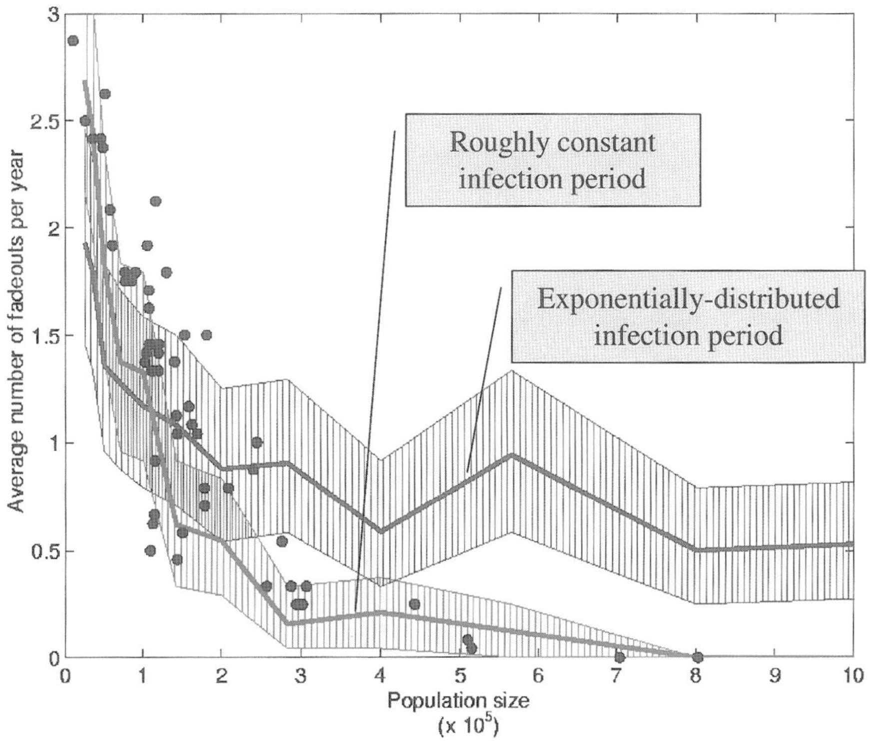

Fig. 5. Observed persistence (inversely measured by the number of local extinctions per year), as a function of population size. The curves (with 95 % confidence limits) show predictions based on different models for the variability of infectious period (for more details, see Keeling & Grenfell, 1997). Reprinted with permission from Keeling & Grenfell (1997). Copyright 1997 American Association for the Advancement of Science.

period generates a proportion of very short infectious periods, which break the chain of transmission in epidemic troughs, causing local extinction. Thus a subtle biological refinement of the SEIR model generates a marked increase in dynamical realism.

Disease persistence in the vaccine era. Standard stochastic theory (Griffiths, 1973) predicts that vaccination of an isolated population should increase the CCS and therefore decrease the probability that the disease will persist locally. Surprisingly, this is not the case. Fig. 1(b) illustrates this: the population threshold for disease persistence remained roughly constant for most of the vaccination era, until incidence became very low in the late 1980s (Bolker & Grenfell, 1996; Grenfell & Harwood, 1997). As discussed below, this effect probably arises from the emergent *metapopulation* dynamics of measles (Grenfell & Harwood, 1997).

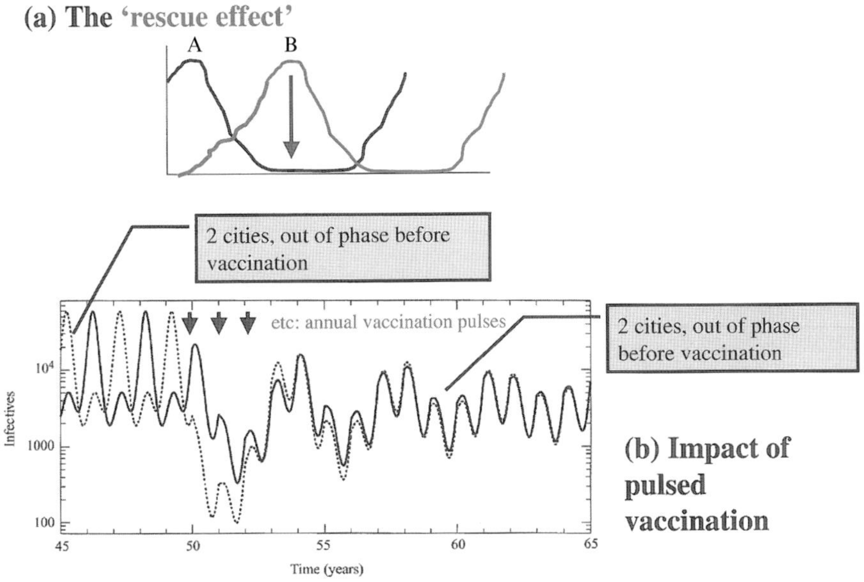

Fig. 6. (a) Schematic illustration of the 'rescue' effect, whereby an epidemic peak in one town might reduce the probability of extinction in the epidemic trough of a neighbouring (out-of-phase) centre. (b) Simulation of the spatial effect of pulsed vaccination. Before vaccination (which starts at time 50), two weakly coupled towns are out-of-phase. Pulses of vaccination at the time shown by the arrows then act to synchronize measles epidemics in the two centres, thus potentially mitigating the 'rescue effect' seen in (a). See Earn *et al.* (1998), from which this figure is reproduced with permission, for more details.

The spatial dynamics of measles is much studied in both epidemiology (see reviews by Anderson & May, 1991; Grenfell & Harwood, 1997; Mollison, 1995) and spatial geography (Cliff *et al.*, 1993). The unexpectedly high persistence of measles in the vaccine era noted above probably occurs because vaccination acts to *decorrelate* measles epidemics spatially, in both the UK (Fig. 1; Bolker & Grenfell, 1996) and the USA (Cliff *et al.*, 1992). In the pre-vaccination era, biennial epidemics of measles are in most cases highly correlated (Fig. 1): even small degrees of epidemiological coupling between large towns can cause epidemics to become synchronized (Bolker & Grenfell, 1996; Lloyd & May, 1996). As we have seen, vaccination produces much more irregular epidemics (Earn *et al.*, 2000), which are more prey to being decorrelated by stochastic fluctuations.

The likely relationship between decorrelation of epidemics and their local persistence is illustrated in Fig. 6(a). Consider two neighbouring towns whose measles epidemics are out of phase – the epidemic trough of one epidemic (A) will then coincide with the epidemic peak of the other (B). Movement of individuals between the two towns can then potentially reduce the probability of extinction in town 'A', compared to the case when the towns are synchronized (for a more systematic exploration of this hypothesis, see Bolker & Grenfell, 1996). This shoring up of the local chain of transmission by spatial diffusion of infection finds parallels in the 'rescue effect', which offsets local species extinction in ecological metapopulations (Earn *et al.*, 1998).

Implications for vaccination strategies. The ultimate aim of any vaccination campaign is to lower the effective reproduction ratio below unity. However, we have seen that, below this sterilizing level of herd immunity, mass vaccination campaigns against measles are likely to be reduced in efficacy by themselves generating local rescue effects, arising from the spatial decorrelation of epidemics. A solution to this problem would be a vaccination strategy which attempted to maintain the correlation of epidemics.

A potential candidate strategy would be to 'pulse' vaccination at regular intervals – each pulse would then tend to synchronize epidemics spatially. Pulse vaccination has already been explored in the theoretical epidemiological literature (Agur *et al.*, 1993; Nokes & Swinton, 1997; Shulgin *et al.*, 1998; Stone *et al.*, 2000), though mainly to optimize vaccine delivery. A preliminary exploration (Fig. 6b) of the spatial effects of pulsed vaccination was performed by Earn *et al.* (1998) – vaccination in pulses can synchronize previous out-of-phase epidemics, thus increasing the regional efficacy of the campaign.

Much more work needs to be done to refine these initial results. For instance, the most risk-averse strategy is likely to involve a combination of pulses and constant mass vaccination of infants; we need to know optimal pulse interval and how trans-national pulses (essentially across a host 'meta-metapopulation') should be applied for regional disease eradication. There is scope for important and exciting theoretical work here.

COMPARATIVE IMPACT AND DYNAMICS OF MICROPARASITES

A very fruitful area for future work is the *comparative dynamics* of disease with different natural histories. For example, recent work shows that pertussis epidemics show markedly different dynamics and response to vaccination to measles, even though the two infections are not hugely different in the overall natural history of transmission (Hethcote, 1997; Rohani *et al.*, 1999). Understanding the dynamics of mumps and rubella also presents challenges, especially given the complexities of protection of at-risk groups by vaccination (Anderson *et al.*, 1987; Anderson & May, 1983).

However, as discussed in the Introduction, acute childhood infections – with their relatively short periods of infection and (seemingly?) relatively stable long-term cross immunity against all strains of the pathogen – only account for part of the epidemiological

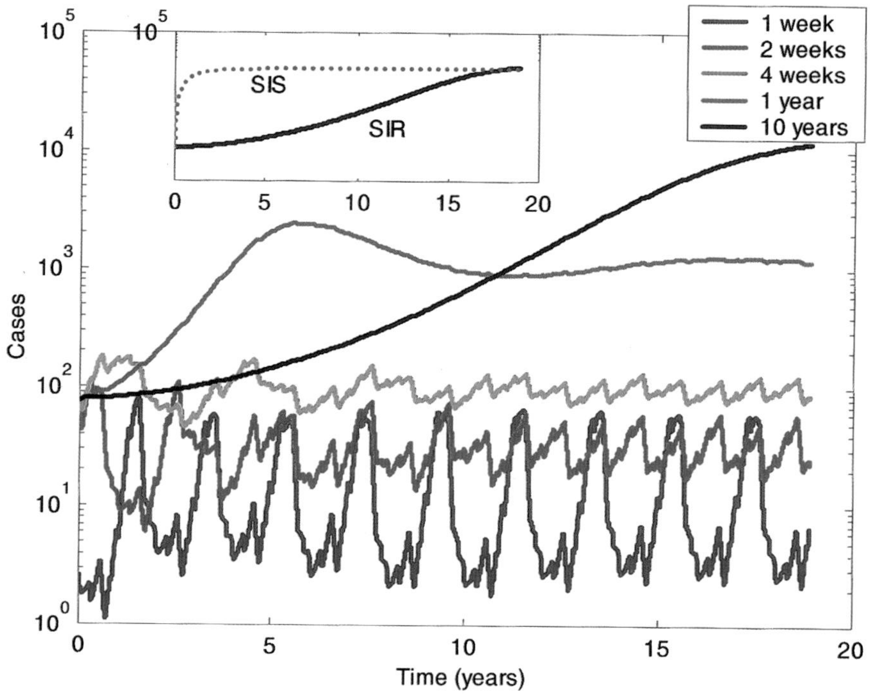

Fig. 7. (Main figure) Numerical solutions of seasonally forced SEIR models (see Fig. 3), showing changes in the dynamics of infection caused by increasing the infectious period (see colour key) while maintaining the basic reproduction ratio of infection at a constant level. To simulate crudely the initial dynamics of a 'novel' infection, the system starts by introducing 6 % infectives into a 20 % susceptible population (a real novel epidemic might be much more violent if everyone is susceptible). (Inset) Comparing the 'slow' dynamics generated by a 10 year infectious period with a (sketched) solution of an SIS model with much faster dynamics (see text).

story. Many infections are much more long-lasting and/or chronic (Anderson & May, 1991) and the dynamical complexities arising from parasite strain dynamics and its interactions with host immunity are a major topic in current analytical epidemiology (Gupta *et al.*, 1996, 1998).

As a bridge with the measles case study, Fig. 7 explores how increasing the infectious period from the measles timescale (a week) to that roughly of human immunodeficiency virus (a decade) affects the deterministic dynamics of infection and its interaction with seasonality in transmission. We assume that a small number of infectious individuals are introduced into a partly susceptible population and that incubation period stays constant. This is a crude and probably unrealistic assumption since incubation period may often scale with infectious period; however, it does not affect the qualitative dynamics of the system. Finally, we assume that R_0 is constant across simulations. Since R_0 is roughly the product of mean *per capita* infection rate and infectious period, this assumption means that:

- a short infectious period implies a relatively high infection rate
- a longer infectious period implies a lower infection rate to give the same R_0.

This aims to mimic a crude evolutionary constraint on the pathogens being compared: pathogens have finite resources which they can devote to some combination of transmission efficiency and maintaining infection in the face of host immunity.

Fig. 7 gives a dynamic illustration of basic steady state results for the SEIR model (Anderson & May, 1991). First, increasing the infectious period greatly augments the mean number of infectious individuals. This is an intuitive result: more individuals are infectious and less are in the recovered class. However, since R_0 is constant, the steady-state proportion of susceptibles will be constant for these different pathogens. Hence, the critical vaccination proportion to eradicate these infections will also be the same, even though the mean incidence is very different.

Dynamically, increasing the infectious period reduces the tendency for cyclical epidemics (May, 1986) – essentially, longer infectious periods 'fill in' troughs following major epidemics. As infectious period increases, we therefore see a transition from seasonally driven biennial epidemics (blue: the 'measles' extreme), via basically annual pre-vaccination epidemics (red green: roughly corresponding to the dynamics of pertussis; Rohani *et al.*, 1999). For long infectious periods, we see slow epidemics with a mild overshoot and trough following the epidemic (purple), and smooth, essentially 'logistic' epidemics with little signature of seasonality for very long infectious periods (black).

Fig. 7 shows how increasing infectious period effects a transition from 'fast' highly cyclical dynamics to 'slow' non-cyclical disease incidence. These differences in behaviour have implications for vaccination policy – even though the deterministic criterion of eradication of these infections is the same, the slow non-cyclic infections are much less prey to stochastic extinction, because the system is never driven into deep troughs by its internal dynamics.

These results illustrate one aspect of the comparative impact of microparasites. However, they give a rather crude picture, for three basic reasons.

(a) At the short infectious period end of the spectrum, the strong interaction between nonlinearity, seasonal forcing and stochasticity can give rise to a complex zoo of dynamical patterns [coexisting attractors, possible chaos, etc. (Earn *et al.*, 2000)]. As illustrated for measles above, the detailed dynamics of a given infection, as well as its response to vaccination, can depend subtly on its exact incubation and infection period and the seasonal pattern of infection rate. For example, the multiannual periodicity of rubella epidemics is still something of a dynamical mystery (Anderson & Grenfell, 1986). The measles case study also illustrates how secular changes in extrinsic variables such as birth rate can affect the behaviour and predictability of epidemics.

(b) The SEIR paradigm is only one possibility for the natural history of infection and immunity. Its assumption of lifelong sterile immunity is at one end of an observed continuum of immunological effectiveness which includes essentially no immune class (e.g. the SIS model for gonorrhoea; Hethcote *et al.*, 1982), carrier states and other complications. Though the basic message of Fig. 7 is correct, the duration of immunity can introduce other dynamical features. For example (Fig. 7, inset), an infection with no immune state generates non-cyclical 'logistic' dynamics similar to our AIDS-like slow infection; however, the timescale of response of the SIS infection can be much quicker. An interesting example of a hybrid between fast and slow dynamics is provided by varicella-zoster – effectively a fast SEIR-type primary infection in children with a slow component from the recrudescence of zoster (Garnett & Grenfell, 1992a, b). Recent work indicates the interesting possibility that zoster recrudescence might dynamically alter varicella incidence, increasing its persistence (Ferguson *et al.*, 1996). If this can be proved to lead to an increase in the CCS for VZV persistence, it would indicate a very unusual combination of fast dynamics (which allows fast invasion of new host populations) and persistence in epidemic troughs.

(c) As noted in the Introduction, the crucial omission from these analyses is the genetic structure of the host and pathogen populations. For many systems, parasite strain structure, and the host immune response to it, will complicate many of the dynamical issues discussed above; however, recent and on-going work on dengue, influenza and other infections shows that the lessons learned from more genetically 'homogeneous' infections about cyclicity, persistence, synchrony, etc., provide a useful toolkit for understanding the added complexities of these systems (Ferguson *et al.*, 1999; Gupta *et al.*, 1998).

CONCLUSION

In summary, our measles case study shows that even one of the simplest and best understood host–microparasite interactions can generate complex and subtle dynamics, depending on how the basic epidemiological oscillator that produces epidemics interacts with stochasticity and various external forces such as seasonality, baby booms and vaccination. Biologically motivated modelling coupled to data analysis is a useful tool for disentangling these issues – apart from their fundamental interest, such models have been very useful in the design of vaccination strategies (Anderson & May, 1991; Babad *et al.*, 1995; Gay *et al.*, 1995).

The major challenge now is to synthesize this population dynamical work with emerging understanding of the genetical dynamics of host–parasite interactions. Such work is of great fundamental interest, but is also important from an applied perspective, given the threat of emerging infections. As illustrated in Fig. 1, the basic dynamical characteristics of a 'novel' epidemic can be crudely understood in terms of relatively few population parameters. However, the subsequent persistence, spread and impact of the pathogen on the host population and the outcome of coevolution of host and parasite will be dependent on a subtle interaction between deterministic and stochastic forces in the host metapopulation, in concert with the population genetic interaction of hosts and parasites. For instance, strains of highly 'epidemic' SEIR-type pathogens could leave a selective impact on local host populations, but die out locally themselves – this would produce more of a skew towards evolutionary changes in the overall host population compared to the pathogen than we would expect from standard spatially aggregated models. There is much scope for exploring this and related questions in host–parasite metapopulation dynamics.

ACKNOWLEDGEMENTS

I thank Matt Keeling and Pej Rohani for helpful discussions and the provision of figures. This work was supported by the Wellcome Trust.

REFERENCES

Agur, Z., Cojocaru, L., Mazor, G., Anderson, R. M. & Danon, Y. L. (1993). Pulse mass measles vaccination across age cohorts. *Proc Natl Acad Sci U S A* **90**, 11698–11702.

Anderson, R. M. & Grenfell, B. T. (1986). Quantitative investigations of different vaccination policies for the control of congenital rubella syndrome (CRS) in the United Kingdom. *J Hyg* **96**, 305–333.

Anderson, R. M. & May, R. M. (1983). Vaccination against rubella and measles: quantitative investigations of different policies. *J Hyg* **90**, 259–325.

Anderson, R. M. & May, R. M. (1991). *Infectious Diseases of Humans: Dynamics and Control*. Oxford: Oxford University Press.

Anderson, R. M., Grenfell, B. T. & May, R. M. (1984). Oscillatory fluctuations in the incidence of infectious disease and the impact of vaccination: time series analysis. *J Hyg* **93**, 587–608.

Anderson, R. M., Crombie, J. A. & Grenfell, B. T. (1987). The epidemiology of mumps in the UK: a preliminary study of virus transmission, herd immunity and the potential impact of immunization. *Epidemiol Infect* **99**, 65–84.

Babad, H. R., Nokes, D. J., Gay, N. J., Miller, E., Morgan-Capner, P. & Anderson, R. M. (1995). Predicting the impact of measles vaccination in England and Wales – model validation and analysis of policy options. *Epidemiol Infect* **114**, 319–344.

Bartlett, M. S. (1956). Deterministic and stochastic models for recurrent epidemics. In *Proceedings of the Third Berkeley Symposium on Mathematical Statistics and Probability*, vol. 4, pp. 81–109. Edited by J. Neyman. Berkeley: University of California Press.

Bartlett, M. S. (1957). Measles periodicity and community size. *J R Stat Soc Ser A* **120**, 48–70.

Bartlett, M. S. (1960a). The critical community size for measles in the U.S. *J R Stat Soc Ser A* **123**, 37–44.

Bartlett, M. S. (1960b). *Stochastic Population Models in Ecology and Epidemiology*. London: Methuen.

Bartlett, M. S. (1966). *An Introduction to Stochastic Processes, with Special Reference to Methods and Applications*. Cambridge: Cambridge University Press.

Black, F. L. (1984). Measles. In *Viral Infections of Humans: Epidemiology and Control*, pp. 397–418. Edited by A. S. Evans. New York: Plenum.

Bolker, B. M. & Grenfell, B. T. (1993). Chaos and biological complexity in measles dynamics. *Proc R Soc Lond B Biol Sci* **251**, 75–81.

Bolker, B. M. & Grenfell, B. T. (1995). Space, persistence and the dynamics of measles epidemics. *Philos Trans R Soc Lond B Biol Sci* **348**, 309–320.

Bolker, B. M. & Grenfell, B. T. (1996). Impact of vaccination on the spatial correlation and dynamics of measles epidemics. *Proc Natl Acad Sci U S A* **93**, 12648–12653.

Cliff, A. D., Haggett, P., Stroup, D. F. & Cheney, E. (1992). The changing geographical coherence of measles morbidity in the United States, 1962–88. *Stat Med* **11**, 1409–1424.

Cliff, A. D., Haggett, P. & Smallman-Raynor, M. (1993). *Measles: an Historical Geography of a Major Human Viral Disease from Global Expansion to Local Retreat, 1840–1990*. Oxford: Blackwell.

Dietz, K. (1976). The incidence of infectious diseases under the influence of seasonal fluctuations. *Lect Notes Biomath* **11**, 1–15.

Earn, D., Rohani, P. & Grenfell, B. T. (1998). Spatial dynamics and persistence in ecology and epidemiology. *Proc R Soc Lond B Biol Sci* **265**, 7–10.

Earn, D. J. D., Rohani, P., Bolker, B. M. & Grenfell, B. T. (2000). A simple model for complex dynamical transitions in epidemics. *Science* **287**, 667–670.

Ellner, S. P., Bailey, B. A., Bobashev, G. V., Gallant, A. R., Grenfell, B. T. & Nychka, D. W. (1998). Noise and nonlinearity in measles epidemics: combining mechanistic and statistical approaches to population modelling. *Am Nat* **151**, 425–440.

Ferguson, N. M., Anderson, R. M. & Garnett, G. P. (1996). Mass vaccination to control chickenpox – the influence of zoster. *Proc Natl Acad Sci U S A* **93**, 7231–7235.

Ferguson, N., Anderson, R. & Gupta, S. (1999). The effect of antibody-dependent enhancement on the transmission dynamics and persistence of multiple-strain pathogens. *Proc Natl Acad Sci U S A* **96**, 790–794.

Fine, P. E. M. & Clarkson, J. A. (1982). Measles in England and Wales – I: an analysis of factors underlying seasonal patterns. *Int J Epidemiol* **11**, 5–15.

Fine, P. E. M. & Clarkson, J. A. (1983). Measles in England and Wales – III: assessing published predictions of the impact of vaccination on incidence. *Int J Epidemiol* **12**, 332–339.

Finkenstädt, B. F. & Grenfell, B. T. (1998). Empirical determinants of measles metapopulation dynamics in England and Wales. *Proc R Soc Lond B Biol Sci* **265**, 211–220.

Finkenstädt, B. F. & Grenfell, B. T. (2000). Time series modelling of childhood diseases: a dynamical systems approach. *J R Stat Soc Ser C* **49**, 187–205.

Finkenstädt, B. F., Keeling, M. J. & Grenfell, B. T. (1998). Patterns of density dependence in measles dynamics. *Proc R Soc Lond B Biol Sci* **265**, 753–762.

Garnett, G. P. & Grenfell, B. T. (1992a). The epidemiology of varicella-zoster virus infections: a mathematical model. *Epidemiol Infect* **108**, 495–511.

Garnett, G. P. & Grenfell, B. T. (1992b). The epidemiology of varicella-zoster virus: the influence of varicella on the prevalence of herpes-zoster. *Epidemiol Infect* **108**, 513–528.

Gay, N. J., Hesketh, L. M., Morgan-Capner, P. & Miller, E. (1995). Interpretation of serological surveillance data for measles using mathematical-models – implications for vaccine strategy. *Epidemiol Infect* **115**, 139–156.

Grenfell, B. T. (1999). Measles as a testbed for characterising nonlinear behaviour in ecology. In *Chaos from Real Data: the Analysis of Non-linear Dynamics in Short Ecological Time Series*. Edited by J. N. Perry & R. Smith. Dordrecht: Kluwer.

Grenfell, B. T. & Bolker, B. M. (1998). Cities and villages: infection hierarchies in a measles metapopulation. *Ecol Lett* **1**, 63–70.

Grenfell, B. T. & Harwood, J. (1997). (Meta)population dynamics of infectious diseases. *Trends Ecol Evol* **12**, 395–399.

Grenfell, B. T., Kleczkowski, A., Ellner, S. P. & Bolker, B. M. (1994). Measles as a case-study in nonlinear forecasting and chaos. *Philos Trans R Soc Lond Ser A Phys Sci Eng* **348**, 515–530.

Grenfell, B. T., Bolker, B. M. & Kleczkowski, A. (1995). Seasonality and extinction in chaotic metapopulations. *Proc R Soc Lond B* **259**, 97–103.

Griffiths, D. A. (1973). The effect of measles vaccination on the incidence of measles in the community. *J R Stat Soc Ser A* **136**, 441–449.

Gupta, S., Maiden, M. C. J., Feavers, I. M., Nee, S., May, R. M. & Anderson, R. M. (1996). The maintenance of strain structure in populations of recombining infectious agents. *Nat Med* **2**, 437–442.

Gupta, S., Ferguson, N. & Anderson, R. (1998). Chaos, persistence, and evolution of strain structure in antigenically diverse infectious agents. *Science* **280**, 912–915.

Hethcote, H. W. (1997). An age-structured model for pertussis transmission. *Math Biosci* **145**, 89–136.

Hethcote, H. W., Yorke, J. A. & Nold, A. (1982). Gonorrhea modeling – a comparison of control methods. *Math Biosci* **58**, 93–109.

Hope Simpson, R. E. (1952). Infectiousness of communicable diseases in the household. *Lancet* ii, 549–554.

Keeling, M. J. & Grenfell, B. T. (1997). Disease extinction and community size: modeling the persistence of measles. *Science* **275**, 65–67.

Keeling, M. J. & Grenfell, B. T. (1998). Effect of variability in infection period on the persistence and spatial spread of infectious diseases. *Math Biosci* **147**, 207–226.

Kermack, W. O. & McKendrick, A. G. (1927). A contribution to the mathematical theory of epidemics. *Proc R Soc Lond Ser A* **115**, 700–721.

Lloyd, A. L. & May, R. M. (1996). Spatial heterogeneity in epidemic models. *J Theor Biol* **179**, 1–11.

Mclean, A. R. & Anderson, R. M. (1988a). Measles in developing countries. Part I. Epidemiological parameters and patterns. *Epidemiol Infect* **100**, 111–133.

Mclean, A. R. & Anderson, R. M. (1988b). Measles in developing countries. Part II. The predicted impact of mass vaccination. *Epidemiol Infect* **100**, 419–442.

May, R. M. (1986). Population biology of microparasitic infections. *Biomathematics* **17**, 405–442.

Miller, E. (1996). Immunisation policies – successes, failures and the future. *J Clin Pathol* **49**, 620–622.

Mollison, D. (editor) (1995). *Epidemic Models: their Structure and Relation to Data*. Cambridge: Cambridge University Press.

Morse, D., Oshea, M., Hamilton, G., Soltanpoor, N., Leece, G., Miller, E. & Brown, D. (1994). Outbreak of measles in a teenage school population – the need to immunize susceptible adolescents. *Epidemiol Infect* **113**, 355–365.

Nokes, D. J. & Swinton, J. (1997). Vaccination in pulses: a strategy for global eradication of measles and polio? *Trends Microbiol* **5**, 14–19.

Rohani, P., Earn, D. J. D. & Grenfell, B. T. (1999). Opposite patterns of synchrony in sympatric disease metapopulations. *Science* **286**, 968–971.

Schaffer, W. M. & Kot, M. (1985). Nearly one dimensional dynamics in an epidemic. *J Theor Biol* **112**, 403–427.

Schenzle, D. (1984). An age-structured model of pre- and post-vaccination measles transmission. *IMA J Math Appl Med Biol* **1**, 169–191.

Schwartz, I. B. & Smith, H. L. (1983). Infinite subharmonic bifurcation in an SEIR model. *J Math Biol* **18**, 233–253.

Shulgin, B., Stone, L. & Agur, Z. (1998). Pulse vaccination strategy in the SIR epidemic model. *Bull Math Biol* **60**, 1123–1148.

Stone, L., Shulgin, B. & Agur, Z. (2000). Theoretical examination of the pulse vaccination policy in the SIR epidemic model. *Math Comput Model* **31**, 207–215.

Sugihara, G. & May, R. M. (1990). Nonlinear forecasting as a way of distinguishing chaos from measurement error in time series. *Nature* **344**, 734–741.

Tait, D. R., Ward, K. N., Brown, D. W. G. & Miller, E. (1996). Measles and rubella misdiagnosed in infants as Exanthem subitum (roseola infantum). *Br Med J* **312**, 101–102.

Van Baalen, M. & Sabelis, M. W. (1995). The dynamics of multiple infection and the evolution of virulence. *Am Nat* **146**, 881–910.

The continuing threat of bunyaviruses and hantaviruses

Richard M. Elliott

Division of Virology, Institute of Biomedical and Life Sciences, University of Glasgow, Church Street, Glasgow G11 5JR, UK

INTRODUCTION

The family *Bunyaviridae* is one of the largest taxonomic groupings of RNA viruses, containing well over 300 named isolates (Elliott *et al.*, 2000). Among the unifying characteristics of these viruses is possession of a tri-segmented single-stranded RNA genome of negative- or ambi-sense polarity which encodes four structural proteins. The three genome segments [called L (large), M (medium) and S (small)] are encapsidated by the nucleocapsid (N) protein and are associated with the viral RNA-dependent RNA polymerase, the L protein, to form ribonucleoprotein complexes (RNPs) termed nucleocapsids. The three RNPs are contained within a lipid envelope into which are embedded two viral glycoproteins, G1 and G2 (Fig. 1). When viewed by cryoelectron microscopy, the viruses are spherical and about 100 nm in diameter, though smaller particles, suggested to contain less than three RNPs, are also seen (Talmon *et al.*, 1987). Virus replication occurs in the cytoplasm of infected cells, and viruses mature primarily by budding at Golgi membranes (see relevant chapters in Elliott, 1996, for review).

The *Bunyaviridae* are classified into five genera – *Bunyavirus*, *Hantavirus*, *Nairovirus*, *Phlebovirus* and *Tospovirus* – on the basis of serological and molecular characteristics (Elliott *et al.*, 2000). Within a genus, viruses show similar patterns in the sizes of their genome segments (Table 1) and structural proteins (Table 2), and whether or not non-structural proteins are also encoded. Viruses which impinge on human health, either directly by causing human disease or indirectly by causing economic losses of domestic animals or crop plants, are found in each of the five genera; some examples are given in

SGM symposium 60: New challenges to health: the threat of virus infection.
Editors G. L. Smith, W. L. Irving, J. W. McCauley, D. J. Rowlands. Cambridge University Press. ISBN 0 521 80614 3 ©SGM 2001.

Table 1. Pattern of *Bunyaviridae* genome segment sizes (kb)

RNA segment	Genus				
	Bunyavirus	*Hantavirus*	*Nairovirus*	*Phlebovirus*	*Tospovirus*
L	6.9	6.5	12.2	6.4	8.8
M	4.5	3.6	4.9	3.2–3.9	4.8
S	0.9	1.7–2.0	1.7	1.7–1.9	2.9

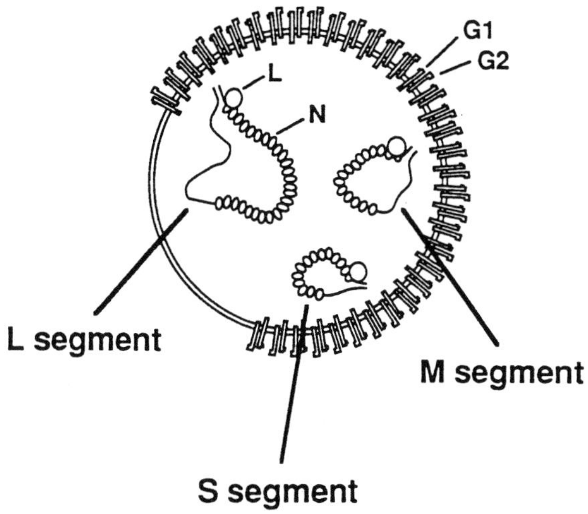

Fig. 1. Schematic of a *Bunyaviridae* particle (courtesy of Dr R. F. Pettersson).

Table 3. For the purpose of this chapter, further discussion is restricted to viruses in the *Bunyavirus* and *Hantavirus* genera.

GENOME CODING AND REPLICATION STRATEGIES

The coding strategies of bunyaviruses and hantaviruses are compared in Fig. 2. For both genera, all three RNA segments encode proteins in a negative-sense manner, and share with the rest of the family the common coding assignment of the structural proteins: L RNA encodes the L protein, M RNA encodes the two glycoproteins and S RNA encodes the N protein. There are differences in the sizes of the glycoproteins and N protein between the genera (Table 2) – for hantavirus, G1 and G2 are fairly similar in size, whereas in bunyaviruses, G1 is considerably larger than G2, and the hantavirus N protein (about 50 kDa) is twice the size of the bunyavirus counterpart. Bunyaviruses encode two non-structural proteins, called NSm and NSs, respectively, in the M and S segments, whereas hantaviruses do not encode any non-structural proteins.

Table 2. Pattern of *Bunyaviridae* protein sizes (kDa)

RNA protein	Genus				
	Bunyavirus	*Hantavirus*	*Nairovirus*	*Phlebovirus*	*Tospovirus*
L segment					
L	259–263	246–247	459	238–241	330–332
M segment					
G1	108–120	68–76	72–84	55–75	72–78
G2	29–41	52–58	30–45	50–70	52–58
NSm*	15–18	None	78–85, 92–115	None or 78	34
S segment					
N	19–25	50–54	48–54	24–30	52
NSs†	10–13	None	None	29–31	29

*NSm, non-structural protein encoded by M segment.

†NSs, non-structural protein encoded by S segment.

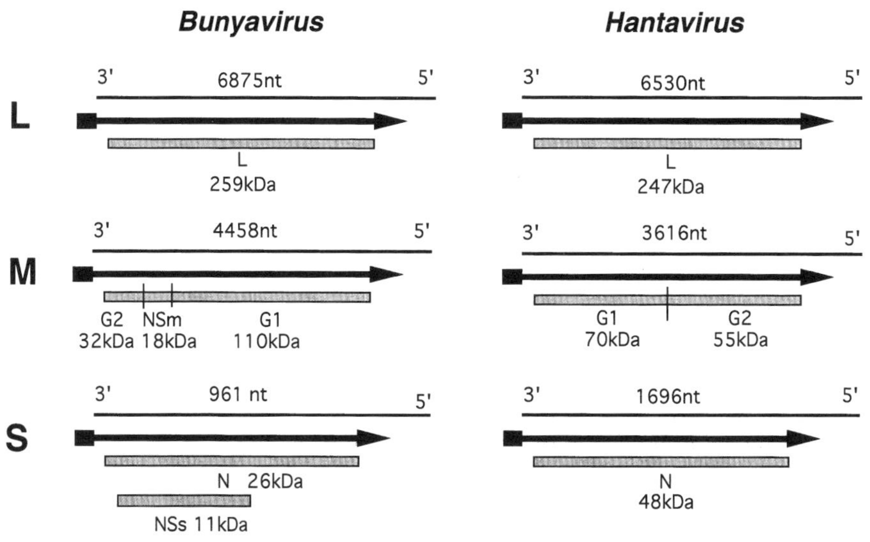

Fig. 2. Comparison of the coding strategies of bunyaviruses and hantaviruses. Genomic RNAs are represented by thin lines (size in nucleotides given above) and mRNAs as thick arrows (■, host-derived primer sequence; ▶, 3′ end). Gene products are shown as stippled boxes, with sizes given for mature proteins. The examples shown are for Bunyamwera and Hantaan viruses.

Table 3. Significant pathogens in the family *Bunyaviridae*

Genus/virus	Host: disease*	Vector	Distribution
Bunyavirus			
Oropouche	Human: fever	Midge	S. America
La Crosse	Human: encephalitis	Mosquito	N. America
Tahyna	Human: fever	Mosquito	Europe
Akabane	Cattle: abortion and congenital defects	Midge	Africa, Asia, Australia
Cache Valley	Sheep, cattle: congenital defects	Mosquito	N. America
Hantavirus			
Hantaan-like viruses	Human: severe haemorrhagic fever with renal syndrome, HFRS (F = 5–15 %)	Field mouse	Eastern Europe, Asia
Seoul-like viruses	Human: moderate HFRS (F = 1 %)	Rat	Worldwide
Puumala-like viruses	Human: mild HFRS, nephropathia epidemica (F = 0.1 %)	Bank vole	Western Europe
Sin Nombre-like viruses	Human: hantavirus pulmonary syndrome (F = 50 %)	Deer mouse	N. and S. America
Nairovirus			
Crimean–Congo haemorrhagic fever	Human: haemorrhagic fever (F = 20–80 %)	Tick	Eastern Europe, Africa, Asia
Nairobi sheep disease	Sheep, goats: fever, haemorrhagic gastroenteritis, abortion	Tick	E. Africa
Phlebovirus			
Rift Valley fever	Human: encephalitis, haemorrhagic fever, retinitis (F = 1–10 %)	Mosquito	Africa
	Domestic ruminants: necrotic hepatitis, haemorrhage, abortion		
Sandfly fever Naples, sandfly fever Sicilian	Human: fever	Phlebotomine fly	Europe, Africa, Asia
Tospovirus			
Tomato spotted wilt virus	>650 plant species: various symptoms	Thrips	Worldwide

*F, Case fatality rate.

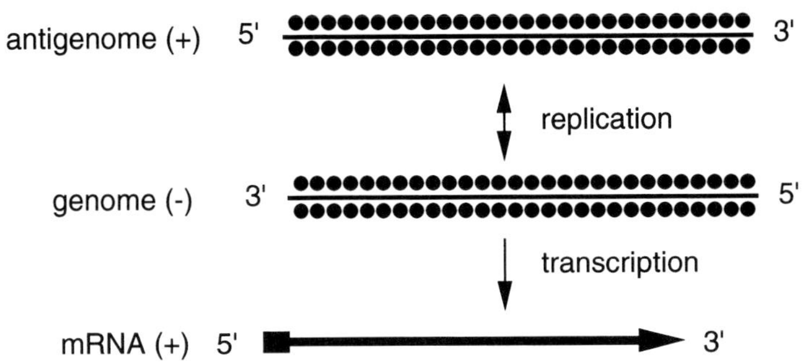

Fig. 3. Transcription and replication. The genome and antigenome are only found encapsidated by N protein (●) in the form of ribonucleocapsid complex. Viral mRNAs contain a host-derived primer sequence (■) at their 5′ ends and are truncated at their 3′ ends (▶) relative to the genomic RNA template; the mRNAs are not polyadenylated.

The glycoproteins are encoded as a precursor protein which is co-translationally cleaved to yield the mature products. In the bunyavirus precursor, NSm is sandwiched between G2 and G1. The designation G1 and G2 is based on the relative electrophoretic mobility of the glycoproteins, the larger protein being called G1. It has been shown that the bunyavirus G2 is functionally equivalent to hantavirus G1 in containing a Golgi targeting/retention signal. Note also that these proteins are located at the N-termini of their respective precursors. G1 and G2 form a heterodimer which is directed to the Golgi by the signal contained in just one of the partners. In the absence of the Golgi-signal-containing glycoprotein, the other glycoprotein is unable to exit the endoplasmic reticulum. The bunyavirus S segment mRNA is bicistronic, and is translated to give N and NSs from alternate AUG initiation codons. The NSs ORF is entirely within the N ORF, but in the +1 frame with respect to N, and therefore there is no amino acid sequence similarity between the two proteins (Elliott, 1996).

Bunyavirus transcription resembles that of influenza virus, although occurring in the cytoplasm, in that mRNA synthesis is primed by cap-containing oligonucleotides derived from host cell mRNAs (Fig. 3). The primers are generated by a viral endonuclease, a function of the L protein, and incorporated into the viral mRNAs, which therefore have non-templated heterogeneous 5′ ends. Viral mRNAs terminate at a signal 50–150 bases before the end of the template. Replication involves the production of a full-length positive-strand RNA, the antigenome, whose synthesis is initiated in a primer-independent manner, and which reads through the transcription termination signal. The antigenome RNA is also encapsidated by N protein in the form of a RNP complex. Hence in infected cells three types of RNA representing each segment are found: full-length negative-sense genomic RNA, full-length positive-sense antigenome

RNA, and positive-sense mRNA which has 5′ non-templated primer sequences (12–18 bases) and is truncated at the 3′ end (for review, see Kolakofsky & Hacker, 1991).

EVOLUTIONARY POTENTIAL OF BUNYAVIRUSES AND HANTAVIRUSES

For segmented genome RNA viruses, the principal modes of evolution are genetic drift, the accumulation of point mutations, insertions or deletions, and genetic shift, the exchange of genome segments (reassortment). Until the emergence of hantavirus pulmonary syndrome (HPS) in 1993, there had been relatively few studies of genetic drift among the *Bunyaviridae*. Studies on field isolates of La Crosse bunyavirus showed that no two isolates had identical RNA oligonucleotide fingerprint patterns (El Said *et al.*, 1979; Klimas *et al.*, 1981). In contrast, nucleotide sequence determination of the M genome segments of three human La Crosse virus isolates, obtained from the brains of children who died in 1960, 1978 and 1993, showed overall only 51 nucleotide changes out of 4526 over the 33-year period (Huang *et al.*, 1995, 1997). The implications of this genetic stability suggest that only some strains of La Crosse virus are virulent for humans.

Following the identification of Sin Nombre virus as a causative agent of HPS, many studies on hantavirus genetic variation and phylogeny have been reported. These studies, together with the few earlier ones, all indicate that genetic drift seems to be the main mechanism of hantavirus evolution, and that hantaviruses co-evolve with their primary rodent host (Nichol *et al.*, 1996; Plyusnin *et al.*, 1996). On this basis, the hantavirus can be grouped into three main clades (Fig. 4) which correspond to the rodent subfamilies that are their primary hosts: Hantaan-, Seoul- and Dobrava-like viruses which are associated with the *Murinae* subfamily (for example, mice and rats); Puumala-, Tula- and Prospect Hill-like viruses with the *Arvicolinae* subfamily (for example, bank voles); and Sin Nombre-like viruses with the *Sigmodontinae* subfamily (for example, deer mice) (Hjelle *et al.*, 1995; Plyusnin *et al.*, 1996).

More dramatic antigenic changes in segmented genome viruses occur following segment reassortment, when progeny viruses from cells infected with two or more different parental viruses contain segments derived from the different parents. This has been most obviously demonstrated with the emergence of pandemic influenza viruses. Bunyavirus reassortment has been extensively studied in the laboratory and has been comprehensively reviewed by Pringle (1996). Evidence for reassortment in nature is accumulating (Akashi *et al.*, 1997). Reassortment is restricted to closely related bunyaviruses; it does not occur between viruses in different serogroups, and even within serogroups some viruses are apparently genetically incompatible. Reassortment can occur in dually infected mosquitoes, and reassortant viruses have been found in both the

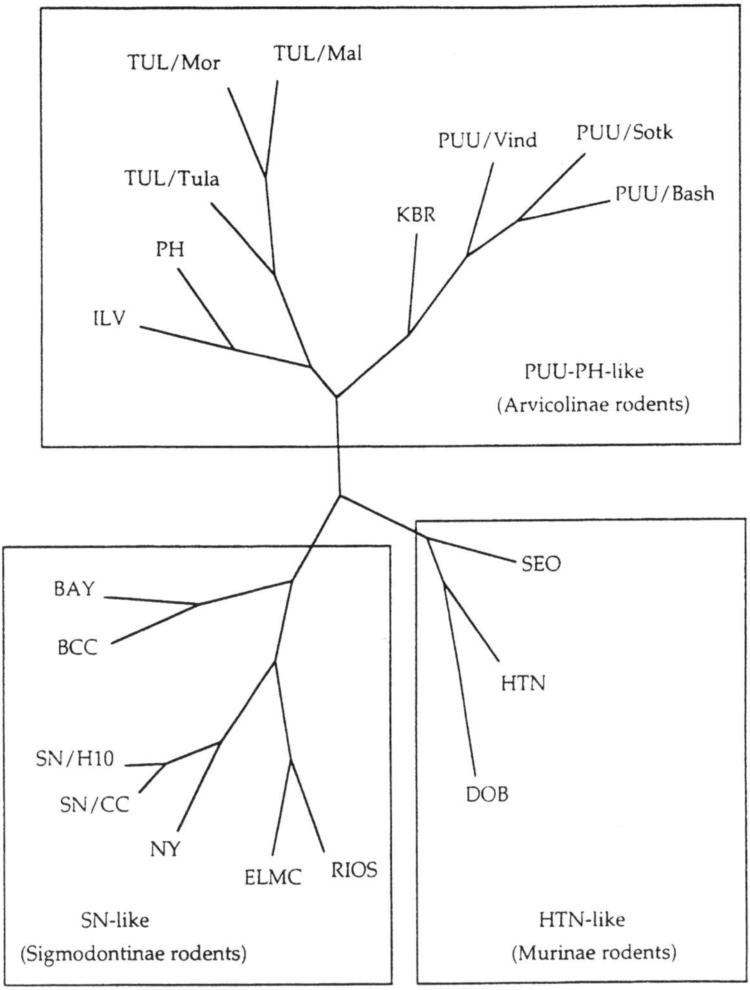

Fig. 4. Consensus phylogenetic tree of hantavirus based on the S segment sequences (coding region). BAY, Bayou virus, strain Louisiana (GenBank accession no. L36929); BCC, Black Creek Canal virus (L39949); DOB, Dobrava virus (L41916); ELMC, El Moro Canyon virus, strain RM-97 (U11427); HTN, Hantaan virus, strain 76118 (M14626); ILV, Isla Vista virus, strain MC-SB-1 (U31534); KBR, Khabarovsk virus, strain MF-43 (U35254); NY, New York virus, strain RI-1 (U09488); PUU/Sotk, Puumala virus, strain Sotkamo (X61035); PUU/Vind, Puumala virus, strain Vindeln/L20Cg/83 (Z48586); PUU/Bash, Puumala virus, strain Bashkiria/CG1820 (M32750); PH, Prospect Hill virus, strain PH-1 (Z49098); SN/H10, Sin Nombre virus, strain H10 (L25784); SN/CC, Sin Nombre virus, strain Convict Creek 107 (L33683); RIOS, Rio Segundo virus, strain RMx-Costa-1 (U18100); SEO, Seoul virus, strain SR-11 (M34882); TUL/Tula, TUL, strain Tula/76Ma/87 (Z30941); TUL/Mal, TUL, strain Malacky/Ma32/94 (Z48235); TUL/Mor, TUL, strain Moravia/5286Ma/94 (Z48573). The PHYLIP program package was used to make 500 bootstrap replicates of the sequence data (Seqboot). Distance matrices were calculated using Kimura's two-parameter model (Dnadist; ratio 2.0) and analysed by the Fitch–Margoliash tree fitting algorithm (Fitch) with global arrangements option set. The CONSENSE program was then used to calculate the consensus tree. Reproduced from Plyusnin *et al.* (1996), with permission.

salivary glands and ovaries. Indeed, Chandler *et al.* (1990) reported that replication and reassortment were enhanced in mosquito ovaries, and transovarial transmission of the reassortant viruses occurred in up to 10% of the progeny. In turn, the reassortant viruses could be transmitted to mice. It is suggested that reassortment in mosquitoes might be the potent driving force for evolution in nature, since a particular mosquito species could feed on multiple vertebrate species which each harboured a different bunyavirus. These laboratory studies are supported by genome analysis of bunyaviruses isolated from mosquitoes trapped in the field, where evidence for naturally occurring reassortants has been obtained (Ushijima *et al.*, 1981).

For hantaviruses, no evidence for reassortment between members of the three clades (Fig. 4) has been obtained. Further, the emergence of HPS-associated viruses was not due to a reassortment event creating a 'new' virus (Spiropoulou *et al.*, 1994; Chizhikov *et al.*, 1995). However, data supporting reassortment between isolates of the Sin Nombre virus group have been reported (Henderson *et al.*, 1995; Li *et al.*, 1995), and it has been suggested that such reassortants have increased virulence for humans (Schmaljohn *et al.*, 1995; Plyusnin *et al.*, 1996). Clearly these viruses have the genetic potential for rapid evolution and warrant continued surveillance.

RNA recombination is a widely observed phenomenon in positive-strand viruses though rare among negative-strand RNA viruses (Pringle, 1996). Evidence has been produced which indicates recombination as another mechanism for evolution of hantaviruses. Two distinct lineages of Tula hantaviruses have been described in Slovakia which show 19 % sequence divergence. Analysis of the S segment sequences of some Tula virus strains indicated that two recombination events had occurred sometime during their evolution, since the sequences contained regions derived from both lineages (Sibold *et al.*, 1999). The significance of these observations awaits further evaluation.

BIOLOGY OF BUNYAVIRUSES AND HANTAVIRUSES

Like the majority of viruses in the family, bunyaviruses are transmitted by arthropod vectors, in their case mosquitoes or midges. Frequently, a particular bunyavirus is associated with one or only a few vector species. Experiments with reassortant bunyaviruses have demonstrated that the principal determinants for permissive replication in a vector species – both to establish a disseminated infection and to be transmitted – map to the M RNA segment gene products, the two virion glycoproteins and NSm. Whether one or all of these gene products is involved has not been determined. The natural life cycles of bunyavirus involve a limited number of warm-blooded vertebrates which can act as amplifying hosts, and may aid in dissemination of virus through migration. Humans are usually regarded as dead hosts, and become infected when they venture into the natural habitat of the vector insect (Beaty & Calisher, 1991).

Viruses in the *Hantavirus* genus are distinct from all other members of the family in that they are not vectored by arthropods. Instead, hantaviruses are maintained in nature as persistent infections of rodents. Normally only a single rodent species harbours a particular hantavirus. Large quantities of virus are found in the lungs, saliva, urine and faeces of infected rodents, and humans become infected by inhaling aerosolized rodent excreta (Lee, 1996).

BUNYAVIRUS AND HANTAVIRUS DISEASE

Few bunyaviruses are known to be associated with human disease and most of those that infect humans cause a self-limiting, febrile illness that is rarely, if ever, fatal. Since these bunyavirus infections frequently occur in regions of the world, particularly Africa, scourged by malaria or other more serious diseases, diagnosis is rarely made and so the true incidence of bunyavirus infections is not known. In Europe, Tahyna bunyavirus is widely distributed – in Germany, Italy, the former Yugoslavia and the Czech Republic – and causes an unpleasant influenza-like illness, sometimes resulting in hospitalization of patients. In the US, La Crosse and a few related bunyaviruses, such as snowshoe hare, Jamestown Canyon, Inkoo and California encephalitis viruses, are a significant cause of paediatric viral encephalitis, often severe and occasionally fatal. However, the overall incidence is low – in the US there are, on average, about 50–75 cases of encephalitis per year, though the number of infections has been estimated as 1000 times higher (Gonzalez-Scarano & Nathanson, 1990).

A bunyavirus of increasing importance is Oropouche virus. The virus was originally recognized in Trinidad in 1955 in a survey for viruses causing human febrile illness (Anderson *et al.*, 1961). In 1962, an epidemic involving 11 000 people in Brazil was reported, and since then over 30 outbreaks in Brazil, Panama and Peru have occurred, involving up to 100 000 patients per epidemic. Clinically the disease is similar to dengue fever, and though not life threatening, is very debilitating and economically significant due to days lost from work. Oropouche virus is rather unusual for a bunyavirus in that it can be transmitted by vectors in different insect families, mosquitoes in the jungle and midges in an urban cycle that includes humans, which can act as an amplifying host. In the forest cycle, monkeys, sloths and perhaps birds are the vertebrate hosts (Le Duc & Pinheiro, 1989). The increase in Oropouche virus epidemics is presumably linked to the increasing insurgence by man into virgin rainforest for agricultural use (that is, clearing the forest) or road construction, thereby exposing humans to bites by infected insects.

Very recently, the disease potential of bunyaviruses has been grimly realized with the isolation of a new virus, termed Garissa virus, from humans with haemorrhagic fever (HF) during an outbreak of Rift Valley fever virus in East Africa. Two HF cases yielded the virus while RT-PCR studies have detected the virus in 14 other HF cases in Kenya and Somalia. What is interesting is that the virus is apparently a reassortant, having the L and S segments from Bunyamwera virus (a relatively innocuous virus for humans) but the M segment from an, as yet, unidentified bunyavirus parent. The L and S segments are almost identical in sequence to those of Bunyamwera whereas the M segment shows 33 % nucleotide and 26 % amino acid divergence from the Bunyamwera virus M segment (S. Nichol, personal communication).

Hantaviruses cause two major types of disease in humans: haemorrhagic fever with renal syndrome (HFRS) and HPS (Table 3). More than 20 antigenically distinct hantaviruses are recognized (Schmaljohn & Hjelle, 1997) and the list continues to grow. Hantaan, Dobrava, Seoul and Puumala viruses are responsible annually for over 200 000 cases of HFRS of varying severity in China and neighbouring countries in the Far East and in Europe (Lee, 1996). Hantavirus infections have frequently been associated with morbidity during military conflict; for example, 10 000 cases of nephropathia epidemica among German troops in Finnish Lapland during World War Two and 3000 cases of Korean HF among UN troops during the Korean War in the 1950s (in fact, the latter was the impetus leading to the isolation of Hantaan virus; Lee *et al.*, 1978). More recently, hantavirus infections have been reported in troops and civilians in war-torn Bosnia–Herzogovina. Dobrava and Puumala viruses cause HFRS in Europe, particularly in the Balkans. Further west, Puumala virus is widely distributed in FennoScandinavia and more than 10 000 cases of nephropathia epidemica are diagnosed annually in Finland (Hedman *et al.*, 1991). Other European hantaviruses such as Tula and Topografov viruses, isolated from voles and lemmings, respectively, are apparently not associated with human disease.

In 1993 in the southwestern United States, hantaviruses showed their ability to arise in a new environment with the dramatic appearance of a virus causing a severe respiratory illness, now called HPS, with about 50 % mortality. This episode is a classic example of an emerging virus, and the rapid identification of the responsible agent a superb illustration of the power of modern diagnostic techniques (Nichol *et al.*, 1993). The virus was eventually named Sin Nombre virus, and sparked considerable further 'virus-hunting' in the area. Many more new hantaviruses have been discovered throughout the Americas, a number of which are associated with HPS. More recently, evidence for direct human-to-human transmission of Andes virus, which causes HPS in Argentina, has been obtained (Wells *et al.*, 1997), adding a further dimension to the disease potential of hantaviruses.

BUNYAVIRUS NON-STRUCTURAL PROTEINS

One of the obvious molecular differences between bunyaviruses and hantaviruses is the lack of non-structural proteins encoded by the latter (Fig. 1; Table 3). One might speculate that the non-structural proteins could be associated with transmission of bunyaviruses by arthropods, though at present there is no experimental evidence to support this hypothesis. However, it appears that both NSm and NSs are accessory proteins that are dispensable for virus growth (at least in the laboratory).

The 16–18 kDa NSm protein of bunyaviruses is a non-glycosylated product which independently localizes to the Golgi of infected cells (Lappin *et al.*, 1994; Nakitare & Elliott, 1993), suggesting a role in virion maturation. However, recent data from analysis of variants of Maguari bunyavirus indicate that NSm may not actually be essential for growth of the virus in tissue culture. Non-temperature-sensitive (ts) revertants of a Maguari virus ts mutant which has the ts lesion in the G1 protein (Pringle, 1996; Murphy & Pringle, 1987) were recovered after passage at non-permissive temperature and shown to synthesize G1 proteins smaller (that is, had a faster electrophoretic mobility) than either the wild-type (wt) or ts parent. Two variants were selected for further study. Subsequent cDNA cloning and sequence determination showed that in one case the region encoding the N-terminus of G1 had an in-frame deletion, removing 238 aa, while in the other case 570 aa were removed, deleting most of NSm and a larger part of G1 (J. Q. Zhao, E. Pollitt & R. M. Elliott, unpublished). These viruses are only slightly impaired for growth in tissue culture; their behaviour in animals has yet to be determined.

The 11 kDa NSs protein encoded by Bunyamwera virus is also dispensable for virus growth but is involved in a number of functions during virus replication. Using the technology developed to recover infectious bunyavirus entirely from cDNAs (Bridgen & Elliott, 1996), a viable virus was produced lacking NSs (Bridgen *et al.*, 2001). This was achieved by introducing five point mutations into the S segment cDNA to knock out the NSs ORF. The transfectant virus, BUNdelNSs, had a small plaque phenotype and grew to levels of about 1 log lower than wt in BHK cells. Whereas wt Bunyamwera virus efficiently shuts off host cell protein synthesis, BUNdelNSs was significantly impaired in this regard, and in addition the mutant virus produced much larger amounts of the N protein. BUNdelNSs was able to kill mice following intracerebral inoculation but more slowly than wt, and immunohistochemical staining of the brains showed a slower spread of the BUNdelNSs virus compared to wt. Bunyamwera virus is highly sensitive to the antiviral state induced by interferon. However, the BUNdelNSs virus was shown to activate an interferon-β promoter, suggesting that one function of NSs is as an interferon antagonist.

Using a reconstituted system it has been shown that only the viral N and L proteins are required to transcribe and replicate a bunyavirus minigenome containing a reporter gene (Dunn *et al.*, 1995). Addition of NSs to the system specifically inhibited the viral polymerase in a dose-dependent manner. This inhibition could also be mediated by the NSs proteins of heterologous bunyaviruses (Weber *et al.*, 2001).

These results indicate the multifunctional nature of this bunyavirus small non-structural protein, and that some of these functions are shared with small non-structural proteins encoded by other negative-strand RNA viruses, for example influenza virus NS1. It is therefore intriguing that hantaviruses do not encode non-structural proteins to perform these activities. It has been suggested that the larger size of the N protein may accommodate extra activities but this remains speculation.

CONCLUSION

Recent outbreaks of human disease associated with members of the *Bunyaviridae* have heightened awareness of the disease potential of this diverse group of viruses, and their vast evolutionary potential. The appearance of HPS in 1993 and rapid identification of Sin Nombre virus played a major role in attracting attention – and funding – to the dramatic effects of emerging viruses on human health. Coupled with the association of a new bunyavirus, Garissa, with HF, these examples surely mandate the need for continued surveillance and further study. The development of a reverse genetic system for Bunyamwera virus (Dunn *et al.*, 1995; Bridgen & Elliott, 1996) has allowed new insights into the role of bunyavirus proteins, and might be exploited in the development of therapeutics and vaccines. Clearly there is an urgent need for similar technology to advance the study and eventual control of hantaviruses, but so far this has proved elusive.

ACKNOWLEDGEMENTS

Work in the author's laboratory is funded by the Medical Research Council and the Wellcome Trust.

REFERENCES

Akashi, H., Kaku, Y., Kong, X. & Pang, H. (1997). Antigenic and genetic comparisons of Japanese and Australian Simbu serogroup viruses: evidence for the recovery of natural virus reassortants. *Virus Res* **50**, 205–213.

Anderson, C. R., Spence, L., Downs, W. G. & Aitken, T. H. G. (1961). Oropouche virus: a new human disease agent from Trinidad, West Indies. *Am J Trop Med Hyg* **10**, 574–578.

Beaty, B. J. & Calisher, C. H. (1991). Bunyaviridae – natural history. *Curr Top Microbiol Immunol* **169**, 27–78.

Bridgen, A. & Elliott, R. M. (1996). Rescue of a segmented negative-strand RNA virus entirely from cloned complementary DNAs. *Proc Natl Acad Sci U S A* **93**, 15400–15404.

Bridgen, A., Weber, F., Fazakerley, J. K. & Elliott, R. M. (2001). Bunyamwera bunyavirus nonstructural protein NSs is a non-essential gene product that contributes to viral pathogenesis. *Proc Natl Acad Sci U S A* (in press).

Chandler, L. J., Beaty, B. J., Baldridge, G. D., Bishop, D. H. L. & Hewlett, M. J. (1990). Heterologous reassortment of bunyaviruses in *Aedes triseriatus* mosquitoes and transovarial and oral transmission of newly evolved genotypes. *J Gen Virol* **71**, 1045–1050.

Chizhikov, V. E., Spiropoulou, C. F., Morzunov, S., Monroe, M. C., Peters, C. J. & Nichol, S. T. (1995). Complete genetic characterisation and analysis of isolation of Sin Nombre virus. *J Virol* **69**, 8132–8136.

Dunn, E. F., Pritlove, D. C., Jin, H. & Elliott, R. M. (1995). Transcription of a recombinant bunyavirus RNA template by transiently expressed bunyavirus proteins. *Virology* **211**, 133–143.

Elliott, R. M. (editor) (1996). *The Bunyaviridae*. New York: Plenum.

Elliott, R. M., Bouloy, M., Calisher, C. H., Goldbach, R., Moyer, J. T., Nichol, S. T., Pettersson, R., Plyusnin, A. & Schmaljohn, C. S. (2000). Bunyaviridae. In *Virus Taxonomy. Seventh Report of the International Committee on Taxonomy of Viruses*, pp. 599–621. Edited by M. H. V. van Regenmortel, C. M. Fauquet, D. H. L. Bishop, E. B. Carsten, M. K. Estes, S. M. Lemon, J. Maniloff, M. A. Mayo, D. J. McGeoch, C. R. Pringle & R. B. Wickner. San Diego: Academic Press.

El Said, L. H., Vorndam, V., Gentsch, J. R., Clewley, J. P., Calisher, C. H., Klimas, R. A., Thompson, W. H., Grayson, M., Trent, D. W. & Bishop, D. H. L. (1979). A comparison of La Crosse virus isolates obtained from different ecological niches and an analysis of the structural components of California encephalitis serogroup viruses and other bunyaviruses. *Am J Trop Med Hyg* **28**, 364–386.

Gonzalez-Scarano, F. & Nathanson, N. (1990). Bunyaviruses. In *Virology*, 2nd edn, pp. 1195–1228. Edited by B. N. Fields & D. M. Knipe. New York: Raven Press.

Hedman, K., Vaheri, A. & Brummer-Korvenkontio, M. (1991). Rapid diagnosis of hantavirus disease with an IgG-avidity assay. *Lancet* **338**, 1353–1356.

Henderson, W. W., Monroe, M. C., St Jeor, S. C., Thayer, W. P., Rowe, J. E., Peters, C. J. & Nichol, S. T. (1995). Naturally occurring Sin Nombre virus genetic reassortants. *Virology* **214**, 602–610.

Hjelle, B., Jenison, S. A., Goade, D. E., Green, W. B., Feddersen, R. M. & Scott, A. A. (1995). Hantaviruses: clinical, microbiologic and epidemiologic aspects. *Crit Rev Clin Lab Sci* **32**, 469–508.

Huang, C., Thompson, W. H. & Campbell, W. P. (1995). Comparison of the M RNA genome segments of two human isolates of La Crosse virus. *Virus Res* **36**, 177–185.

Huang, C., Thompson, W. H., Karabatsos, N., Grady, L. & Campbell, W. P. (1997). Evidence that fatal human infections with La Crosse virus may be associated with a narrow range of genotypes. *Virus Res* **48**, 143–148.

Klimas, R. A., Thompson, W. A., Calisher, C. H., Clark, G. G., Grimstad, P. R. & Bishop, D. H. L. (1981). Genotypic varieties of La Crosse virus isolated from different geographic regions of the continental United States and evidence for a naturally occurring intertypic recombinant La Crosse virus. *Am J Epidemiol* **114**, 112–131.

Kolakofsky, D. & Hacker, D. (1991). Bunyavirus RNA synthesis: genome transcription and replication. *Curr Top Microbiol Immunol* **169**, 143–159.

Lappin, D. F., Nakitare, G. W., Palfreyman, J. W. & Elliott, R. M. (1994). Localisation of Bunyamwera bunyavirus G1 glycoprotein to the Golgi requires association with G2 but not NSm. *J Gen Virol* **75**, 3441–3451.

Le Duc, J. W. & Pinheiro, F. P. (1989). Oropouche fever. In *The Arboviruses: Epidemiology and Ecology*, vol. 4, pp. 1–14. Edited by T. P. Monath. Boca Raton, FL: CRC Press.

Lee, H. W. (1996). Epidemiology and pathogenesis of haemorrhagic fever with renal syndrome. In *The Bunyaviridae*, pp. 253–267. Edited by R. M. Elliott. New York: Plenum.

Lee, H. W., Lee, P. W. & Johnson, K. J. (1978). Isolation of the etiologic agent of Korean hemorrhagic fever. *J Infect Dis* **137**, 298–308.

Li, D., Schmaljohn, A. L., Anderson, K. & Schmaljohn, C. S. (1995). Complete nucleotide sequences of the M and S segments of two hantavirus isolates from California: evidence for reassortment in nature among viruses related to hantavirus pulmonary syndrome. *Virology* **206**, 973–983.

Murphy, J. & Pringle, C. R. (1987). Bunyavirus mutants: reassortment group assignment and G1 protein variants. In *The Biology of Negative Strand Viruses*, pp. 357–362. Edited by B. W. J. Mahy & D. Kolakofsky. Amsterdam: Elsevier.

Nakitare, G. W. & Elliott, R. M. (1993). Expression of the Bunyamwera virus M genome segment and intracellular localisation of NSm. *Virology* **195**, 511–520.

Nichol, S. T., Spiropoulou, C. F., Morzunov, S., Rollin, P. E., Ksiazek, T. G., Feldman, H., Sanchez, A., Childs, J., Zaki, S. & Peters, C. J. (1993). Genetic identification of a hantavirus associated with an outbreak of acute respiratory illness. *Science* **262**, 914–917.

Nichol, S. T., Ksiazek, T. G., Rollin, P. E. & Peters, C. J. (1996). Hantavirus pulmonary syndrome and newly described hantaviruses in the United States. In *The Bunyaviridae*, pp. 269–280. Edited by R. M. Elliott. New York: Plenum.

Plyusnin, A., Vapalahti, O. & Vaheri, A. (1996). Hantaviruses: genome structure, expression and evolution. *J Gen Virol* **77**, 2677–2687.

Pringle, C. R. (1996). Genetics and genome segment reassortment. In *The Bunyaviridae*, pp. 189–226. Edited by R. M. Elliott. New York: Plenum.

Schmaljohn, C. S. & Hjelle, B. (1997). Hantaviruses: a global disease problem. *Emerg Infect Dis* **3**, 95–104.

Schmaljohn, A. L., Li, D., Negleg, D. L., Bressler, D. S., Turell, M. J., Korch, G. W., Ascher, M. S. & Schmaljohn, C. S. (1995). Isolation of a newfound hantavirus from California. *Virology* **206**, 963–972.

Sibold, C., Meisel, H., Kruger, D. H., Labuda, M., Lysy, J., Kozuch, O., Pejooch, M., Vaheri, A. & Plyusnin, A. (1999). Recombination in Tula hantavirus evolution: analysis of genetic lineages for Slovakia. *J Virol* **73**, 667–675.

Spiropoulou, C. F., Morzunov, S., Feldman, H., Sachez, A., Peters, C. J. & Nichol, S. T. (1994). Genome structure and variability of a virus causing hantavirus pulmonary syndrome. *Virology* **200**, 715–723.

Talmon, Y., Prasad, B. V. V., Clerx, J. P. M., Wang, G.-J., Wah, C. & Hewlett, M. (1987). Electron microscopy of vitrified-hydrated La Crosse virus. *J Virol* **61**, 2319–2321.

Ushijima, H., Clerx-Van-Haaster, C. M. & Bishop, D. H. L. (1981). Analyses of Patois group bunyaviruses: evidence for naturally occurring recombinant bunyaviruses and existence of immune precipitable and nonprecipitable nonvirion proteins encoded in bunyavirus-infected cells. *Virology* **110**, 318–332.

Weber, F., Dunn, E. F. & Elliott, R. M. (2001). The Bunyamwera virus nonstructural protein NSs inhibits viral RNA synthesis in a minireplicon system. *Virology* (in press).

Wells, R. M., Estani, S. S., Yadon, Z. E., Enria, D., Padula, P., Pini, N., Mills, J. N., Peters, C. J., Segura, E. L. & The Hantavirus Pulmonary Syndrome Study Group for Patagonia (1997). An unusual hantavirus outbreak in southern Argentina: person-to-person transmission? *Emerg Infect Dis* **3**, 171–174.

Calicivirus, myxoma virus and the wild rabbit in Australia: a tale of three invasions

B. J. Richardson

Centre for Biostructural and Biomolecular Research, University of Western Sydney Hawkesbury, Richmond, NSW 2783, Australia

INTRODUCTION

The history, ecology and genetics of the European rabbit and the two viruses, myxoma virus and calicivirus, used to control the rabbit in Australia offer a useful set of parallels and insights into the behaviour of such systems. In this review, the process of invasion of Australia by the rabbit and the consequent invasions of the Australian rabbit population by myxoma virus and a calicivirus are described.

THE RABBIT

The European rabbit (*Oryctolagus cuniculus* L.) evolved in what is now the Iberian Peninsula and southern France in Europe. During classic antiquity, mariners moved rabbits to islands in the Mediterranean. The species was then spread by human agencies across much of western Europe and, finally, during the age of European colonization of other parts of the world, populations were established on hundreds of islands and several continents (Thompson & King, 1994). The rabbit arrived in Australia from England with the first European settlers in 1788. Over the next 70 years, rabbits were repeatedly brought to Australia and on occasion either escaped or were released into the wild. While some of these populations survived and a few spread slowly, they did not thrive (Rolls, 1984; Stodart & Parer, 1988).

Invasion

In 1859, this pattern changed with the arrival in Australia of a shipment of 13 rabbits from near Glastonbury in England. These were taken to Barwon Park, a property near Geelong in Victoria, where they were cosseted as their numbers built up. Eventually

SGM symposium 60: New challenges to health: the threat of virus infection.
Editors G. L. Smith, W. L. Irving, J. W. McCauley, D. J. Rowlands. Cambridge University Press. ISBN 0 521 80614 3 ©SGM 2001.

they escaped or were released from their enclosures. They were also distributed as gifts by the owner to his friends and associates. This release, which included wild rabbits, not just semi-domesticated hutch rabbits, was spectacularly successful, unlike previous introductions. In 1867, only 9 years after their arrival, 14 253 rabbits were shot on Barwon Park (Rolls, 1984). From Geelong, rabbits spread across Victoria and ultimately across the southern two-thirds of Australia (Fig. 1a). The first mistake was the false set of assumptions made, on the basis of previous experience, regarding the capacity of the rabbit to maintain itself in the Australian environment. Secondly, there was a failure in containment procedures that allowed it to escape in an uncontrolled manner. The combined effects of these two errors was catastrophic.

The rate of spread of rabbits varied from 15 km per year to over 100 km per year and they reached the Indian Ocean coast of Western Australia, over 2500 km from the release site, within 50 years. This scourge, which must have exceeded half a billion animals at its height, created environmental and agriculture disasters wherever they went. A description by a landowner of the arrival of the rabbit is available: '*Last Friday morning soon after sunrise, I met a swarm coming from the hills. I never saw such a thing before. The ground was scarcely to be seen for about a mile in length. Five weeks since, I could not find a rabbit on my land, but since then we have killed thousands*' (quoted in Coman, 1999). They conquered all the environments of southern Australia from the snowfields of the Australian Alps to the central deserts. Its range extension was only halted by ecological factors (oceans and temperature effects on its breeding in the tropics). A third observation can be made: the extravagant behaviour of the rabbit in the new context of Australian environments could not have been simply predicted from its ecology in northern Europe.

The response of Australians to the invasion by the rabbit followed classic patterns. In the early stages, it consisted of attempts to isolate infected areas from uninfected areas by building rabbit-proof fences across the countryside (Fig. 1b). Some of these were immense undertakings and were often over a thousand kilometres in length. The fences were unsuccessful as the rabbits were beyond the fence lines before the fences could be completed. The next move was undertaken by local landowners, who fenced their properties with rabbit-proof wire and then systematically destroyed the rabbits and warrens on their land. Experience showed that all the rabbits had to be destroyed or the procedure was fruitless. Vigilance was needed to detect and destroy foci of reinfection. Such an approach was very expensive and never-ending. Elsewhere in the settled areas of Australia, poisoning was used to reduce the numbers for short periods. In the vast, almost uninhabited, arid zone, nothing could be done (Williams *et al.*, 1995; Coman, 1999). The damage done by the rabbit is difficult to overestimate. Even after myxomatosis had reduced the population to a small proportion of its previous levels, loss of

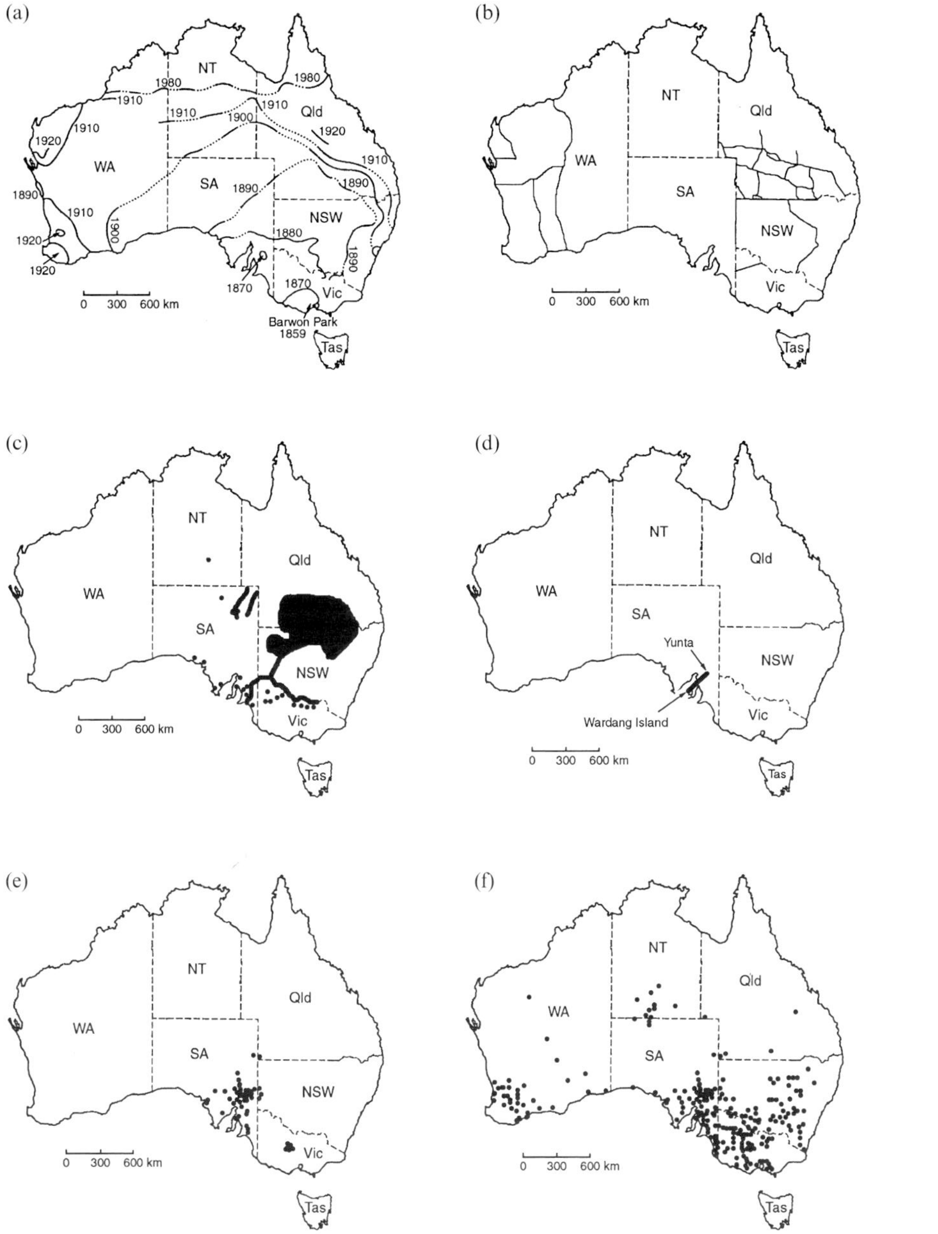

Fig. 1. (a) Historic spread of the rabbit across Australia (after Stodart & Parer, 1988). (b) Examples of rabbit-proof fences built in an attempt to stop the spread of the rabbit. (c) Distribution of myxomatosis by the end of autumn 5 months after its escape (after Fenner & Fantini, 1999). (d) Trajectory of spread of RHDV from Wardang Island predicted for the meteorological conditions present on the 13 October 1995. (e) Distribution of RHDV 5 months after its escape from Wardang Island. (f) Distribution of RHDV 12 months after its escape and prior to the start of inoculation campaigns (after Kovaliski, 1998).

agricultural production was estimated at 600 million Australian dollars per year, or 3 % of Australia's agricultural production (Wilson, 1995; Coman, 1999). The environmental damage, particularly the effective halt to regeneration of the native trees of inland Australia for almost a century, is incalculable. Needless to say, Australians were desperate to control this plague by any means possible.

Evolution of the invader

What then have been the effects on the rabbit of its invasion and accommodation to Australia? Firstly, there was little loss of genetic variation in the transfer from Europe to Australia. A study of variation in seven microsatellite loci showed a small reduction in the number of rare alleles from 1.4 ± 0.2 in France to 0.9 ± 0.2 in Australia. There was no change in mean heterozygosity (0.64 to 0.67). Similarly, there was no divergence in the level of mean heterozygosity in different parts of Australia (A.-M. Vachot, personal communication; Mougel, 1997). Similar results are found for allozyme and immunoglobulin variation (Richardson *et al.*, 1980; van der Loo *et al.*, 1987). There is, however, evidence of spatial structuring at the 3–10 km scale in allozymes and mitochondrial DNA in the more temperate parts of Australia (Richardson, 1980; Zenger, 1996). Genetic structuring at this scale, however, seems to be absent in the more arid parts of the continent (Fuller *et al.*, 1996; Zenger, 1996). Spatial structuring includes changes as great as 0.3 in allele frequencies within a few kilometres (Fig. 2). Examination of the morphology of the rabbit also shows some evidence of local adaptation to environmental conditions, with darker hair and shorter limbs in colder and wetter environments and the reverse in hotter, more arid environments. These changes have occurred within 80 generations of the release of the rabbit (Stodart, 1965; Myers *et al.*, 1989). Some of this divergence is genetically based and the consequence of selection, while other aspects are a direct response to conditions during development. Aspects of the ecology of the rabbit also vary with conditions, with high fertility (28 kittens per doe per year) and mortality in drier areas and lower fertility (14 kittens per doe per year) and mortality in colder and wetter areas. This variation seems to be ecological rather than genetically based. There does not seem to be, however, any innate regulatory mechanism on numbers in the rabbit. In the drier parts of the country particularly, the populations rapidly build up to plague proportions in good years and eat everything, before changing behaviour and moving in a massive wave, perhaps reminiscent of the original spread of the rabbit, across the country in a desperate effort to find food before dying in massive numbers (Myers *et al.*, 1989).

Lessons

In summary then, the rabbit has lost little genetic variation despite undergoing a series of population bottlenecks, it rapidly adapted to different environmental conditions in Australia and, after the original spread, is relatively sedentary in temperate areas,

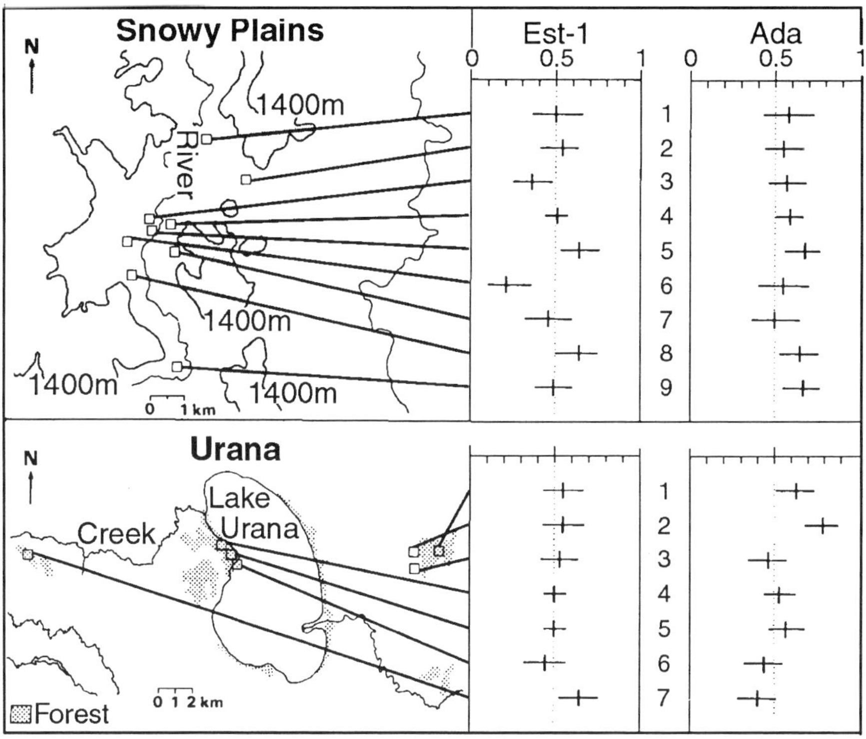

Fig. 2. Microdistribution of gene frequencies in two different regions of Australia (after Richardson, 1980).

allowing genetic divergence (probably by both selection and genetic drift) to develop between areas only a few kilometres apart.

THE MYXOMA VIRUS

The disease myxomatosis was first reported by Sanarelli in laboratory rabbits in 1898 in Montevideo and the virus was illustrated by Aragão in 1927. It is a poxvirus placed in the genus *Leporipoxvirus*. Though spontaneous cases of the disease in domestic rabbits were reported in several South American countries, it was many years before the reservoir host was identified. This proved to be a tropical forest rabbit, *Sylvilagus brasiliensis*, a South American relative of the European rabbit. In *Sylvilagus*, an infection is characterized by the development of a small localized dermal tumour. This is in marked contrast to the nature of the disease in the European rabbit, where it is characterized by the production of mucinous tumours. In the original virus strain used in Australia, a lump is found at the original site of intradermal inoculation after 3 days and secondary skin lesions are apparent from the sixth day after inoculation. The eyes are closed by day 9 and this leads on to the production of copious opalescent discharges and secondary infections. There is oedematous swelling of the perineum and of the head, especially at the base of the ears. Death occurs in >99 % of animals in a mean time of 10.8 days (Fenner & Fantini, 1999). Like the changes in the ecology of the rabbit when it reached Australia, the symptomatology or virulence of myxomatosis in the European rabbit could not be predicted from its behaviour in its original host, *Sylvilagus*.

Dr Aragão, who was studying the disease in Brazil, had determined that it could be spread by contact and was host-specific. He realized the potential of the virus as a biological control agent and wrote to Australian colleagues in 1919. He also sent them virus cultures. The idea was passed on to government officers, who were unenthusiastic, and the cultures were placed in quarantine. Over the next 30 years, the argument for and against the use of myxomatosis raged and several series of laboratory and yard trials were undertaken, with less than impressive results. Because of the concerns of the Chief Quarantine Officer, the Australian trials took place in drier areas where there were few biting insects. Some of the trials were carried out in Britain, Denmark and Sweden, where the quarantine regulations were not so strict. The intent of those involved in these countries was to establish the disease for rabbit control. The problem was that the disease did not seem to be highly infectious, as infections faded away after a few generations of disease. During this period, the virus was also tested against a range of Australian native species and its host specificity was confirmed (Fenner & Fantini, 1999).

Invasion

In 1950, trials were undertaken in field pen populations at four sites in southern New South Wales in higher rainfall areas. Like previous trials, the results were an apparently unsuccessful disease outbreak. For example, in one area almost 100 rabbits in a population of 700 animals were inoculated during spring with the virus. Over the 3 months leading up to Christmas, the number of cases gradually decreased and the population increased to 1000 animals. Few mosquitoes were seen and contact infection clearly was not effective. Then, after work had closed down, a phone call reported that many sick rabbits were present in the yards. Shortly thereafter, sick rabbits were reported from other sites many kilometres from the study pens. Within a month, myxomatosis had spread along the rivers for over 600 km from the study sites and by mid-February was to be found throughout the river systems of inland Australia (Fig. 1c). Those involved quickly realized that flying insects were acting as the vector and these later were shown to be mosquitoes (Fenner & Fantini, 1999).

At this point, one of the most fascinating human events in Australian virology occurred. The weather conditions that led to the large number of mosquitoes needed to set off the myxomatosis outbreak also supported, in the same area, an outbreak of a human disease only seen once before (in 1917/8). This disease, called Murray Valley encephalitis, occurred in the same areas and at the same time as the spread of myxomatosis and created consternation in the local population, who feared it was due to myxoma virus. The virus causing the human disease was recovered and reassuring press releases were made. The public was not convinced, however, and the chairman of one of the local hospital boards challenged the scientists to demonstrate their commitment to their belief that the encephalitis was not caused by the myxoma virus, and that the myxomatosis virus was harmless to humans, by injecting themselves with live myxoma virus. After discussion and careful consideration, Sir Macfarlen Burnet and Dr Frank Fenner injected one another with 100 rabbit infectious doses. On hearing what they had done, the head of CSIRO, Dr Ian Clunies Ross, insisted on also being injected. Other than a slight reddening at the injection site there were no symptoms and no antibodies were detected. The relevant Federal Minister announced the results of the demonstration in Parliament and the public concern of a link between the diseases was set to rest (Burnet, 1968; Fenner & Fantini, 1999). It is interesting to speculate what virologists would do today if faced with the same public concern.

Though the myxoma virus killed more than 90 % of the rabbits in areas where it occurred, it was restricted to sites close to the rivers. After dying down in the following winter, it broke out again in the spring of 1952. Over the following 9 months, the disease swept through the populations of eastern Australia, aided by inoculation campaigns. Through natural spread and the inoculation campaigns, by the winter of 1955 the virus was effectively coextensive with the rabbit across Australia. The effects of the arrival of the disease on a local rabbit population can be seen in data from Lake Urana from 1951 to 1953 (Fig. 3; Myers *et al.*, 1954). Detailed studies of the number and distribution of biting insects clearly demonstrated that mosquitoes, particularly *Culex annulirostris*, were the usual vectors, though biting flies and fleas also could be important in suitable circumstances. Transmission was mechanical and the work by Fenner (1953) led him to conclude that mosquitoes could be considered 'flying pins', as there was no vector specificity, no extrinsic incubation period and virus concentration in the cells of skin lesions rather than the general level of viraemia was critical in contaminating biting insects.

Evolution of the invader

Attenuated strains of the virus were recovered in the field as early as 1952 and the distribution of virulence in virus recovered from field populations is summarized in Table 1 (Fenner & Fantini, 1999). It is clear that the original highly virulent Grade I virus used in the original study and again used in the follow-up inoculation campaigns was quick to adapt to the conditions it found in the field. By 1952/3, Grade III virulence was

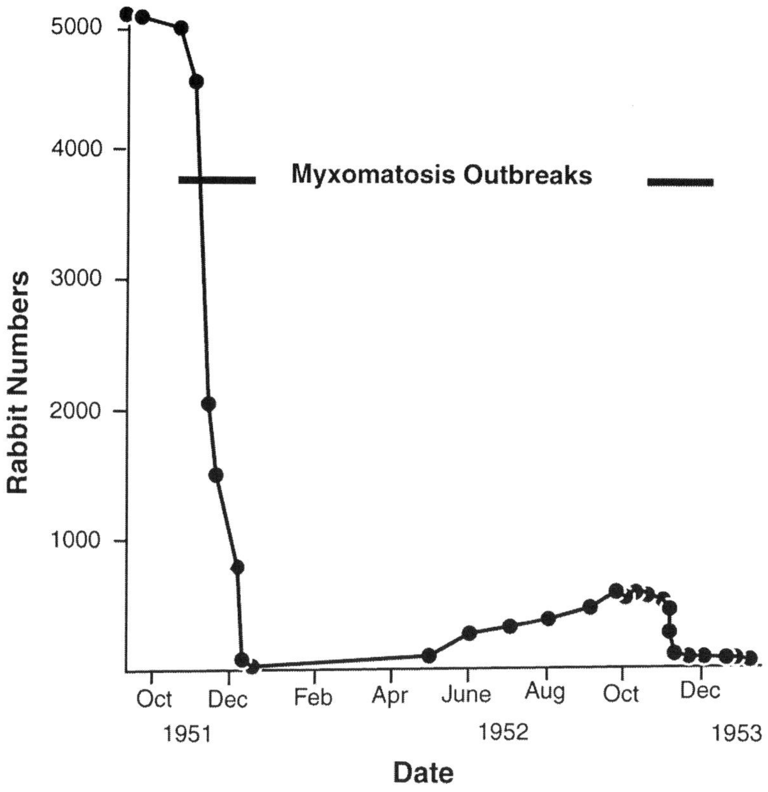

Fig. 3. Effect of myxoma virus on the size of the rabbit population at Urana during the first 3 years after its arrival (Myers *et al.*, 1954).

the norm in field populations. The strong competitive advantage of the less virulent virus can be seen in a study carried out at Lake Urana in 1954, where Grade I virus was released into the population by inoculation of rabbits. A field strain broke out shortly after the releases and quickly replaced the more virulent virus (Table 2). It is noteworthy that strains of similarly reduced virulence appeared at many locations across Australia. The pattern is unlikely to be due to the spread of the same strain and it is safe to assume that reduced virulence developed in many different localities shortly after the virus appeared in each population.

At the same time as the virus was adapting to its new host and field situation, studies of the rabbits showed that the field population was also responding to the challenge of the virus with increased genetic resistance. It was shown that the virus causes less mortality

Table 1. Attributes of the different virulence grades of myxoma virus when tested against unselected, antibody-negative, animals and the percentage of different grades recovered from the field during the first 7 years after escape (Fenner & Fantini, 1999)

	Virulence grade				
	I	**II**	**III**	**IV**	**V**
Survival time (days)	<13	13–16	17–28	29–50	∝
Mortality rate	99.5	95–99	70–95	50–70	<50
Year:					
1950/1	>99	0	0	0	0
1951/2	33	50	17	0	0
1952/3	4	13	74	9	0
1953/4	16	25	50	9	0
1954/5	16	16	42	26	0
1955/6	0	3	55	25	17
1956/7	0	6	55	24	15

Table 2. Competitive replacement of a virulent strain of myxoma virus (Grade I) inoculated into a field population by field strain (Grade III) at Lake Urana (Fenner *et al.*, 1957)

The population was inoculated with virulent virus (242 rabbits inoculated) during November and early December. Figures show the number of each form of virus recovered from sick rabbits in the field and the percentage of the total population showing symptoms.

Month	Virulent strain	Field strain	Disease incidence (%)
November	7	1	10
December	10	16	24
January	1	9	10
February	0	2	2

in animals younger than 2 months, through protection by maternal antibodies. At high ambient temperatures, survival rates were also higher. The reduction in the virulence of the virus in the field to Grade II or III allowed 10 % or more of infected animals to survive and provided ideal selection conditions for genetic resistance to evolve in the host. To test for the development of resistance, each year young rabbits not previously challenged with virus were collected in the field and housed in mosquito-proof animal houses until they were about 4 months of age, when they were challenged with a standard Grade I virus. At the same time, control strains of rabbits that had never been challenged with myxomatosis were tested. The results for rabbits from Lake Urana are

Table 3. Development of resistance in Australian rabbits to 'Standard Laboratory Strain' (Grade I) myxoma virus

Year	Mortality rate (%)
1950	>99
1956/7	87
1961	73
1961/6	68
1967/71	66
1972/5	67
1976/81	60

Table 4. Effect of myxomatosis on rabbit numbers after 20 years at the four original study sites (Myers, 1970)

	No. of rabbits	
Site	**Before myxomatosis**	**20 years later**
Rutherglen (6 ha)	1000	3
Coreen (8 ha)	400	2
Balldale (8 ha)	300	20
Urana (154 ha)	5000	11

shown in Table 3 (Marshall & Fenner, 1958). Clearly, susceptibility was reduced in this rabbit population over the few years following the original release. By 1957, more than 10 % of rabbits were surviving challenge with the Grade I virus that originally killed >99 % of animals (Marshall & Douglas, 1961).

A question of critical concern in Australia was whether, with increased rabbit resistance and virus attenuation, the virus was still effective in limiting the number of rabbits found, compared to the pre-myxomatosis situation. Myers measured the densities of rabbits at four of his early study sites in 1970 and found the densities were much lower than those obtained pre-myxomatosis (Myers, 1970) (Table 4). Parer *et al.* (1985) also tested the effectiveness of myxoma virus in suppressing rabbit numbers by infecting young animals in two natural populations with an attenuated strain that caused very few deaths but immunized the survivors. They showed that the populations protected against the local field strain in this fashion increased eightfold and tenfold in size in 2 years. Clearly, the field strain of the virus was still a key mortality factor in the population ecology of the rabbit in Australia.

Table 5. Interaction between viruses obtained from different sites and rabbits from different sites (as % survival ±SE) (Parer *et al.*, 1994)

The unselected rabbits were derived from wild rabbits from New Zealand.

	Virus from:		
Rabbits from:	**Urana**	**Yathong**	**France**
Urana	41 ± 8	20 ± 4	2 ± 2
Yathong	60 ± 8	45 ± 6	46 ± 8
Cape Naturaliste	45 ± 15	0 ± 0	0 ± 0
Unselected	4 ± 3	0 ± 0	0 ± 0

Evolution of the host

By the early 1960s, the situation seemed to have stabilized with 60–70 % of myxoma virus strains isolated in the field having a virulence grade of III. After 1970, however, strains with increased virulence (Grades I and II) were increasingly obtained (when tested against genetically unselected laboratory rabbits). The few tests carried out during the 1990s show that virus field strains (when tested against unselected laboratory rabbits) were now of Grade I virulence with attributes similar to the strain originally released. In wild rabbits from different areas, these Grade I viruses killed between 0 and 50 % of the animals tested, compared with the effectively 100 % mortality observed 40 years earlier (Kerr in Fenner & Fantini, 1999). The net effect has been increased virulence in the virus to offset the increased resistance in the rabbit.

A comparison of the resistance level in rabbits collected from various natural populations with the virulence of virus from the same or different populations showed that the nature of the resistance and the virus' response to these changes differed from location to location (Table 5; Parer *et al.*, 1994). Clearly, the nature of the resistance/myxoma virus response pattern differs from place to place and it is likely that the range of potential sources of genetic resistance in the rabbit is far greater than that found at any one location. The capacity for rabbit populations to diverge, seen in the allozyme and mitochondrial DNA data sets, is also apparent in the resistance data. The overall effect in the field is an apparent balance with similar numbers of rabbits killed by the local strain of myxoma virus in each area. The myxoma virus that first spread across Australia was clearly 'too virulent' and less virulent strains evolved and replaced the original strain within a few months. As the rabbits developed resistance, field strains of the virus presumably were 'too weak' and evolution favoured increased virulence against unselected rabbits. It is clear that an evolutionary interplay is occurring between the rabbit and the virus. The observed balance is established differently at different localities, and at one locality at different times, as different genetic resources are brought into play by the rabbit and the virus, but what could the selective basis of this balance be?

It is often assumed that a virus will coevolve with its host to reduce the adverse effects of the virus, particularly death rates. This, however, is not always so, as can be seen by the maintenance of virulence and, in the face of increasing resistance, increased virulence in myxoma virus in Australia. What then are the factors affecting transmission of the myxoma virus in Australia? Firstly, the nature and ecology of the vector involved is critical. In Australia, the virus was originally spread by mosquitoes, with rabbit fleas becoming important once they were introduced. Secondly, the likelihood of a mosquito becoming infected with the virus increases the longer the host remains infectious, that is, over what time periods are high levels of virus available in the skin of infected rabbits? This varies with the intensity of the infection. More virulent viruses tend to give higher titres but death intervenes relatively quickly; low virulence viruses on the other hand do not lead to early death but then neither do they give high levels of virus (Table 1). The optimum is an intermediate condition and how this optimum is reached depends on the genetic resources underlying the virulence of the virus and the resistance of the host.

Recently, myxoma virus has been sequenced (Cameron *et al.*, 1999) and through genetic engineering is used experimentally in studies of mechanisms of pathogenesis (e.g. Hnatick *et al.*, 1999; Lalani *et al.*, 1999) and in the development of recombinants for vaccination or immunocontraception (e.g. Barcena *et al.*, 1999; Holland & Jackson, 1994).

Lessons

In summary, the history of myxomatosis in Australia shows many similarities to the history of the rabbit: it escaped rather than was released (though in each case there was a clear intention to ultimately release it), the pattern of invasion did not match that expected as the mode of transmission was not as expected, the speed of the invasion was much faster than expected and was effectively unstoppable, and the virus evolved rapidly to local conditions in each part of the country and continues to evolve.

CALICIVIRUS

Rabbit haemorrhagic disease (RHD) was first observed in 1984 in domestic rabbits in China that had been imported from East Germany (Lui *et al.*, 1984). The disease was usually fatal and highly infectious, and outbreaks occurred across parts of China and Korea over the next 9 months. It was detected in Italy in 1986 and then across Europe. It spread rapidly to rabbits in many parts of the world, including the Middle East, Africa, Mexico, the United States and India (Morisse *et al.*, 1991). The causative agent was identified as a calicivirus and called rabbit haemorrhagic discasc virus (RHDV; Ohlinger & Thiel, 1991). Unfortunately, in much of the Australian literature, it is called rabbit calicivirus or RCV, which is the acronym for a related virus (see below).

After infection, domestic rabbits die within 30–36 h, while Australian wild rabbits die within 20–24 h (Lenghaus *et al.*, 1994), though in the field, death occurred 21–48 h after infection in wild rabbits (Cooke in Fenner & Fantini, 1999). Consistently the disease is characterized by the presence of necrotizing hepatitis and, commonly, necrosis of the spleen. While severe haemorrhages are seen, more generally the disease presents with disseminating intravascular coagulation in most organs, but particularly in the lungs, heart and kidneys (Marcato *et al.*, 1991; Plassiart *et al.*, 1992). Mortality rates in adult rabbits are high (>99 %), though the rates are much lower in immature rabbits, with all rabbits less than 10 days old surviving. Maternal immunity also protects young animals for 2–3 months.

The origin of the virus is unknown; however, there are two related forms, one in the European brown hare, where it is responsible for the European brown hare syndrome (EBHS), and a second, benign, calicivirus in the European rabbit (RCV). The single-stranded RNA genomes of RHDV and EBHSV have been sequenced and are 7437 and 7442 bases in length [excluding the poly(A) tail] (Gould *et al.*, 1997). There is 0–11 % divergence between strains of RHDV collected at different places in Europe, and 40–47 % divergence between RHDV and EBHSV strains for the capsid protein gene (Nowotny *et al.*, 1997). RHDV and RCV are more similar, however, with RCV providing immunological protection against infection by RHDV (Capucci *et al.*, 1996). It is possible, therefore, that RHDV was derived from RCV. Another alternative is that the virus is derived from another, more distantly related, host species. The similarity between the RCV and RHDV sequences compared to that with other caliciviruses, and the presence of antibodies to, presumably, RCV in Czech rabbit sera collected 12 years before RHDV was first reported (Rodak *et al.*, 1990) and from up to 20 % of rabbits at an Australian site each year since 1992 (B. Richardson & A. Philbey, unpublished), make the origin of RHDV from a more distantly related form unlikely.

At the time of the first outbreak of RHDV in wild rabbits in Spain, an Australian scientist, Dr B. Cooke, was investigating the possibility of using Spanish fleas adapted to drier conditions to improve the effectiveness of myxomatosis in the arid zone of Australia, where rabbits were still causing massive environmental damage. On return to Australia, Dr Cooke raised the possibility of using RHDV to control rabbit numbers. As the virus was effective in arid areas in Europe, host-specific, highly infectious and highly lethal, it was an attractive possible biological control agent (Fenner & Fantini, 1999).

Invasion

In 1989, the Australian and New Zealand governments jointly began the process of assessing the potential of and risks related to the use of RHDV as a biological control agent. The virus had been inoculated into 26 species in other countries without any evidence of disease. Also, in the years that the virus had been widespread in Europe, there were no reports of illness or disease in humans or stock (Bureau of Resource Sciences, 1996). The first step in Australia then was to test the virus against Australian native species in the high security facility of the Australian Animal Health Laboratory. Tests were carried out on four to six individuals of each of 15 introduced and 15 native vertebrate species (Bureau of Resource Sciences, 1996). Tests included the use of PCR, which only works with DNA, to detect replication of this RNA virus in these species. No replication was detected (Gould *et al.*, 1997). In 1994, the governments agreed to move to the next stage of the project with a projected release date, if all was well, of 1998. This stage included the preparation of an Environmental Impact Assessment, an assessment by the Biological Control Authority (which includes public submissions) and field trials of the virus on an island.

Wardang Island in South Australia (Fig. 1d) was chosen for the field trials with the approval of the Australian Quarantine and Inspection Service. The pens, each of which contained a warren, were double-fenced and then the entire study area was double-fenced again. All rabbits in the warrens were killed and the warrens treated to remove fleas before rabbits were reintroduced. Surrounding swampy areas were treated to kill mosquito larvae and all carcasses of dead animals were removed as soon as possible after death to reduce the chance of scavengers coming in contact with the virus. Fly traps were used to reduce the fly population as much as possible. All clothing and footwear was either changed or treated when each barrier fence was crossed. It was assumed that animal-to-animal contact was the usual method of transmission. Vector control was precautionary.

A series of studies over the first 6 months showed that the virus could be spread by contact infection within a social group. The level of contact spread, however, was low with only 13 cases of contact infection following 14 inoculations. The speedy removal of the carcasses of all animals that died, whether above or below ground, meant that infection from dead bodies was not likely. Restocking of pens 7 weeks after an experiment never led to infected animals, so it seemed probable that the virus did not survive in the pens for this length of time (at least in the absence of carcasses) (Cooke, 1999).

Safety procedures included the monitoring of sentinel rabbits in control pens and in warrens elsewhere on the island. In September 1995, sentinel rabbits in the pens and in warrens 900 metres outside the enclosure areas were found dead with RHD. Immediately, the contingency plan in place was followed, with all animals killed in both areas and the infected warrens outside the pens destroyed. Nevertheless, cases continued to appear across the island through to mid-October. Surveys of warrens on the mainland opposite the island led to the discovery of a dead RHDV-infected animal 5 km from the enclosures. Two weeks later, the first cases from Yunta 360 km away and from Blinman 390 km away were reported. These two outbreaks probably were initiated at the same time, and this was also at about the same time as the virus jumped to the mainland. Study of the meteorological conditions during the critical period showed that weather conditions were such as to provide suitable wind trajectories on the 13th October (Fig. 1d; Wardhaugh & Rochester, 1996; Cooke, 1999). The particular areas where the distant infections started are areas of high rabbit density. In other areas along this trajectory, densities were much lower, reducing the likelihood that infections would be established.

By 5 months after the escape, cases had been detected from sites over much of South Australia as well as in Victoria. Within 12 months, the virus had been detected at hundreds of sites across the range of the rabbit (Fig. 1e, f; Kovaliski, 1998). The rate of spread varied from 9 km per month in summer to 414 km per month in spring. At this point in time, approval was given for inoculations of wild populations to be undertaken by rabbit control authorities. Over the next year, the virus was distributed across the range of the rabbit in Australia.

The epidemiological pattern shown by the virus proved to be different to that seen with myxomatosis. The virus has been most effective in the drier parts of Australia (<200 mm rainfall), with regular annual outbreaks reducing population sizes to 10–20 % of the pre-RHD levels. In slightly wetter areas, the population size has fallen but to 20–50 % of previous levels with annual outbreaks. As rainfall increases, the effectiveness of the virus decreases, and outbreaks become patchy, occurring intermittently. In some coastal areas and along the northern fringe of the rabbit's range, the virus does not occur naturally and when released does not spread. Outbreaks occur much less commonly in summer than in other seasons. In one studied population in Victoria, however, the pattern is different. At Bacchus Marsh in Victoria, the original reduction in population size was only 40 % compared to the 70–80 % expected. This population recovered in the breeding season following the arrival of the virus and has continued to maintain its numbers ever since (S. McPhee in Cooke, 1999). The pattern of the disease in most areas is one of recurrence of the disease in seasons other than summer whenever sufficient numbers of susceptible animals (i.e. young animals no longer covered by maternal immunity) are available. The outbreaks may occur in late spring or in the following autumn or winter (Cooke, 1999).

The rate of spread of the virus in England was estimated as 200–300 metres per week (R. Trout in Cooke, 1999) and was assumed to be due to contact transmission between animals. The situation in Australia was quite different with natural rates of spread of over 90 km per week in remote areas (Kovaliski, 1998) where human interference is unlikely. It is assumed that such a rate is maintained by the transfer of the virus from location to location by flying insects. In marginal areas, the virus reappears intermittently and affects effectively all the animals in some warrens but none in other warrens only a few metres away (e.g. Moriarty *et al.*, 2000). In high rainfall areas, for example around Sydney, inoculation programmes lead to either no spread of the virus or only a slow spread for a few hundred metres before the infection dies out. It is assumed that this is due to the absence of suitable vectors to support long or short distance transmission between social groups (B. J. Richardson, unpublished).

In areas where the virus is active, samples of field-collected insects have been tested for the presence of RHDV using reverse transcriptase PCR (Asgari *et al.*, 1998). The most commonly contaminated insects were large flies, particularly those of the genus *Calliphora*. Members of this genus are found in all seasons but are less common in summer. This pattern matched the onset of outbreaks, which are less common in summer. The flies can carry virus for up to 9 days after feeding on an infected liver. Single specimens of fly faeces or regurgita are sufficient to infect a rabbit (Asgari *et al.*, 1998). While transmission by insects is clearly mechanical, the process by which transmission by insects occurs is unclear: is the virus deposited on the rabbit (for example on eye membranes, or on the fur where it is ingested during grooming), or are fly faeces deposited on grass stems later eaten by the rabbits? Given their role in the transmission of myxoma virus, it might be predicted that mosquitoes might also be important vectors in the transmission of RHDV; however, there is no clear evidence of such a role, though they may be implicated in particular situations (Cooke, 1999).

Rabbit fleas (*Spilopsyllus cuniculi*) also carry the virus and it is possible that they are a key factor in the spread of the virus between members of the same social group. It is known that rabbit fleas move, not only from dead animals to live animals, but between living animals and from the burrow floor to live rabbits (Mead-Briggs, 1964). Alternatively, contact transmission may be the most important factor within social groups. Such a mode of transmission could be based on droplet spread or through virus-contaminated urine, etc. Such mechanisms were not effective, however, even if present, in the Wardang trials. The relative importance of different mechanisms of transmission of the virus between adjacent social groups, which seems to be the point at which local transmission fails, is uncertain. The modes of transmission and any critical vector species at each geographic scale are still unclear.

In areas where the virus shows an annual epidemic, it is still unclear as to where the virus 'hides' between outbreaks. Is it maintained in 'carriers' or in dead carcasses in warrens, or through some other means?

Evolution of virus and host

One of the notable aspects of the interaction between myxoma virus and the rabbit was the rapid appearance of attenuated virus and of genetic resistance in the rabbit. No evidence of attenuated RHDV has been obtained in either Europe or Australia. Studies of the nucleotide sequence of the virus show that up to 2 % sequence variation may be observed in the field (Asgari *et al.*, 1999). A non-pathogenic RHDV-like virus discovered in domestic rabbit colonies (RCV) is not considered to be derived from RHDV but to be a distinct species (Capucci *et al.*, 1996). In the same fashion, the survival of significant numbers of infected rabbits in the field would lead one to expect the evolution of genetic resistance. While no tests have been carried out to look for changes in pathogenicity, as they were for myxomatosis (Tables 1 and 3), there is no evidence of recovery in rabbit numbers in the field except at Bacchus Marsh (see above). This evidence, however, is negative and effective evolution of the virus and its host may be occurring unnoticed. If the virus requires the death of rabbits for effective transmission, for example dead carcasses to provide a contaminated food source for flies or as a source of virus for a new outbreak after several months, then attenuated virus will not be found. Similarly, any genetic resistance developed by the rabbit is likely to be covered by increased virulence of the virus.

Lessons

A number of parallels to the history of the rabbit and myxoma virus in Australia are apparent. Firstly, the virus escaped rather than was released. Assumptions about the nature of the mechanism of transmission and therefore the precautions needed were incorrect. Once released the virus spread very rapidly and was unstoppable by human efforts. The range of the virus was ultimately limited by ecological factors.

SIMILARITIES AND DIFFERENCES IN THE THREE INVASIONS

The marked differences between the co-evolution of the three systems, Australian environment/rabbit, rabbit/myxoma virus and rabbit/RHDV, warrant serious consideration. In each of the systems there is a dynamic and evolving balance between transmissibility, virulence and genetic resistance in the host. In the rabbit, there is no evidence at this time of any change in the reproductive capacity or mobility of the rabbit that could be interpreted as a change in transmissibility. The usual mode of control, by reducing population size by poisoning and then allowing the population to recover before repoisoning, that is, by the use of a highly r selected system, is not likely to change the adaptive situation. As a consequence, there is no evidence of 'attenuation' of the effect (virulence) of the rabbit on its 'host', the Australian environment.

For optimum transmission, myxoma virus requires high virus titre in the skin of the rabbit for as long as possible and this is found at intermediate levels of virulence. Increased resistance in the host leads to selection on the virus to restore the level and availability of virus, and this is reflected at present in increased virulence. Dead rabbits, however, are not a necessary condition for transmission and, ultimately, death and transmission capacity may become uncoupled.

The situation with RHDV is different, as in most areas in Australia it is probable that the virus requires dead rabbits to act as a source of food/contamination for adult flies and for maggots if long-distance transmission (at least) is to be effective. One might predict that no matter what the rabbit does in an attempt to increase resistance, the virus will counter. One possible way out for the rabbit (anthropomorphically speaking) is for it to change the mode of transmission so that death is not needed; this would then change the selection conditions faced by the virus. In such circumstances, attenuation may well occur. The situation at Bacchus Marsh may provide the opportunity to examine this prediction by measuring the rate of change in genetic structure of the virus in this area compared to that found elsewhere in Australia. At another level, it may be that the change in mode of transmission to one requiring the death of the host provided the driving force for the evolution of RHDV in the first place.

The speed of range extension in all three systems was shockingly fast. Both myxoma virus and RHDV repeatedly moved hundreds of kilometres in a few weeks and were halted only by ecological barriers. It is clear that insect-spread, highly infectious, viruses will be very difficult to stop. The unexpected behaviour of both viruses in Australia places a serious question mark against our capacity to predict behaviour and protect populations against invading organisms.

It is also worth noting that the consecutive invasion of a vertebrate host by two highly infectious viruses, each known to kill >99 % of infected individuals, led to a reduction in the total host population, but not to anything like the level needed to exterminate the species. This may be seen as good news for the human species in the face of human viruses, but as bad news for the ideal of destroying a pest species by the use of biological control procedures.

REFERENCES

Asgari, S., Hardy, J. R. E., Sinclair, R. G. & Cooke, B. D. (1998). Field evidence of mechanical transmission of rabbit haemorrhagic disease virus (RHDV) by flies (Diptera: Calliphoridae) among wild rabbits in Australia. *Virus Res* **54**, 123–132.

Asgari, S., Hardy, J. R. E. & Cooke, B. D. (1999). Sequence analysis of rabbit haemorrhagic disease virus (RHDV) in Australia: alterations after its release. *Arch Virol* **144**, 135–145.

Barcena, J., Morales, M., Vazquez, B., Boga, J. A., Parra, F., Lucientes, J., Pages-Mante, A., Sanchez-Vizcaino, J. M., Blasco, R. & Torres, J. M. (1999). Horizontal transmissible protection against myxomatosis and rabbit hemorrhagic disease by using a recombinant myxoma virus. *J Virol* **74**, 1114–1123.

Bureau of Resource Sciences (1996). *Rabbit Calicivirus Disease: a Report under the Biological Control Act 1984*. Canberra: Bureau of Resource Sciences.

Burnet, M. (1968). *Changing Patterns: an Atypical Autobiograpy*. Melbourne: Heinemann.

Cameron, C., Hota-Mitchell, S., Chen, L., Barrett, J., Cao, J. X., Macaulay, C., Willer, D., Evans, D. & McFadden, G. (1999). The complete DNA sequence of myxoma virus. *Virology* **264**, 298–318.

Capucci, L., Fuss, P., Lavazza, A., Pacciarni, M. L. & Rossi, C. (1996). Detection and preliminary characterisation of a new calicivirus related to rabbit haemorrhagic disease virus but nonpathogenic. *J Virol* **70**, 8614–8623.

Coman, B. (1999). *Tooth and Nail: the Story of the Rabbit in Australia*. Melbourne: Text Publishing.

Cooke, B. D. (1999). *Rabbit Calicivirus Disease Program Report 2: Epidemiology, Spread and Release in Wild Rabbit Populations in Australia*. Canberra: Bureau of Rural Resources.

Fenner, F. (1953). Host parasite relationships in myxomatosis of the Australian wild rabbit. *Cold Spring Harbor Symp Quant Biol* **18**, 291–294.

Fenner, F. & Fantini, B. (1999). *Biological Control of Vertebrate Pests: the History of Myxomatosis – an Experiment in Evolution*. Wallingford: CABI.

Fenner, F., Poole, W. E., Marshall, I. D. & Dyce, A. L. (1957). Studies on the epidemiology of infectious myxomatosis of rabbits. VI. The experimental introduction of the European strain of myxoma virus into Australian wild rabbit populations. *J Hyg* **55**, 192–206.

Fuller, S. J., Mather, P. B. & Wilson, J. C. (1996). Limited genetic differentiation among wild *Oryctolagus cuniculus* L (rabbit) populations in arid eastern Australia. *Heredity* **77**, 138–145.

Gould, A. R., Kattenbelt, J. A., Lenghaus, C., Morrissy, C., Chamberlain, T., Collins, B. J. & Westbury, H. A. (1997). The complete nucleotide sequence of rabbit haemorrhagic disease virus (Czech strain V351) – use of the polymerase chain reaction to detect replication in Australian vertebrates and analysis of viral population sequence variation. *Virus Res* **47**, 7–17.

Hnatick, S., Barry, M., Zeng, W., Liu, L., Lucas, A., Percy, D. & McFadden, G. (1999). Role of the C-terminal RDEL motif of the myxoma virus M-T4 protein in terms of apoptosis regulation and viral pathogenesis. *Virology* **263**, 290–306.

Holland, M. K. & Jackson, R. J. (1994). Virus-vectored immunocontraception for control of wild rabbit: identification of target antigens and construction of recombinant viruses. *Reprod Fertil Dev* **6**, 631–642.

Kovaliski, J. (1998). Monitoring the spread of rabbit haemorrhagic disease virus as a new biological agent for the control of European rabbits in Australia. *J Wildl Dis* **34**, 421–428.

Lalani, A. S., Masters, J., Graham, K., Liu, L., Lucas, A. & McFadden, G. (1999). Role of the myxoma virus soluble CC-chemokine inhibitor glycoprotein, M-T1, during myxoma virus pathogenesis. *Virology* **256**, 233–245.

Lenghaus, C., Westbury, H., Collins, B., Ratnamohan, M. & Morrissy, C. (1994). Overview of the RHD project in the Australian Animal Health Laboratory. In *Rabbit Haemorrhagic Disease: Issues in Assessment for Biological Control*, pp. 104–129. Edited by R. K. Munro & R. T. Williams. Canberra: AGPS.

van der Loo, W., Arthur, C. P., Richardson, B., Wallace, M. & Hamers, R. (1987). Nonrandom allele associations between unlinked protein loci: are the allotypes of the immunoglobulin constant regions adaptive? *Proc Natl Acad Sci U S A* **84**, 3047–3079.

Lui, S. J., Xue, H. P., Pu, B. Q. & Quian, N. H. (1984). A new viral disease of rabbits. *Anim Husb Vet Med* **16**, 352–355 (in Chinese).

Marcato, P. S., Benazzi, C., Vecchi, G., Galeotti, M., Della Salda, L., Sarli, G. & Lucidi, P. (1991). Clinical and pathological features of viral haemorrhagic disease of rabbits and the European brown hare syndrome. *Rev Sci Tech Off Int Épizoot* **10**, 371–392.

Marshall, I. D. & Douglas, G. W. (1961). Studies in the epidemiology of infectious myxomatosis of rabbits. VIII. Further studies on the innate resistance of Australian wild rabbits exposed to myxomatosis. *J Hyg* **59**, 117–122.

Marshall, I. D. & Fenner, F. (1958). Studies in the epidemiology of infectious myxomatosis of rabbits. V. Changes in the innate resistance of Australian wild rabbits exposed to myxomatosis. *J Hyg* **56**, 288–302.

Mead-Briggs, A. R. (1964). Some experiments concerning the interchange of rabbit fleas, *Spilopsyllus cuniculi* (Dale), between living rabbits. *J Anim Ecol* **33**, 13–26.

Moriarty, A., Saunders, G. & Richardson, B. J. (2000). Mortality factors acting on adult rabbit population in central-western NSW. *Wildl Res* **27**, 613–619.

Morisse, J. P., le Gall, G. & Boilletot, E. (1991). Hepatitis of viral origin in Leporidae: introduction and aetiological hypothesis. *Rev Sci Tech Off Int Épizoot* **10**, 283–295.

Mougel, F. (1997). *Variation de trois types de marqueurs génétiques dans l'évolution de l'e-specè Oryctolagus cuniculus*. Thesis, University of Paris XI Orsay.

Myers, K. (1970). The rabbit in Australia. In *Dynamics of Populations*, pp. 478–506. Edited by P. J. den Boer & G. R. Gradwell. Wageninen: Centre for Agricultural Publishing and Documentation.

Myers, K., Marshall, I. D. & Fenner, F. (1954). Studies on the epidemiology of myxomatosis of rabbits. III. Observations on two successive epizootics in Australian wild rabbits on the reverine plain of south-eastern Australia 1951–1953. *J Hyg* **53**, 337–360.

Myers, K., Parer, I. & Richardson, B. J. (1989). Leporidae. In *Fauna of Australia*, vol. 1B, *Mammalia*, pp. 917–931. Edited by D. W. Walton & B. J. Richardson. Canberra: AGPS.

Nowotny, N., Bascunana, C. R., Ballagipordany, A., Gavierwiden, D., Uhlen, M. & Belak, S. (1997). Phylogenetic analysis of rabbit haemorrhagic disease and European brown hare syndrome viruses by comparison of sequences from the capsid protein gene. *Arch Virol* **142**, 657–673.

Ohlinger, V. F. & Thiel, H.-J. (1991). Identification of the viral haemorrhagic disease virus of rabbits as a calicivirus. *Rev Sci Tech Off Int Épizoot* **10**, 311–323.

Parer, I., Conolly, D. & Sobey, W. R. (1985). Myxomatosis: the effects of annual introduction of an immunizing strain and a highly virulent strain of myxoma virus into rabbit populations at Urana, N.S.W. *Aust Wildl Res* **12**, 347–423.

Parer, I., Sobey, W. R., Conolly, D. & Morton, R. (1994). Virulence of strains of myxoma virus and the resistance of wild rabbits, *Oryctolagus cuniculus* (L.), from different locations in Australia. *Aust J Zool* **42**, 347–362.

Plassiart, G., Guelfi, J. F., Ganière, J. P., Wang, B., André-Fontaine, G. & Wyers, M. (1992). Hematological parameters and visceral lesions relationships in rabbit viral haemorrhagic disease. *J Vet Med B* **38**, 443–453.

Richardson, B. J. (1980). Ecological genetics of the wild rabbit in Australia. III. Comparisons of the microgeographical distribution of alleles in two different environments. *Aust J Biol Sci* **33**, 385–391.

Richardson, B. J., Rogers, P. M. & Hewitt, G. M. (1980). Ecological genetics of the wild rabbit in Australia. II. Protein variation in British, French and Australian rabbits and the geographical distribution of the variation in Australia. *Aust J Biol Sci* **33**, 371–383.

Rodak, L., Smid, B., Valicek, L., Vesely, T., Stepanek, J., Hampl, J. & Jurak, E. (1990). Enzyme-linked immunosorbent assay of antibodies to rabbit haemorrhagic disease virus and determination of its major structural proteins. *J Gen Virol* **71**, 1075–1080.

Rolls, E. C. (1984). *They all Ran Wild: the Animals and Plants that Plague Australia*. Sydney: Angus and Robertson.

Stodart, E. (1965). A study of the biology of the wild rabbit in climatically different regions in eastern Australia. III. Some data on the evolution of coat colour. *CSIRO Wildl Res* **10**, 73–82.

Stodart, E. & Parer, I. (1988). Colonisation of Australia by the rabbit *Oryctolagus cuniculus*. *Division of Wildlife and Ecology, Project Report No. 6*.

Thompson, H. V. & King, C. M. (editors) (1994). *The European Rabbit. The History and Biology of a Successful Colonizer*. Oxford: Oxford University Press.

Wardhaugh, K. & Rochester, W. (1996). *Wardang Island. A Retrospective Analysis of Weather Conditions in Respect to Insect Activity and Displacement*. Unpublished Report to the Meart Research Corporation.

Williams, C. K., Parer, I., Coman, B., Burley, J. & Braysher, M. (1995). *Managing Vertebrate Pests: Rabbits*. Canberra: AGPS.

Wilson, G. (1995). *The Economic Impact of Rabbits on Agricultural Production: a Preliminary Assessment*. Canberra: ACIL Economics & Policy.

Zenger, K. R. (1996). *The genetic variation and evolution of the mitochondrial DNA non-coding region in Australian wild rabbit populations (Oryctolagus cuniculus (L))*. Thesis, University of Western Sydney.

Potential of influenza A viruses to cause pandemics

Alan J. Hay

Division of Virology, National Institute for Medical Research, The Ridgeway, Mill Hill, London NW7 1AA, UK

INTRODUCTION

The last century ended with heightened awareness of an impending influenza pandemic, i.e. a major worldwide epidemic caused by the emergence of a novel influenza A subtype against which the population has little or no immunity. The 'bird 'flu' (H5N1) outbreak in Hong Kong in 1997, which claimed the lives of six of its 18 victims, was terminated at the end of the year by the slaughter of the poultry in the live bird markets, which removed the source of infection (Claas *et al.*, 1998; Subbarao *et al.*, 1998; Bender *et al.*, 1999). Shortly after, in late 1998 in southern China and in early 1999 in Hong Kong, cases of mild influenza caused by another influenza A subtype, H9N2, indicated that the repopulated bird markets still harboured potential human pathogens (Peiris *et al.*, 1999; Lin *et al.*, 2000). Most of these infections resulted from avian-to-human transmission and there was little evidence of human-to-human spread; neither virus became established (as yet) in the human population (Bridges *et al.*, 2000; Katz *et al.*, 1999). The high mortality associated with influenza A H5N1 infection (of both chickens and people) raised the spectre of a recurrence of a 1918–1919-like pandemic which is estimated to have claimed in excess of 40 million lives worldwide (Crosby, 1989). A major concern was the possibility that co-infection of an individual by one of these viruses and a contemporary human strain might generate, by genetic reassortment (recombination), a virus with novel antigens and a capacity for efficient human-to-human transmission, akin to the viruses which were responsible for the 1957–1958 H2N2 Asian 'flu and 1968–1969 H3N2 Hong Kong 'flu pandemics (Scholtissek *et al.*, 1978; Schafer *et al.*, 1993). It is over 30 years since this last pandemic and there is general concern that the next is imminent. However, whereas recurrent epidemics of

SGM symposium 60: New challenges to health: the threat of virus infection.
Editors G. L. Smith, W. L. Irving, J. W. McCauley, D. J. Rowlands. Cambridge University Press. ISBN 0 521 80614 3

influenza are fairly predictable annual events, the future occurrence of a pandemic is unpredictable both as to its timing and the identity of the prospective virus. We do not know whether these or related viruses circulating in poultry in S.E. Asia pose a real threat or whether the cases of human infection represent false alarms. As yet we have little knowledge of the genetic factors important for limiting or facilitating the emergence and adaptation of a novel human virus. These incidents have, however, provided an added stimulus to studies of host specificity, interspecies transmission, adaptation and pathogenicity of the viruses and preparations for future pandemics.

INFLUENZA VIRUSES

Three types of influenza viruses infect the human population: influenza A, B and C viruses. The influenza C viruses cause mild infections more akin to the common cold and do not contribute to typical epidemics of influenza, which are due to A and B viruses. The latter viruses have a similar genetic make-up; their genomes comprise eight segments of single-stranded negative-sense RNA which encode a similar set of 10 or 11 protein products, respectively (reviewed by Lamb & Krug, 1996). Their haemagglutinins (HAs), responsible for attachment to and infection of cells, have similar receptor specificities and bind to sialic acid (*N*-acetylneuraminic acid)-containing glycoproteins and glycolipids on the cell surface; their receptor-destroying enzymes are neuraminidases (sialidases; NAs). The genome of influenza C viruses, in contrast, is composed of seven segments, one of which encodes a haemagglutinin–esterase-fusion (HEF) protein which possesses both the receptor-binding and membrane fusion activities of the haemagglutinins of A and B viruses and the receptor-destroying activity corresponding to the neuraminidases (Herrler & Klenk, 1991). Furthermore, the receptor-binding specificity for 9-*O*-acetyl *N*-acetylneuraminic acid and acetyl-esterase (receptor-destroying) activity of the HEF protein differ from those of the HA and NA of influenza A and B viruses (Herrler *et al.*, 1985; Rogers *et al.*, 1986). The coding strategies of influenza A and B viruses differ in certain important respects. Whereas the M2 proton channel of influenza A viruses is translated from a spliced mRNA of the M gene (Lamb *et al.*, 1981), the functionally equivalent NB channel of influenza B is translated from a bicistronic mRNA of the NA gene (Shaw *et al.*, 1983), the extra BM2 protein product of the spliced mRNA of the influenza B virus M gene having a different function (Odagiri *et al.*, 1999). How these differences and the differences in ion channel activity of M2 and NB (Hay, 1998) relate to the principal difference between A and B viruses, their host range, is not known. Whereas influenza B (and C) viruses are largely restricted to man, influenza A viruses are predominantly viruses of avian species. Furthermore, influenza A viruses exhibit greater genetic and antigenic diversity and comprise a variety of antigenically distinct subtypes, including various combinations of 15 HA(H) and 9 NA(N) subtypes (Webster *et al.*, 1992). The diversity of influenza A viruses is further enhanced by the high frequency of genetic reassortment (recombination) between the segmented genomes of two viruses following dual infection of a cell. All subtypes (defined by the combination of H and N antigens) are perpetuated in wild aquatic bird populations, including ducks, shorebirds and gulls, which act as a natural reservoir for influenza A viruses. The viruses are well adapted to these avian hosts. In ducks, infections are largely asymptomatic; the virus replicates preferentially in cells lining the intestine and large concentrations of virus are excreted in the faeces (Webster *et al.*, 1978), providing an efficient means of transmission to domestic as well as other feral avian species. Although periodically these viruses may infect mammalian species, sometimes causing severe disease, few subtypes have become established; these include three in man (H1N1, H2N2 and H3N2), three in pigs (H1N1, H1N2 and H3N2) and two in horses (H7N7 and H3N8) during the 20th century. There is therefore a large barrier which restricts transmission of 'novel' subtypes from the avian reservoir to immunologically naïve mammals.

The susceptibility of pigs to infection by both avian and human viruses led to the notion that pigs may have a role as an intermediate host, not only in transmission of avian viruses to the human population, but also as a 'mixing vessel' to promote genetic reassortment between avian and human virus genomes to facilitate the emergence of a potential pandemic strain (Scholtissek *et al.*, 1985). For example, the novel H2N2 and H3N2 subtypes which emerged to cause the pandemics of 1957–1958 and 1968–1969, respectively, were the result of genetic reassortment whereby the circulating human H1N1 or H2N2 virus acquired three genes (H2, N2 and PB1) or two genes (H3 and PB1), respectively, from avian viruses (Kawaoka *et al.*, 1989). Recent examples of the emergence in pigs of reassortant viruses which contain genes from human, swine and avian viruses, such as the H1N2 subtype in the UK (Brown *et al.*, 1998) and H3N2 viruses in the USA (Zhou *et al.*, 1999), serve to emphasize this potential role of the pig in zoonotic disease. Sporadic human infections by swine viruses of both H1N1 and H3N2 subtypes, some of which have caused serious disease, have not resulted in significant person-to-person spread and those viruses did not become established in the human population (Goldfield *et al.*, 1977; Patriarca *et al.*, 1984; Wells *et al.*, 1991; Claas *et al.*, 1994; V. Gregory, W. Lim, K. Cameron, M. Bennett, A. Klimov, N. Cox, A. Hay & Y. Lin, unpublished). In particular, swine viruses which arose by genetic reassortment between 'avian' H1N1 and 'human' H3N2 viruses co-circulating in Italian pigs (Castrucci *et al.*, 1993) have subsequently been responsible for human infection (Claas *et al.*, 1994; V. Gregory, W. Lim, K. Cameron, M. Bennett, A. Klimov, N. Cox, A. Hay & Y. Lin, unpublished).

HOST RANGE RESTRICTION

Influenza viruses in general replicate efficiently in the natural host but are restricted in their replication in other host species. Thus, as demonstrated experimentally, avian influenza viruses replicate poorly in the respiratory tract of human and non-human primates, such as squirrel monkeys, as well as other animals such as ferrets and hamsters, and human viruses replicate poorly in birds (Murphy *et al.*, 1982; Beare & Webster, 1991). Studies of genetic reassortants between avian and human (or animal) viruses have shown that many genetic factors may influence host range, including those which affect infection of the host, efficient replication in susceptible cells and spread to other individuals in the host population, the predominant factors depending on the virus and host species under study.

As regards the first of these, receptor specificity of the virus haemagglutinin (HA) is an important correlate of host range. Although all influenza A viruses recognize oligosaccharides with terminal sialic acid, they differ in their preference for two major types of sialyloligosaccharides, distinguished by the glycosidic linkage, either α2,3 or α2,6, between the terminal sialic acid and the penultimate galactose residue. Whereas avian and equine viruses exhibit a preference for attachment to sialic acid (SA) linked α2,3 to galactose (Gal), SAα2,3Gal, human viruses have a preference for the SAα2,6Gal disaccharides (Rogers & Paulson, 1983; Connor *et al.*, 1994). The importance of this difference is emphasized by the correlation between receptor specificity and the receptors present on the respective host cells at the site of infection. Thus the predominance of SAα2,3Gal oligosaccharides on cells of the duck intestine correlates with the specificity of avian viruses for these receptors (Ito *et al.*, 1998). Conversely, human viruses infect ciliated epithelial cells of the human trachea, which have predominately SAα2,6Gal oligosaccharides, but not mucin-producing goblet cells, which have primarily SAα2,3Gal oligosaccharides (Baum & Paulson, 1990). Furthermore, specificity for SAα2,6Gal avoids potential inhibition by SAα2,3Gal-containing mucins. The presence of SAα2,3Gal, and not SAα2,6Gal, on cells lining the lungs of seals and whales is consistent with the apparent frequent infection of these sea mammals by avian viruses of different subtypes (Ito *et al.*, 1999).

An explanation for the susceptibility of pigs to infection by both avian and human viruses, and their potential role as a 'mixing vessel' for generating avian–human reassortant viruses, is provided by the presence of both SAα2,3Gal and SAα2,6Gal receptors on cells of the pig trachea (Ito *et al.*, 1998). Furthermore, the observation that the 'avian-like' H1N1 viruses, which entered the pig population in Europe in the late 1970s, underwent a change in receptor specificity from recognition of both α2,3 and α2,6 sialic acid–galactose linkages to specificity for SAα2,6Gal, about 1984, emphasizes the potential of the pig to act as an intermediate host in the adaptation of avian viruses to a mammalian, potentially human, host. For example, the HAs of the H2N2 and H3N2 pandemic viruses, which entered the human population in 1957 and 1968, respectively, were derived from avian viruses and had apparently acquired the binding specificity for SAα2,6Gal, typical of human viruses. However, the H5N1 viruses isolated from fatal (and non-fatal) cases of human infection in Hong Kong in 1997 were shown to retain the preference for binding SAα2,3Gal of the chicken viruses from which they derived (Matrosovich *et al.*, 1999). It is apparent, therefore, that, in this instance, 'avian' receptor specificity did not prevent avian-to-human transmission or efficient replication of the viruses in the human respiratory tract; whether it reduced the potential for human-to-human transmission is not known. The ability of H9N2 viruses, circulating in poultry in Hong Kong, to infect pigs (K. Shortridge & D. Markwell, personal communication) as well as humans may be related to recognition by the HA of both α2,6 and α2,3 linkages (Y. Ha, D. J. Stevens, J. J. Skehel & D. C. Wiley, unpublished). It remains to be seen whether this property will enable these viruses to acquire a greater capacity to spread within human or pig populations.

Studies of mutants with altered receptor-binding specificity identified the importance of residue 226, in the receptor-binding site, in specifying preference for α2,3 or α2,6 linkages by H3 HAs. Leucine 226 correlated with selective binding to α2,6 whereas glutamine 226 correlated with α2,3 specificity (Rogers *et al.*, 1983). Comparisons of receptor-binding specificities and amino acid sequences of a large number of human, animal and avian H2 and H3 HAs broadened this correlation to include residue 228. The conservation of leucine 226 and serine 228 in human H2 and H3 HAs correlated with specificity for SAα2,6Gal, whereas glutamine 226 and glycine 228 in HAs of avian and equine viruses correlated with specificity for SAα2,3Gal receptors (Connor *et al.*, 1994). Recent X-ray crystallographic studies of complexes between H3, H5 and H9 HAs, with differing receptor specificities, and oligosaccharides with terminal sialic acid in an α2,3 or α2,6 linkage have provided an explanation of the structural basis of specificity. Although the sialic acid receptor is bound in the same orientation by the sites on the different HA subtypes, the affinity of binding is influenced by the interaction of leucine or glutamine 226 with different conformations, *cis* or *trans*, respectively, of the α2,6 or α2,3 linkages of sialic acid to galactose and the width of the receptor-binding site (Eisen *et al.*, 1997; Y. Ha, D. J. Stevens, J. J. Skehel & D. C. Wiley, unpublished). It is suggested that the combination of leucine 226 and glycine 228 of the H9 HA represents an intermediate in the evolution to leucine 226/serine 228 which is optimal for binding to α2,6 receptors.

Differences in amino acids close to the receptor-binding site, including some which alter glycosylation of the HA, may also influence the affinity and/or specificity of attachment (Ohuchi *et al.*, 1997; Gambaryan *et al.*, 1998). Several studies of mutant viruses selected in different host cells, of avian or mammalian origin, have shown that attachment of additional oligosaccharides close to the receptor-binding site causes reduced receptor binding, presumably due to steric interference with HA–receptor interaction.

Such changes may upset the important balance between the receptor-binding and receptor-destroying activities of the HA and NA, respectively.

The role of the neuraminidase is to cleave the terminal sialic acid receptor from glycoconjugates on virus-infected cells and the glycoproteins of progeny virus to facilitate virus release and prevent virus aggregation, and to remove potential intercellular inhibitors, so as to promote spread of virus to other cells and susceptible individuals. The complementary nature of the receptor-binding and receptor-destroying activities of the HA and NA, respectively, is reflected in the specificities of sialic acid recognition. Thus the preferential hydrolysis of SAα2,3Gal by avian virus NAs contrasts with the broader specificity, for SAα2,3Gal and SAα2,6Gal oligosaccharides, of the N2 neuraminidases of recent human H3N2 (and H2N2) viruses. The latter 'dual' specificity corresponds to the dual role of NA in cleaving the SAα2,6Gal disaccharides on host cells and removal of sialic acid linked α2,3 to galactose in potentially inhibitory mucins, facilitating spread of virus. The changing specificity during evolution of the N2 of H2N2 viruses between 1957 and 1968 and subsequently of H3N2 viruses after 1968 from initial high specificity for SAα2,3Gal of the avian virus derived N2 to the broader recognition of α2,3 and α2,6 linkages after 1972 emphasizes the significance of the specificity of substrate binding (Baum & Paulson, 1991). An amino acid change at position 275, associated with this change in specificity, is located close to the highly conserved catalytic site and appears to affect indirectly the affinity of binding of different substrates (Kobasa *et al.*, 1999).

Correlation between the specificities of HA and NA has also been observed in the preference of swine viruses for *N*-glycolylneuraminic acid (Neu5Gc) over *N*-acetylneuraminic acid (Neu5Ac). The relative increase in activity of NA against Neu5Gc-containing gangliosides during evolution of influenza A viruses in the swine population correlates with the preference of their HA for binding Neu5Gc and the predominance of oligosaccharides with this terminal residue in pig respiratory epithelium (Xu *et al.*, 1995; Suzuki *et al.*, 1997). The inability of human viruses to recognize Neu5Gc reflects the absence of Neu5Gc-containing oligosaccharides in the human respiratory tract. Differences in other features of NA structure have been associated with host specificity. For example, it has been suggested that shortening of the stalk of NA, due to amino acid deletions, together with reduced enzymic activity is associated with adaptation of the NA following transmission of H5N1 viruses from ducks to domestic chickens (Matrosovich *et al.*, 1999).

In addition to the surface HA and NA spikes, involved in transmitting virus between cells of susceptible hosts, various other gene products, necessary for efficient replication, have been shown to be important determinants of host range. Results of studies in a number of *in vitro* and *in vivo* virus–host systems have emphasized the importance of, for example, the PB2 polymerase component, the nucleoprotein (NP) and the products of the M gene, the major structural M1 protein and the M2 proton channel, in modulating the extent of replication following infection. Single gene reassortants in which the M or NP genes of the avian virus A/Mallard/New York/6750/78 (H2N2) replaced the corresponding gene in the human virus A/Udorn/72 (H3N2) were restricted in their replication in the respiratory tract of squirrel monkeys, comparable to that of the parental avian virus (Tian *et al.*, 1985). Replacement of the PB2 gene of A/Los Angeles/2/87 by that of A/Mallard/New York/78 restricted replication in human volunteers as well as squirrel monkeys (Clements *et al.*, 1992). In this case, the determinant of host range was traced to a single amino acid, residue 627. Host range mutants in which glutamic acid 627, typical of avian viruses, was substituted by lysine, typical of human viruses, had a phenotype similar to that of the parental human virus (Subbarao *et al.*, 1993).

PATHOGENICITY

In addition to the immune status of the population and its susceptibility to infection, the impact of influenza depends on the pathogenic potential of the virus, exacerbation by secondary infections, as well as environmental factors.

Virus infectivity is dependent on proteolytic cleavage of the haemagglutinin precursor HA0 to HA1 and HA2 to generate the N-terminal 'fusion peptide' of HA2, which is directly involved in promoting membrane fusion between the virus and endosome membranes during virus infection (Maeda & Ohnishi, 1980; Skehel *et al.*, 1982). Thus the extent of virus infection and spread of infection from the primary site to other tissues or organs in the body depends on the susceptibility of the HA to cleavage by intracellular and/or extracellular proteases and the tissue distribution of the activating proteases. For the majority of viruses, including those which infect humans, cleavage occurs at a single arginine residue, which separates HA1 and HA2, by extracellular trypsin-like endoproteases. For example, tryptase Clara, secreted by Clara cells of the bronchiolar epithelium in the respiratory tract of rats, and enzymes with similar activities in human and pig respiratory tracts have been implicated in the activation of infectivity (Kido *et al.*, 1992). Proteases produced by co-infecting bacteria or mycoplasmas may also participate and have been shown to be responsible for promoting virus replication in the lower respiratory tract and exacerbating the severity of disease (Tashiro *et al.*, 1987; Webster & Rott, 1987).

Susceptibility of the HA to cleavage activation by intracellular proteases is a prime determinant of the high pathogenicity of 'Fowl Plague' viruses of H5 and H7 subtypes, which cause systemic infection and rapid death in domestic poultry, including chickens and turkeys. The broad tissue tropism of these viruses, in contrast to the more localized respiratory or cloacal infection by low pathogenicity strains, is primarily due to the presence of a sequence of basic amino acids at the proteolytic cleavage site, e.g. RERRRKKR↓G (arrow denotes point of cleavage to generate the N-terminal glycine of HA2) of the pathogenic H5N1 Hong Kong viruses, such as A/Hong Kong/156/97. This multibasic sequence renders the HA susceptible to cleavage activation by ubiquitously distributed furin-like enzymes during transport through the *trans* Golgi network to the plasma membrane of the infected cell, prior to assembly into and release of progeny virus (Stieneke-Grober *et al.*, 1992; Horimoto *et al.*, 1994).

Introduction of highly pathogenic viruses into domestic poultry may occur directly from wild birds in which they cause no symptomatic disease (Kawaoka *et al.*, 1987). Alternatively, they may emerge during adaptation of a non-pathogenic strain in the new host. During an outbreak of avian H5N2 influenza in Mexico in 1994–1995, viruses which initially caused mild disease acquired high lethality for poultry. This phenotypic change correlated with changes in the amino acid sequence at the cleavage site, QRETR↓G→Q**RK**RKTR↓G (changes indicated in bold), including the insertion of two additional basic amino acids by sequence duplication (Garcia *et al.*, 1996; Perdue *et al.*, 1997). Similar genetic/phenotypic changes have been observed during experimental adaptation *in ovo* and *in vivo*. During an earlier outbreak of disease among chickens in Pennsylvania in 1983, emergence of a highly pathogenic variant, which caused 80 % mortality, was due to the loss of an oligosaccharide side chain attached to asparagine 11 of the HA of the non-virulent parent virus, which increases accessibility of the cleavage site to protease (Kawaoka *et al.*, 1984).

There have been few instances of human disease caused by pathogenic avian viruses. However, the Hong Kong incident in 1997 illustrated what might happen were a similar pathogenic avian virus, with a highly cleavable HA1/HA2 site, to acquire the ability to spread within the human population. Serological studies of household and social contacts showed that, in this instance, although person-to-person transmission may have occurred, most infections resulted from direct faecal–oral transmission from infected poultry and the H5N1 virus did not acquire a capacity for efficient spread within the human population (Katz *et al.*, 1999). The severity of disease ranged from mild respiratory symptoms to fatality in six of the 18 cases. Although it is not known whether, in the latter cases, the viruses caused systemic infections, differences in virulence for mice correlated well with differences in the outcomes of human infection. The viruses separated into two groups, the more virulent causing systemic infections in mice, while replication of the less virulent viruses was restricted to the respiratory tract (Gao *et al.*, 1999). It is evident, therefore, that the presence of the sequence of basic amino acids is not sufficient in itself to determine virulence in mammalian hosts and that other factors are important. Analyses of the influence of other genes identified the PB2 polymerase component as an important factor and that an amino acid difference at 627, glutamic acid, typical of avian viruses, or lysine, typical of human viruses, may be an important determinant of the replication capacity of these viruses in mammals (Gao *et al.*, 1999; see above).

Another mechanism for potentiating proteolytic cleavage of the HA was discovered from studies of the neurovirulent variant A/WSN/33, derived by passage in mouse brain of the original human influenza A virus isolate, A/WS/33. Two structural features of the neuraminidase were shown to be important: the absence of a conserved oligosaccharide attached to asparagine 130 of the N1 neuraminidase and the presence of a lysine at the C-terminus (Li *et al.*, 1993; Goto & Kawaoka, 1998). By analogy with the role of the C-terminal lysine of plasminogen-binding proteins, it is proposed that binding of plasminogen by the terminal lysine of the NA increases the local concentration of plasmin, following activation of the proenzyme, in the proximity of the HA, thus enhancing activation of the HA, which possesses a single arginine at the cleavage site, in a variety of tissues including the brain.

Relatively little is known about the factors which determine the virulence of human viruses. Various genes have been shown to influence the replicative capacity of different viruses and would consequently modulate the severity of the febrile response. In view of the enormous impact of the 1918–1919 pandemic, one of the principal reasons for identifying and studying the virus responsible is to gain an understanding of why it caused such high mortality, in particular among young adults. The first wave of the epidemic in the spring of 1918 was relatively mild, but was followed in the autumn by a particularly lethal form of the disease which engulfed the world during the following 6 months. Although the majority of fatalities were due to secondary bacterial pneumonia, large numbers of patients succumbed to viral pneumonia, the pathological consequences of which were confined to the respiratory tract. Sequence analyses of DNA amplified by PCR from fixed or frozen lung tissue obtained from three victims, two US soldiers who died in September 1918 and an Alaskan Inuit woman buried in the permafrost in November, established that they had been infected with an H1N1 subtype virus (Taubenberger *et al.*, 1997). The HA and NA components of the three viruses were very closely related and had properties intermediate between those of subsequently isolated avian and mammalian viruses (Reid *et al.*, 1999, 2000). Phylogenetic comparisons with the corresponding genes of avian, human and swine viruses place the 1918 sequences near the root of the mammalian clade, indicating that the genes had been introduced from an avian source just prior to the pandemic and that the virus was an ancestor of the human and swine H1N1 viruses isolated after 1930. Examination of the structural features of the HA and NA has, however, presented no clues as to why the viruses were so virulent, since these components possessed features similar to those of other human viruses and none of those discussed above, which are known to contribute to the pathogenicity of avian and mammalian viruses.

AVIAN-TO-MAMMALIAN TRANSMISSION

Since information on the human viruses preceding the 1918 pandemic is limited to serological evidence of their antigenic properties, we have no information on the genetic relationships or whether the 1918 virus was, like the novel subtypes of 1957 and 1968, the result of genetic reassortment between the circulating human virus and an avian strain or represents the introduction of a novel avian virus *in toto*. The ability of avian viruses circulating in domestic poultry to infect people has been clearly demonstrated by recent events in Hong Kong – infections by the virulent chicken H5N1 virus in 1997 (Claas *et al.*, 1998) and by the avirulent quail H9N2 virus in 1999 (Lin *et al.*, 2000). As regards factors which may be important for avian-to-human transmission, it is of interest that, despite co-circulation of genetically distinct viruses of the same subtypes, six of the genes (with the exception of those encoding the surface antigens) of the chicken H5N1 and quail H9N2 viruses which caused human infections were common to the two subtypes. These viruses were thus the result of genetic reassortment involving a 'common' precursor (Guan *et al.*, 1999; Hoffman *et al.*, 2000). In the absence of any obvious similarities to the corresponding genes, and their products, of human viruses, there are no clues as to which if any of these 'avian' genes were important for human infection.

On the other hand, apparent subtype specificity in avian-to-mammalian transmission emphasizes the importance of the HA and NA surface components. For example, an avian H3N8 virus which caused a serious outbreak of disease among horses in north-east China in 1989 was of the same subtype as the contemporary viruses circulating in horse populations in other parts of the world, but was genetically distinct, being more closely related to avian than to the equine viruses (Guo *et al.*, 1992). Furthermore, several examples of the introduction of genetically distinguishable avian viruses of the H1N1 subtype into pig populations in different parts of the world stress their susceptibility to infection by viruses of this subtype (Pensaert *et al.*, 1981; Guan *et al.*, 1996). These observations lend support to the notion that the human population may also be more susceptible to infection by certain subtypes, resulting in the cyclical re-emergence of a few subtypes. Thus H1N1 viruses, which had been replaced by the H2N2 subtype in 1957, reappeared in 1977 and have since co-circulated with the H3N2 subtype, introduced in 1968. Serological studies suggesting that H2 and H3 subtype viruses were responsible for the pandemics which began in 1889 and 1898, respectively, provide evidence for recycling of these subtypes (Davenport *et al.*, 1969), although doubt has been cast on the interpretations (Dowdle, 1999). On this basis, faced with an increasing immunologically naïve population, a favourite candidate to cause the next pandemic is therefore a H2 virus, which has been 'absent' from the human population since 1968. However, as for the timing of the next pandemic, it is not possible to predict which subtype of influenza A will be responsible.

SURVEILLANCE

The hallmark of influenza viruses is their ability to change and circumvent the immunity of the host to previous infection. During interpandemic periods, gradual changes in the antigenic characteristics of A and B influenza viruses result in recurrent annual epidemics of disease.

Despite the greater impact of pandemics following the introduction of a novel antigenic subtype, the cumulative impact of annual epidemics exceeds that of pandemics. Of prime importance in immunity is antibody to HA, which neutralizes virus infectivity by blocking HA binding to its receptor, although antibody to the neuraminidase also contributes by reducing virus dissemination. Thus the accumulation of amino acid changes in both components may contribute to reduced immunity. Since significant antigenic changes, in particular among H3N2 viruses, can occur in the space of 1 or 2 years, it is necessary to monitor continually the viruses circulating in the community so as to update effectively the composition of influenza vaccines to reflect such changes.

The main objectives of the World Health Organization Global Influenza Surveillance Network, which was established in 1948, are, therefore, the early detection of novel viruses which enter the population and may have the potential to spread and cause a pandemic, and the identification of immunologically significant antigenic variants among circulating viruses, so that the most representative, up-to-date variants of A and B viruses are included in the vaccine. The recent events in Hong Kong have reminded us of the need for constant vigilance. The circulation in poultry in other parts of Asia of H9N2 viruses similar to the human isolates (Cameron *et al.*, 2000) stresses the importance of increasing surveillance of human viruses in some parts of the world, currently poorly served by the global programme, and integrating increased surveillance of animal and avian viruses.

REFERENCES

Baum, L. G. & Paulson, J. C. (1990). Sialyloligosaccharides of the respiratory epithelium in the selection of human influenza virus receptor specificity. *Acta Histochem Suppl* **40**, 35–38.

Baum, L. G. & Paulson, J. C. (1991). The N2 neuraminidase of human influenza virus has acquired a substrate specificity complementary to the hemagglutinin receptor specificity. *Virology* **180**, 10–15.

Beare, A. S. & Webster, R. G. (1991). Replication of avian influenza viruses in humans. *Arch Virol* **119**, 37–42.

Bender, C., Hall, H., Huang, J., Klimov, A., Cox, N., Hay, A., Gregory, V., Cameron, K., Lim, W. & Subbarao, K. (1999). Characterization of the surface proteins of influenza A (H5N1) viruses isolated from humans in 1997–1998. *Virology* **254**, 115–123.

Bridges, C. B., Katz, J. M., Seto, W. H. & 17 other authors (2000). Risk of influenza A (H5N1) infection among health care workers exposed to patients with influenza A (H5N1), Hong Kong. *J Infect Dis* **181**, 344–348.

Brown, I. H., Harris, P. A., McCauley, J. W. & Alexander, D. J. (1998). Multiple genetic reassortment of avian and human influenza A viruses in European pigs, resulting in the emergence of an H1N2 virus of novel genotype. *J Gen Virol* **79**, 2947–2955.

Cameron, K. R., Gregory, V., Banks, J., Brown, I. H., Alexander, D. J., Hay, A. J. & Lin, Y. P. (2000). H9N2 subtype influenza A viruses in poultry in Pakistan are closely related to the H9N2 viruses responsible for human infection in Hong Kong. *Virology* **276**, 36–41.

Castrucci, M. R., Donatelli, I., Sidoli, L., Barigazzi, G., Kawaoka, Y. & Webster, R. G. (1993). Genetic reassortment between avian and human influenza A viruses in Italian pigs. *Virology* **193**, 503–506.

Claas, E. C. J., Kawaoka, Y., De Jong, J. C., Masurel, N. & Webster, R. G. (1994). Infection of children with avian-human reassortant influenza virus from pigs in Europe. *Virology* **204**, 453–457.

Claas, E. C. J., Osterhaus, A. D. M. E., van Beek, R., De Jong, J. C., Rimmelzwaan, G. F., Senne, D. A., Krauss, S., Shortridge, K. F. & Webster, R. G. (1998). Human influenza A H5N1 virus related to a highly pathogenic avian influenza virus. *Lancet* **351**, 472–477.

Clements, M. L., Subbarao, E. K., Fries, L. F., Karron, R. A., London, W. T. & Murphy, B. R. (1992). Use of single-gene reassortant viruses to study the role of avian influenza A virus genes in attenuation of wild-type human influenza A virus for squirrel monkeys and adult human volunteers. *J Clin Microbiol* **30**, 655–662.

Connor, R. J., Kawaoka, Y., Webster, R. G. & Paulson, J. C. (1994). Receptor specificity in human, avian, and equine H2 and H3 influenza virus isolates. *Virology* **205**, 17–23.

Crosby, A. (1989). *America's Forgotten Pandemic*. Cambridge: Cambridge University Press.

Davenport, F. M., Minuse, E., Hennessy, A. V. & Francis, T., Jr (1969). Interpretations of influenza antibody patterns of man. *Bull WHO* **41**, 453–460.

Dowdle, W. R. (1999). Influenza A virus recycling revisited. *Bull WHO* **77**, 820–828.

Eisen, M. B., Sabesan, S., Skehel, J. J. & Wiley, D. C. (1997). Binding of the influenza A virus to cell-surface receptors: structure of five hemagglutinin-sialyloligosaccharide complexes determined by X-ray crystallography. *Virology* **232**, 19–31.

Gambaryan, A. S., Marinina, V. P., Tuzikov, A. B., Bovin, N. V., Rudneva, I. A., Sinitsyn, B. V., Shilov, A. A. & Matrosovich, M. N. (1998). Effects of host-dependent glycosylation of hemagglutinin on receptor-binding properties of H1N1 human influenza A virus grown in MDCK cells and in embryonated eggs. *Virology* **247**, 170–177.

Gao, P., Watanabe, S., Ito, T., Goto, H., Wells, K., McGregor, M., Cooley, A. J. & Kawaoka, Y. (1999). Biological heterogeneity, including systemic replication in mice, of H5N1 influenza A virus isolates from humans in Hong Kong. *J Virol* **73**, 3184–3189.

Garcia, M., Crawford, J. M., Latimer, J. W., Rivera-Cruz, E. & Perdue, M. L. (1996). Heterogeneity in the haemagglutinin gene and emergence of the highly pathogenic phenotype among recent H5N2 avian influenza viruses from Mexico. *J Gen Virol* **77**, 1493–1504.

Goldfield, M., Bartley, J. D., Pizzuti, W., Black, H. C., Altman, R. & Halperin, W. E. (1977). Influenza in New Jersey in 1976: isolations of influenza A/New Jersey/76 virus at Fort Dix. *J Infect Dis* **136**, S347–S355.

Goto, H. & Kawaoka, Y. (1998). A novel mechanism for the acquisition of virulence by a human influenza A virus. *Proc Natl Acad Sci U S A* **95**, 10224–10228.

Guan, Y., Shortridge, K. F., Krauss, S., Li, P. H., Kawaoka, Y. & Webster, R. G. (1996). Emergence of avian H1N1 influenza viruses in pigs in China. *J Virol* **70**, 8041–8046.

Guan, Y., Shortridge, K. F., Krauss, S. & Webster, R. G. (1999). Molecular characterization of H9N2 influenza viruses: were they the donors of the "internal" genes of H5N1 viruses in Hong Kong? *Proc Natl Acad Sci U S A* **96**, 9363–9367.

Guo, Y., Wang, M., Kawaoka, Y., Gorman, O., Ito, T., Saito, T. & Webster, R. G. (1992). Characterization of a new avian-like influenza A virus from horses in China. *Virology* **188**, 245–255.

Hay, A. J. (1998). Functional properties of the virus ion channels. In *Textbook of Influenza*, pp. 74–81. Edited by K. G. Nicholson, R. G. Webster & A. J. Hay. London: Blackwell Science.

Herrler, G. & Klenk, H. D. (1991). Structure and function of the HEF glycoprotein of influenza C virus. *Adv Virus Res* **40**, 213–234.

Herrler, G., Rott, R., Klenk, H.-D., Müller, H.-P., Shukla, A. K. & Schauer, R. (1985). The receptor-destroying enzyme of influenza C virus is neuraminate-O-acetylesterase. *EMBO J* **4**, 1503–1506.

Hoffman, E., Stech, J., Leneva, I., Krauss, S., Scholtissek, C., Chin, P. S., Peiris, M., Shortridge, K. F. & Webster, R. G. (2000). Characterization of the influenza A virus gene pool in avian species in Southern China: was H6N1 a derivative or a precursor of H5N1? *J Virol* **74**, 6309–6315.

Horimoto, T., Nakayama, K., Smeekens, S. P. & Kawaoka, Y. (1994). Proprotein-processing endoprotease PC6 and furin both activate hemagglutinin of virulent avian influenza viruses. *J Virol* **68**, 6074–6078.

Ito, T., Nelson, J., Couceiro, S. S. & 9 other authors (1998). Molecular basis for the generation in pigs of influenza A viruses with pandemic potential. *J Virol* **72**, 7367–7373.

Ito, T., Kawaoka, Y., Nomura, A. & Otsuki, K. (1999). Receptor specificity of influenza A viruses from sea mammals correlates with lung sialyloligosaccharides in these animals. *J Vet Med Sci* **61**, 955–958.

Katz, J. M., Lim, W., Bridges, C. B. & 13 other authors (1999). Antibody response in individuals infected with avian influenza A (H5N1) viruses and detection of anti-H5 antibody among household and social contacts. *J Infect Dis* **180**, 1763–1770.

Kawaoka, Y., Neave, C. W. & Webster, R. G. (1984). Is virulence of H5N2 influenza viruses in chickens associated with loss of carbohydrate from the hemagglutinin? *Virology* **139**, 303–316.

Kawaoka, Y., Nestorowicz, A., Alexander, D. J. & Webster, R. G. (1987). Molecular analyses of the hemagglutinin genes of H5 influenza viruses: origin of a virulent turkey strain. *Virology* **158**, 218–227.

Kawaoka, Y., Krauss, S. & Webster, R. G. (1989). Avian-to-human transmission of the PB1 gene of influenza A viruses in the 1957 and 1968 pandemics. *J Virol* **63**, 4603–4608.

Kido, H., Yokogoshi, Y., Sakai, K., Tashiro, M., Kishino, Y., Fukutomi, A. & Katunuma, N. (1992). Isolation and characterization of a novel trypsin-like protease found in rat bronchiolar epithelial Clara cells. A possible activator of the viral fusion glycoprotein. *J Biol Chem* **267**, 13573–13579.

Kobasa, D., Kodihalli, S., Luo, M., Castrucci, M. R., Donatelli, I., Suzuki, Y., Suzuki, T. & Kawaoka, Y. (1999). Amino acid residues contributing to the substrate specificity of the influenza A virus neuraminidase. *J Virol* **73**, 6743–6751.

Lamb, R. A. & Krug, R. M. (1996). *Orthomyxoviridae*: the viruses and their replication. In *Fields Virology*, pp. 1353–1596. Edited by B. N. Fields, D. M. Knipe & P. M. Howley. Philadelphia: Lippincott–Raven.

Lamb, R. A., Lai, C.-J. & Choppin, P. W. (1981). Sequences of mRNAs derived from genome RNA segment 7 of influenza virus: colinear and interrupted mRNAs code for overlapping proteins. *Proc Natl Acad Sci U S A* **78**, 4170–4174.

Li, S., Schulman, J., Itamura, S. & Palese, P. (1993). Glycosylation of neuraminidase determines the neurovirulence of influenza A/WSN/33 virus. *J Virol* **67**, 6667–6673.

Lin, Y. P., Shaw, M., Gregory, V. & 10 other authors (2000). Avian-to-human transmission of H9N2 subtype influenza A viruses – relationship between H9N2 and H5N1 human isolates. *Proc Natl Acad Sci U S A* **97**, 9654–9658.

Maeda, T. & Ohnishi, S. (1980). Activation of influenza virus by acidic media causes hemolysis and fusion of erthrocytes. *FEBS Lett* **122**, 283–287.

Matrosovich, M., Zhou, N., Kawaoka, Y. & Webster, R. (1999). The surface glycoproteins of H5 influenza viruses isolated from humans, chickens, and wild aquatic birds have distinguishable properties. *J Virol* **73**, 1146–1155.

Murphy, B. R., Hinshaw, V. S., Sly, D. L., London, W. T., Hosier, N. T., Wood, F. T., Webster, R. G. & Chanock, R. M. (1982). Virulence of avian influenza A viruses for squirrel monkeys. *Infect Immun* **37**, 1119–1126.

Odagiri, T., Hong, J. & Ohara, Y. (1999). The BM2 protein of influenza B virus is synthesized in the late phase of infection and incorporated into virions as a subviral component. *J Gen Virol* **80**, 2573–2581.

Ohuchi, M., Ohuchi, R., Feldmann, A. & Klenk, H.-D. (1997). Regulation of receptor binding affinity of influenza virus hemagglutinin by its carbohydrate moiety. *J Virol* **71**, 8377–8384.

Patriarca, P. A., Kendal, A. P., Zakowski, P. C., Cox, N. J., Trautman, M. S., Cherry, J. D., Auerbach, D. M., McCusker, J., Belliveau, R. R. & Kappus, K. D. (1984). Lack of significant person-to-person spread of swine influenza-like virus following fatal infection in an immunocompromised child. *Am J Epidemiol* **119**, 152–158.

Peiris, M., Yuen, K. Y., Leung, C. W., Chan, K. H., Ip, P. L. S., Lai, R. W. M., Orr, W. K. & Shortridge, K. F. (1999). Human infection with influenza H9N2. *Lancet* **354**, 916–917.

Pensaert, M., Ottis, K., Vandeputte, J., Kaplan, M. M. & Bachmann, P. A. (1981). Evidence for the natural transmission of influenza A virus from wild ducks to swine and its potential importance for man. *Bull WHO* **59**, 75–78.

Perdue, M. L., Garcia, M., Senne, D. & Fraire, M. (1997). Virulence-associated sequence duplication at the hemagglutinin cleavage site of avian influenza viruses. *Virus Res* **49**, 173–186.

Reid, A. H., Fanning, T. G., Hultin, J. V. & Taubenberger, J. K. (1999). Origin and evolution of the 1918 "Spanish" influenza virus hemagglutinin gene. *Proc Natl Acad Sci U S A* **96**, 1651–1656.

Reid, A. H., Fanning, T. G., Janczewski, T. A. & Taubenberger, J. K. (2000). Characterization of the 1918 "Spanish" influenza virus neuraminidase gene. *Proc Natl Acad Sci U S A* **97**, 6785–6790.

Rogers, G. N. & Paulson, J. C. (1983). Receptor determinants of human and animal influenza virus isolates: differences in receptor specificity of the H3 hemagglutinin based on species of origin. *Virology* **127**, 361–373.

Rogers, G. N., Paulson, J. C., Daniels, R. S., Skehel, J. J., Wilson, I. A. & Wiley, D. C. (1983). Single amino acid substitutions in influenza haemagglutinin change receptor binding specificity. *Nature* **304**, 76–78.

Rogers, G. N., Herrler, G., Paulson, J. C. & Klenk, H. D. (1986). Influenza C virus uses 9-O-acetyl-*N*-acetylneuraminic acid as a high affinity receptor determinant for attachment to cells. *J Biol Chem* **261**, 5947–5951.

Schafer, J. R., Kawaoka, Y., Bean, W. J., Suss, J., Senne, D. & Webster, R. G. (1993). Origin of the pandemic 1957 H2 influenza A virus and the persistence of its possible progenitors in the avian reservoir. *Virology* **194**, 781–788.

Scholtissek, C., Rhode, W., von Hoyningen, V. & Rott, R. (1978). On the origin of the human influenza virus subtypes H2N2 and H3N2. *Virology* **87**, 13–20.

Scholtissek, C., Burger, H., Kistner, O. & Shortridge, K. F. (1985). The nucleoprotein as a possible major factor in determining host specificity of influenza H3N2 viruses. *Virology* **147**, 287–294.

Shaw, M. W., Choppin, P. W. & Lamb, R. W. (1983). A previously unrecognised influenza B virus glycoprotein from a bicistronic mRNA that also encodes the viral neuraminidase. *Proc Natl Acad Sci U S A* **80**, 4879–4883.

Skehel, J. J., Bayley, P. M., Brown, E. B., Martin, S. R., Waterfield, M. D., White, J. M., Wilson, I. A. & Wiley, D. C. (1982). Changes in the conformation of influenza virus hemagglutinin at the pH optimum of virus-mediated membrane fusion. *Proc Natl Acad Sci U S A* **79**, 968–972.

Stieneke-Gröber, A., Vey, M., Angliker, H., Shaw, E., Thomas, G., Roberts, C., Klenk, H.-D. & Garten, W. (1992). Influenza virus hemagglutinin with multibasic cleavage site is activated by furin, a subtilisin-like endoprotease. *EMBO J* **11**, 2407–2414.

Subbarao, E. K., London, W. & Murphy, B. R. (1993). A single amino acid in the PB2 gene of influenza A virus is a determinant of host range. *J Virol* **67**, 1761–1764.

Subbarao, K., Klimov, A., Katz, J. & 13 other authors (1998). Characterization of an avian influenza (H5N1) virus isolated from a child with a fatal respiratory illness. *Science* **279**, 393–396.

Suzuki, T., Horiike, G., Yamazaki, Y. & 10 other authors (1997). Swine influenza virus strains recognise sialylsugar chains containing molecular species of sialic acid predominantly present in the swine tracheal epithelium. *FEBS Lett* **404**, 192–196.

Tashiro, M., Ciborowski, P., Klenk, H.-D., Pulverer, G. & Rott, R. (1987). Role of *Staphylococcus* protease in the development of influenza pneumonia. *Nature* **325**, 536–537.

Taubenberger, J. K., Reid, A. H., Krafft, A. E., Bijwaard, K. E. & Fanning, T. G. (1997). Initial genetic characterization of the 1918 "Spanish" influenza virus. *Science* **275**, 1793–1796.

Tian, S.-F., Buckler-White, A. J., London, W. T., Reck, L. J., Chanock, R. M. & Murphy, B. R. (1985). Nucleoprotein and membrane protein genes are associated with restriction of replication of influenza A/Mallard/NY/78 virus and its reassortants in squirrel monkey respiratory tract. *J Virol* **53**, 771–775.

Webster, R. G. & Rott, R. (1987). Influenza virus A pathogenicity: the pivotal role of hemagglutinin. *Cell* **50**, 665–666.

Webster, R. G., Yakhno, M., Hinshaw, V. S., Bean, W. J. & Murti, K. G. (1978). Intestinal influenza: replication and characterization of influenza viruses in ducks. *Virology* **84**, 268–278.

Webster, R. G., Bean, W. J., Gorman, O. T., Chambers, T. M. & Kawaoka, Y. (1992). Evolution and ecology of influenza A viruses. *Microbiol Rev* **56**, 152–179.

Wells, D. L., Hopfensperger, D. J., Arden, N. H., Harmon, M. W., Davis, J. P., Tipple, M. A. & Schonberger, L. B. (1991). Swine influenza virus infections: transmission from ill pigs to humans at a Wisconsin Agricultural Fair and subsequent probable person-to-person transmission. *JAMA* **265**, 478–481.

Xu, G., Suzuki, T., Maejima, Y., Mizoguchi, T., Tsuchiya, M., Kiso, M., Hasegawa, A. & Suzuki, Y. (1995). Sialidase of swine influenza A viruses: variation of the recognition specificities for sialyl linkages and for the molecular species of sialic acid with the year of isolation. *Glycoconj J* **12**, 156–161.

Zhou, N. N., Senne, D. A., Landgraf, J. S., Swenson, S. L., Erickson, G., Rossow, K., Liu, L., Yoon, K. J., Krauss, S. & Webster, R. G. (1999). Genetic reassortment of avian, swine, and human influenza A viruses in American pigs. *J Virol* **73**, 8851–8856.

The hepatitis viruses as emerging agents of infectious diseases

Stanley M. Lemon

Departments of Microbiology & Immunology and Internal Medicine, The University of Texas Medical Branch, Galveston, TX 77555-1019, USA

INTRODUCTION

In the context of emerging viral infections, the hepatitis viruses present a fascinating contrast of the old and the new. Although the physicians of antiquity were almost certainly aware of viral hepatitis as a disease entity associated with icterus, it is only in the past century that this has been recognized clearly to be an infectious process. Furthermore, it was almost the mid-part of the twentieth century before it was shown unequivocally that hepatitis was not a single disease, and that very similar disease processes could result from infection by very different viral agents that are incapable of eliciting cross-protective immunity. Even more striking, however, it is only in the past decade that we have gained a full appreciation of the number and diversity of the hepatitis viruses.

The era of modern hepatitis virology began with the discovery of the 'Australia antigen' by Blumberg and associates in 1965 and the subsequent association of this antigen with hepatitis B virus (HBV) (Blumberg *et al.*, 1965; Prince, 1968). The identification of hepatitis A virus (HAV) came almost 8 years later, with the immune electron microscopic demonstration of HAV particles in human faecal material by Feinstone and colleagues (Feinstone *et al.*, 1973), and the hepatitis delta virus (HDV) was discovered by Rizzetto and coworkers at the end of the 1970s (Rizzetto *et al.*, 1980). With each of these discoveries came new serological tests and subsequent clinical studies that have redefined the epidemiological range of these infections and on more than one occasion demonstrated the existence of yet undiscovered hepatitis agent(s). By the mid-1970s, the realization that most cases of transfusion-transmitted hepatitis could be attributed to

SGM symposium 60: New challenges to health: the threat of virus infection.
Editors G. L. Smith, W. L. Irving, J. W. McCauley, D. J. Rowlands. Cambridge University Press. ISBN 0 521 80614 3 ©SGM 2001.

neither HAV nor HBV had led to the recognition that there was at least one other infectious agent responsible for post-transfusion 'non-A, non-B' hepatitis (Prince *et al.*, 1974; Feinstone *et al.*, 1975). A decade and a half long search for this agent culminated eventually in the discovery of hepatitis C virus (HCV) at the end of the 1980s (Choo *et al.*, 1989). Finally, the molecular cloning of hepatitis E virus (HEV) (Reyes *et al.*, 1990), the cause of water-borne or enteric 'non-A, non-B' hepatitis, brought this chapter in the annals of hepatitis viruses to an end with the discovery of the last of the five hepatitis viruses.

Present evidence suggests that these viruses account for almost all cases of acute viral hepatitis. The vast majority of persons presenting with this disease can be shown to have infection with one or more of these agents, providing scant evidence for another yet undiscovered hepatitis virus (Alter *et al.*, 1997). There are few compelling data to support the claim that infection with GB virus C (GBV-B) (mistakenly labelled 'hepatitis G virus') is frequently associated with liver injury (Simons *et al.*, 1995; Linnen *et al.*, 1996). There is even less evidence to suggest a role for the more recently discovered TT virus (TTV) (Okamoto *et al.*, 1999) in the causation of liver disease. Nonetheless, both GBV-B and TTV appear to be commonly transmitted by blood transfusion.

THE FIVE VIRAL AGENTS OF HEPATITIS

These five very different viruses, HAV, HBV, HCV, HDV and HEV, thus comprise the aetiologic agents responsible for either acute or chronic viral hepatitis in humans. [Other systemic virus infections may be associated with liver disease. Yellow fever virus may cause severe, often fatal, necrotic liver disease, at times mimicking the clinical picture of fulminant viral hepatitis. Similarly, primary infections with Epstein–Barr virus and cytomegalovirus are typically associated with mild hepatitis, although this rarely dominates the clinical picture.] For the most part, they share only a common tropism for the liver, with the hepatocyte representing the dominant site of viral replication and either acute or chronic forms of hepatitis representing the major clinical manifestations associated with infection. The convenient alphabetical classification of these viruses reflects neither phylogenetic nor taxonomic relationships between these viruses, but has evolved with the recognition of each form of viral hepatitis as a unique infectious disease and as much from happy coincidence as rational foresight.

With one notable exception described in greater detail below (HBV and HDV), these viruses share no antigenic specificities and infection with one does not confer protection against another. However, these viruses do have distinct but overlapping epidemiological characteristics that allow them to be categorized into two major groups that correlate well with the presence or absence of a viral envelope, a fundamental aspect of viral structure. Thus those viruses that lack a lipid-containing outer viral envelope (HAV and

HEV) share a number of clinical and epidemiological features that distinguish them from the enveloped viruses (HBV, HCV and HDV). This correlation of the clinical and epidemiological features of these viruses with their structural characteristics may relate to the fact that the absence of a lipid envelope in HAV and HEV confers stability on these viruses when they are secreted from infected hepatocytes into the bile. Both of these viruses gain entry to the intestinal tract via this route, and their spread via faeces dominates their epidemiology.

In contrast, HBV, HCV and HDV all possess lipid envelopes and are thus likely to be rapidly inactivated by bile if they were secreted into the biliary canaliculi from infected hepatocytes. These viruses are not found in faeces as infectious particles in biologically significant quantities, and their transmission is thus dependent upon several other routes, most often involving virus shed from a mucosal surface or by direct percutaneous exposure to blood. In contrast to HAV and HEV, each of these three enveloped viruses is capable of causing long-term, persistent infection, and each has been shown to be an important cause of chronic viral hepatitis and cirrhosis. It is interesting to speculate that the ability to initiate persistent infection may have arisen as an adaptive response to the relatively inefficient transmission of these viruses between individuals. For example, persistence would increase the chance of transmission of a virus that is spread by sexual contact, but may not be necessary to sustain the transmission of a virus spread by the faecal–oral route.

THE DISEASE BURDEN ASSOCIATED WITH HEPATITIS VIRUS INFECTION

In aggregate, these viruses are responsible for a considerable disease burden worldwide. Within the United States alone, more than 45 000 cases of acute hepatitis, mostly due to HAV, HBV and HCV, are reported annually (Centers for Disease Control and Prevention, 1995). These represent only a fraction of all cases, however, because less than 1 out of every 5–10 cases appears likely to be reported to public health authorities. Fulminant hepatic failure and death occur in a small proportion of patients with acute hepatitis A or B (the latter much more likely with coincident HDV infection), but these clinical end points are rarely associated with acute HCV infection in the United States (Farci *et al.*, 1996).

As dramatic as acute viral hepatitis may be, the major disease burden associated with hepatitis virus infections arises from the development of cirrhosis, liver failure and hepatocellular carcinoma in individuals with long-term persistence of HBV (with or without HDV) or HCV. The majority of persons who are initially infected with HCV fail to clear the virus and resolve the infection. Most eventually develop at least chemical and histological evidence of chronic liver injury (Alter *et al.*, 1992), even though the large majority of these individuals remain free of symptoms for years. HBV has an intermediate tendency to establish persistence, and chronic hepatitis B is largely confined to infants infected at birth and immunologically impaired adults (Seeff *et al.*, 1987; Cerny & Chisari, 1999).

In most persons, chronic infections with either of these viruses are relatively well tolerated, with little evidence of disease save elevated serum activities of liver-derived alanine and aspartate aminotransferase enzymes (Beasley *et al.*, 1981; Seeff, 1997). However, a small proportion of those who are infected go on to develop life-threatening liver failure or malignancy. There is little understanding of what selects these unfortunate individuals for either the insidiously progressive fibrotic reaction that characterizes the development of hepatic cirrhosis, or the malignant transformation of infected hepatocytes that leads to hepatocellular carcinoma.

THE HEPATITIS VIRUSES AS EMERGING AGENTS OF INFECTION

Several of the hepatitis viruses can be considered to be 'emerging' infectious agents in every sense of the word. The epidemiological studies that followed the discovery of new viruses have helped redefine the natural history of the diseases caused by these agents, while at the same time increasing our awareness of the magnitude of their contribution to human disease. In modern times, however, several of these viruses are undergoing significant changes in their epidemiology, and, in the case of HCV in particular, these changes have led to impressive attendant increases in disease incidence and infection prevalence. Although vaccines are available for protection against HAV as well as HBV, only the prevalence of hepatitis B has yet been impacted on by immunization and only in certain regions. At the same time, recently recognized HBV mutants are deserving of additional study. HAV remains a major and perhaps increasing threat to health in some developing countries, while HCV continues to emerge as an agent with considerable morbid potential.

HEPATITIS A VIRUS

Virology

HAV is a positive-sense, single-stranded RNA virus classified within the genus *Hepatovirus* of the family *Picornaviridae*. Unlike other hepatitis viruses, it can be propagated in conventional cell cultures. Cell culture isolation of virus is not a useful approach to diagnosis, however, because wild-type virus typically replicates poorly in cell culture. Several cell-culture-adapted HAV variants have been shown to be highly attenuated in their ability to cause disease in otherwise susceptible primates. However, these strains have been most useful in the production of formalin-inactivated HAV vaccines. While expensive, these vaccines are both highly effective and very safe (Lemon &

Thomas, 1997). Unfortunately, the relatively high cost of these vaccines has limited their use in the general population.

The changing threat of hepatitis A

As indicated above, infection usually occurs by the faecal–oral route of transmission, and is associated with extensive shedding of the virus in faeces during the 3–5 week incubation period and extending into the early days of the illness. Consistent with this mode of transmission, the prevalence of anti-HAV is clearly related to age as well as a number of socio-economic factors (Szmuness *et al.*, 1976, 1977). However, the age-related nature of antibody prevalence in many Western countries appears to be due largely to a cohort effect created by a recent decline in the overall incidence of HAV infection (Frosner *et al.*, 1978). In contrast to well-developed Western nations, infection still occurs during early childhood in many developing countries, often at an early age when specific symptoms of hepatitis A may be minimal or absent.

This situation may be changing, however, as public health sanitation continues to improve in these countries, leading to increased susceptibility to the virus among young adults. The potential impact of this effect is best typified by an epidemic of hepatitis A in Shanghai, which reportedly involved over 300 000 persons in early 1988 (Halliday *et al.*, 1991). However, the emerging nature of hepatitis A is also evident in reports of school-centred disease outbreaks in countries such as China in which virtually all children were probably asymptomatically infected at an early age previously. Unfortunately, the relatively high cost and lack of availability of hepatitis A vaccines has precluded a significant impact of immunization on hepatitis A rates in any nation to date.

Although the incidence of hepatitis A appears to be declining over the past several decades in the United States (Bell *et al.*, 1998), as it is in most developed nations, the globalization of the economy and the increasing importation of unprocessed foods from Mexico and other emerging markets has resulted in occasionally dramatic, multi-state outbreaks of hepatitis A due to contaminated produce (Hutin *et al.*, 1999). Such events are likely to continue on a sporadic basis as long as the majority of the adult population has been neither naturally infected nor immunized, and thus remains susceptible to the infection.

Also impressive over the past few decades has been the frequency with which large outbreaks of hepatitis A have occurred among urban homosexual men. In one study from Seattle (Corey & Holmes, 1980), an annual infection rate of 22 % was observed when seronegative homosexual men were followed prospectively. Men reporting frequent oral–anal exposure were found to be at significantly increased risk of becoming infected with HAV. Similar epidemics of hepatitis A have been described among urban homosexual men in Copenhagen, Amsterdam, Melbourne, London and other large, cosmopolitan Western cities in recent years. Risk factors for the acquisition of hepatitis A in these studies have generally paralleled those determined in the Seattle outbreak (Henning *et al.*, 1995).

Blood-borne hepatitis A is a rare but potential problem when units of blood or plasma are collected from infected donors who are in an early stage of the infection, prior to the onset of symptoms. Viraemia is present throughout most of the incubation period (Lemon *et al.*, 1990), but its potential for contributing to the parenteral transmission of HAV has been generally under-appreciated. If persons incubating the disease donate blood or plasma, or share needles with others for the injection of illicit drugs, the virus can be readily transmitted. Several outbreaks of hepatitis A among haemophilic patients receiving contaminated, solvent–detergent-inactivated factor VIII preparations have called special attention to the potential for parenteral transmission of this virus (Mannucci *et al.*, 1994). Solvent–detergent methods that were introduced for the inactivation of lipid-enveloped viruses contaminating such blood products have no effect on the infectivity of HAV (Lemon *et al.*, 1994). Other factors contributing to the spread of HAV by such products included the very large numbers of donor units in the plasma pools used for their manufacture, as well as the high purity of the product, which effectively excluded potentially protective immunoglobulins from being co-administered with the clotting factor.

Although transfusion-transmitted cases of hepatitis A remain relatively rare despite the absence of specific screening, numerous outbreaks of hepatitis A have been reported among users of illicit injected drugs (Widell *et al.*, 1983; Shaw *et al.*, 1999). Interestingly, the incidence of hepatitis A cases in illicit drug users peaked in the late 1980s within the United States, and has since declined in parallel with decreases in the incidence of HBV and HCV, two viruses which are well documented to be transmitted by exposure to contaminated blood. This argues that HAV may be similarly spread by parenteral means, rather than by faecal–oral transmission among drug users as often suspected previously.

HEPATITIS E VIRUS

Like HAV, HEV is a single-stranded, positive-sense RNA virus. Its phylogenetic relationship to other viruses is uncertain. Although the organization of the genome of HEV closely resembles that of the caliciviruses, its proteins show a much closer relatedness to those of the alphaviruses. Reliable propagation of HEV in cell culture has not been achieved, and much less is known about the biology of HEV compared to HAV. Strains of HEV recovered from patients in various geographic regions may differ significantly with respect to their nucleotide sequence, but these differences are not sufficiently large to suggest the existence of different serotypes.

In many ways, however, the course of HEV infection mimics the course of HAV infection, with extensive faecal shedding of the virus during the latter part of the 4–6-week-long incubation period. Fulminant hepatitis is a more common complication than with HAV, however, particularly in pregnant women, in whom the overall mortality rate may be as high as 20 % (Khuroo, 1980). HEV has been identified in developing nations of both hemispheres, where transmission generally occurs via the faecal–oral route and is frequently associated with contaminated drinking water. Outbreaks at times have been enormous, causing tens of thousands of cases, such as that which occurred in Delhi in 1955–1956 due to breakdown in the sanitary water supply (Wong *et al.*, 1980). Sporadic cases have also been described in well-developed countries, including the United States.

Given the sporadic nature of hepatitis E, even in regions with very poor public health sanitation, a nonhuman reservoir for this infection has long been suspected. Much attention has been focused in recent years on closely related viruses that commonly infect domestic swine (Meng *et al.*, 1997). Young pigs that were naturally infected by the swine virus had microscopic evidence of hepatitis, and were viraemic prior to seroconversion. The putative capsid gene (ORF2) of the swine virus shares about 90–92 % identity with human HEV strains at the amino acid level. Thus swine HEV is closely related to, but distinct from, human HEV. It is interesting to note that infection with swine HEV is common among swine herds in the United States, despite the virtual absence of indigenously acquired hepatitis E in humans. Interestingly, 50 % or more of wild rats trapped in the United States also have been shown to have serological evidence of infection with HEV (Kabrane-Lazizi *et al.*, 1999). The relationship of these zoonotic viruses to disease in humans will clearly require further study.

Vaccines based on the expression of the major HEV capsid protein from recombinant DNA appear promising, and are in early stages of clinical evaluation.

HEPATITIS B VIRUS

Molecular virology

HBV, an hepadnavirus, is the only hepatitis virus with a DNA genome. However, this virus has a unique replication cycle which involves an RNA intermediate and a reverse transcription step (Chisari, 1992). The HBV particle (sometimes called the Dane particle) is a spherical double-shelled structure approximately 47 nm in diameter. The outer shell is a lipid envelope containing the surface antigen, HBsAg. The envelope can be removed by mild detergent treatment to reveal a stable nucleocapsid with a distinct antigenic specificity (core antigen or HBcAg). This nucleocapsid contains the genomic DNA, which in the mature virus particle consists of two complementary linear strands of DNA that are base-paired with each other to form a partially double-stranded, circular molecule of about 3.2 kb. Also present in the nucleocapsid is a virally encoded reverse transcriptase/DNA polymerase. Several overlapping reading frames in the HBV genome encode the three carboxy-coterminal S proteins, the core protein, the reverse transcriptase/polymerase and a transactivator protein, 'X'. Alternate transcriptional start sites lead to the synthesis of two different RNA messages containing the core protein open reading frame (ORF), and different translation initiation codons in these two mRNAs result in the expression of the HBcAg found in the nucleocapsid as well as an amino-terminally extended 'precore' protein. The precore protein contains a signal sequence which directs it to the endoplasmic reticulum, where it is processed to a soluble form known as 'e' antigen (HBeAg). Although the HBeAg and HBcAg molecules contain overlapping amino acid sequences, they represent very different antigenic specificities.

Natural history and the host immune response

HBV is most commonly transmitted sexually between adults, but can also be transmitted parenterally following inapparent percutaneous exposure to contaminated blood. Horizontal transmission occurs between children, and vertical transmission occurs between mothers (especially if HBeAg-positive) and their newborn infants. Infection with HBV usually entails a lengthy incubation period (2–6 months) between exposure and the onset of liver disease (Lee, 1997). Viraemia is typically present throughout much of this incubation period, but virus is also shed from mucous membranes, resulting in relatively efficient sexual transmission. Symptomatic hepatocellular disease probably occurs in most immunocompetent infected adults, the vast majority of whom (probably over 99 %) successfully clear the infection and develop permanent immunity (Norman *et al.*, 1993). Following such acute infection with HBV, anti-HBs, anti-HBc and anti-HBe antibodies all appear, with immunity largely dependent upon the anti-HBs response. Newborn infants and immunocompromised adults who are infected with HBV are very likely to become persistently infected. Such individuals have chronic viraemia (associated with circulating HBsAg) and usually substantial anti-HBc (IgG) but not anti-HBs antibody responses. Persistent carriers of HBV may or may not have associated hepatocellular injury.

The emerging issue of natural and drug-selected HBV mutants

Immunization is the mainstay of hepatitis B control and is recommended universally for newborn infants in many countries. Vaccination is generally accomplished with recombinant vaccines that contain the small S protein expressed in yeast. Immunization programmes have been aggressively promoted by the World Health Organization and other agencies, resulting in impressive regional decreases in the incidence and prevalence of hepatitis B infection and its attendant complications (Lemon & Thomas, 1997). No significant serotypic differences have been described among HBV strains from various regions of the world, although minor antigenic variation is common (antigenic subtypes). However, mutations in the S protein have been suggested to be involved in uncommon cases of apparent vaccine failure (Waters *et al.*, 1992). These mutants have been recognized particularly in infected infants who have failed perinatal prophylaxis with vaccines. Nonetheless, the standard, licensed recombinant hepatitis B vaccine has been shown to provide protection against such variants in a chimpanzee model (Ogata *et al.*, 1999). Furthermore, even if such virus mutants were completely resistant to vaccine-induced antibodies, mathematical modelling suggests that it would take decades for a vaccine-resistant S mutant to emerge as the dominant HBV strain due to the generally slow spread of HBV within human populations (Wilson *et al.*, 2000).

A different type of HBV variant with mutations within the precore coding region has been described by a number of investigators (Hasegawa *et al.*, 1991; Liang *et al.*, 1994). These precore mutants are unable to express HBeAg, generally due to the presence of a mutation within the precore region, but they retain the ability to replicate and express the core protein (HBcAg). These viruses have appeared in the course of persistent HBV infection, generally in association with an HBeAg to anti-HBe seroconversion, despite the continuing presence of circulating HBV DNA. Alternatively, they may be the cause of new infections. In some situations, particularly following liver transplantation, these precore mutant HBVs have been associated with clinically aggressive forms of liver disease. However, the impact of these mutations on the natural history of chronic hepatitis B is not well characterized and requires further study (Blum, 1997).

Yet a third type of HBV mutant that has been recognized in recent years is represented by lamivudine-resistant variants with amino acid substitutions involving the YMDD motif within the viral polymerase. The antiviral drug lamivudine has been increasingly used for suppression of HBV replication in patients with hepatitis B, but only rarely is its use associated with complete resolution of the infection. Drug-resistant variants emerge in as many as 50 % of patients with chronic hepatitis B who are treated with the antiviral for periods as long as 24 months (Lau *et al.*, 2000; Liaw *et al.*, 2000). Consistent with *in vitro* assays showing the resistance of such variants to lamivudine, patients in whom these variants appear usually have a return in the magnitude of their HBV viraemia to levels approaching those pretherapy. The polymerase mutations in these virus variants appear to compromise their replication capacity, however, and the variants are replaced by wild-type virus upon cessation of drug therapy (Sakugawa *et al.*, 1999). There is as yet no evidence that the emergence of these drug-resistant variants carries any adverse prognostic implications.

HEPATITIS D VIRUS

HDV is a defective subviral satellite of HBV that is a cause of severe and often fatal liver disease in persons infected with HBV (Taylor, 1999). The HDV particle is approximately 35 nm in diameter, with an outer lipid envelope containing predominantly the small S protein of HBV. The envelope contains a poorly organized ribonucleoprotein structure, comprised of a 1.7 kb single-stranded, minus-sense, circular RNA genome in association with an HDV-encoded protein, the hepatitis delta antigen (HDAg). The RNA is catalytically active, and both minus and plus strands are capable of self-cleavage and ligation in the absence of proteins.

HDV is dependent upon a coinfecting hepadnavirus for provision of its envelope, and thus cannot replicate in the absence of HBV infection. Infection with HDV can occur either simultaneously with HBV (acute coinfection) or in previously infected chronic carriers of HBV (HDV superinfection). HDV infection is diagnosed by demonstration of antibodies to HDAg, with superinfections distinguished from acute coinfections by the absence of IgM antibodies to HBcAg. Severe and even fulminant hepatitis is common in both coinfections and superinfections, with appreciable mortality rates in both settings (De Cock *et al.*, 1986). Typically, HBV carriers who develop persistent HDV infection will experience an acceleration in the progression of their disease, with an increased risk of death due to cirrhosis and/or liver cancer.

In the United States, multiply transfused individuals (e.g. haemophilic patients) and users of illicit parenteral drugs have historically been at highest risk of acquiring HDV infection. Transmission appears to be predominantly by percutaneous exposure to contaminated blood, but there are data supporting the sexual transmission of this virus. Infection with HDV has been increasingly recognized among hepatitis B carriers in Asia (Taiwan and Japan). However, infection with Asian genotypes may lead to a milder course of infection than reported previously with European and South American strains of the virus (Sakugawa *et al.*, 1999). Although a potentially serious infection, there is little evidence to suggest a significant emerging role for this hepatitis agent at present.

HEPATITIS C VIRUS

Molecular virology

HCV is a single-stranded, positive-sense RNA virus with a genome length of approximately 9.6 kb (Houghton *et al.*, 1991; Major & Feinstone, 1997). It is currently classified within a separate genus of the family *Flaviviridae*, the genus *Hepacivirus*. The

HCV genome contains a single large ORF that follows a relatively lengthy 5′ nontranslated region of approximately 342 bases containing an internal ribosome entry segment (IRES) directing cap-independent initiation of viral translation. The large ORF encodes a polyprotein which undergoes post-translational cleavage, under control of cellular and viral proteinases. This yields a series of structural proteins which includes a core or nucleocapsid protein, two envelope glycoproteins, E1 and E2, and at least six nonstructural replicative proteins. These include NS2 (which with the adjacent NS3 sequence demonstrates *cis*-active metalloproteinase activity at the NS2/NS3 cleavage site), NS3 (a serine proteinase/NTPase/RNA helicase), NS4A (serine proteinase accessory factor), NS4B, NS5A and NS5B (RNA-dependent RNA polymerase).

With the exception of the 5′ nontranslated region, there is substantial genetic heterogeneity among different stains of HCV. This has led to the classification of HCV strains into a series of genetically distinct 'genotypes'. The genetic distance between some of these genotypes is large enough to suggest that there may be biologically significant serotypic differences as well. There is little understanding of the extent to which infection with a virus of any one genotype might confer protection against viruses of a different genotype. However, there appears to be only low-level homologous protection when chimpanzees are challenged twice with the same strain of HCV (Farci *et al.*, 1992).

Natural history and disease pathogenesis

One of the most striking features of HCV, yet to be explained, is its proclivity to establish persistent infections. This occurs in the majority of infected persons, and is typically associated with serum aminotransferase activities that are elevated to levels that are indicative of liver injury (Alter *et al.*, 1992). The histopathological hallmarks of chronic hepatitis C are portal inflammation, centrilobular, predominantly microvesicular steatosis, and progressive hepatic fibrosis. This culminates in cirrhosis in about a third of patients after 20–30 or more years of infection, but it may be present as early as 5 years after infection. Infected individuals with well-compensated, early cirrhosis are at greatly increased risk for developing hepatocellular carcinoma (Nishiguchi *et al.*, 1995). Interestingly, in the United States, most deaths related to HCV infection occur as a result of cirrhosis and liver failure, while a much smaller fraction occur as a result of liver cancer. For reasons that remain completely obscure, these proportions are reversed in Japan, with most HCV-related deaths due to liver cancer and a smaller fraction due to cirrhosis.

Recent work in Galveston involving transgenic animal models of hepatitis C and hepatocyte-derived cell lines with inducible expression of HCV proteins suggests that the accumulation of intracellular reactive oxygen species (ROS) plays a major role in this disease process (M. Okuda and others, unpublished). ROS appear to be generated as a result of core-protein-mediated mitochondrial injury. This mitochondrial injury not only leads directly to the accumulation of ROS, but also renders the hepatocyte more vulnerable to the additional oxidative stress generated by the inflammatory response to HCV infection or even moderate levels of ethanol consumption. This may explain the apparent synergy between HCV infection and ethanol consumption in enhancing the progression of the liver injury. Furthermore, ROS can serve as a stimulus for activation of stellate cells, the cell type in the liver that is responsible for the abnormal production of collagen and that plays a key role in the fibrosis accompanying chronic hepatitis C. It is also possible that nonspecific ROS-induced cellular DNA damage, coupled with accelerated hepatocyte turnover, may be the process that leads ultimately to liver cancer. This hypothesis is supported by recent microarray evidence of substantial stochastic gene dysregulation in transgenic mice expressing the structural proteins of HCV (M. Beard and others, unpublished).

Hepatitis C as an emerging infection

The single most common risk factor for hepatitis C is illicit injection drug use. This is a practice that has become increasingly prevalent in recent decades in many societies, with resultant increases in HCV transmission. Active HCV infections are highly prevalent among drug users, and exposure to HCV generally occurs soon after a user begins illicit drug use. Within the United States, at least two-thirds of all community-acquired HCV infections are related to illicit injection drug use (Alter *et al.*, 1990). Approximately a third of persons with acute hepatitis C admit to such drug use within the 6 months prior to their illness, while a somewhat greater proportion have other potential indicators of injection drug use. In contrast to HBV, the sexual transmission of HCV appears to be relatively inefficient, although specific sexual practices or numbers of sex partners have been related to the risk of infection. Maternal–infant transmission of HCV also occurs at much lower rates than with HBV, but it is increased when the mother is infected with human immunodeficiency virus.

In the 1980s, the annual incidence of HCV infection in the United States was approximately 15/100 000. Since then, it has declined significantly due to changes in illicit drug practices. The results of a large survey carried out at the beginning of the last decade indicate that there are approximately 3.9 million infected individuals within the United States, representing about 1.8 % of the general population (Alter *et al.*, 1999). The prevalence in men is generally higher than in women, and increased among racial minorities in comparison to Caucasian Americans. The highest infection prevalence was found among persons 30–50 years of age, presumably representing an upsurge of new infections associated with illicit injection drug use in young adults beginning in the 1960s.

Although the prevalence of HCV is remarkably similar to that found in the United States in many parts of the world, there are distinct geographic regions where the prevalence is much higher. In Egypt, for example, HCV infection is present in over 10 % of the general population (Abdel-Wahab *et al.*, 1994; Darwish *et al.*, 1996). Localized regions with similarly high rates of infection have been reported also in some European and Asian countries (Nakashima *et al.*, 1995; Chiaramonte *et al.*, 1996). In such areas, HCV infection is generally more prevalent among persons over 40 years of age, suggesting that transmission may have been due to conditions that no longer exist, such as traditional folk remedies or the reuse of nonsterile needles for injection. For example, in a study carried out in an isolated region of Japan, almost half of the individuals over 40 years of age had a serological profile indicative of HCV infection (Kiyosawa *et al.*, 1994). Traditional healing practices involving skin puncture and the drawing of blood with nonsterile instruments were suggested to have facilitated transmission of the virus. On the other hand, in Egypt, it was suggested that extensive use of nonsterile needles for administration of parenteral antischistosomal therapy may have contributed to the spread of the infection (Frank *et al.*, 2000).

In Japan, hepatitis C now exceeds hepatitis B in contributing to the development of hepatocellular carcinoma, one of the most common types of cancer worldwide (Kiyosawa & Furuta, 1994). In this regard, Japan appears to be unique among Asian countries. Hepatocellular carcinoma has been an increasing problem in Japan, with current mortality rates of more than 20 deaths per 10^5 persons per year, up from ~8 deaths per 10^5 persons per year prior to 1980. This increase is largely due to an increase in HCV-associated cancer. HBV and HCV presently account for 17 % and 76 % of cases, respectively, a change from 1970 when >60 % of cases of hepatocellular carcinoma could be related to HBV infection. This increase in the mortality rate and shift to association with HCV is likely due to the spread of HCV among the Japanese due to widespread use of injection amphetamines just after World War II (Lemon *et al.*, 2000).

A 1990 Ministry of Health and Welfare report indicated that as many as 5 % of young adult Japanese (ages 15–25) used oral amphetamines between 1945 and 1951, and that the illicit use of injected amphetamines became common between 1951 and 1954 (Lemon *et al.*, 2000). In 1954, there were 55 000 drug-related arrests and an estimated 200 000–500 000 active users in Japan. As occurred in the United States approximately 25–35 years later, the shared use of nonsterile needles for injection of illicit drugs is likely to have promoted the spread of HCV during this period. Illicit amphetamine use subsided by the late 1960s, but a second phase of HCV dissemination may have resulted from blood transfusions prior to the advent of specific viral screening. Phylogenetic analyses of HCV strains recovered in Japan suggest a genetic dispersion of the virus during the 1950s, whereas similar analyses in the United States suggest dispersion of the virus during the late 1960s and early 1970s (Lemon *et al.*, 2000).

The societal events in Japan that appear to explain the sharp increases that occurred in the incidence of HCV-related hepatocellular carcinoma during the 1980s in Japan may foretell similar increases in liver cancer (and cirrhosis) within the United States population. Indeed, this picture is consistent with prevalence of HCV infection in the United States being highest among those 30–50 years of age, and the notion that end-stage cirrhosis and cancer generally develop after an interval of 30 or more years or persistent infection. At present, within the United States, it is estimated that there are between 8000 and 10 000 deaths due to HCV infection (Alter, 1997). The age-specific prevalence of infection, taken in the context of what is known of the natural history of the disease, suggests that this number may double or triple over the next 10–20 years.

That this may be happening already is indicated by a recent survey of hepatocellular carcinoma in the United States showing that the incidence of histologically proved cancer increased from 1.4 cases per 100 000 population during 1976–1980 to 2.4 per 100 000 in 1991–1995 (El-Serag & Mason, 1999). During the latter period, the incidence of cancer among black men was over twice that among Caucasian men, similar to the approximately twofold difference in the prevalence of HCV infection in these ethnic groups. While certainly not proven, it is reasonable to suspect that any increase in the incidence of liver cancer may be due to earlier increases in the prevalence of HCV infection in the American population. A recent report describing the *in situ* detection of HCV in liver tissues from over 40 % of recent cases of liver cancer in the United States provides support for that view (Abe *et al.*, 1998). Should this trend continue, and particularly if it is mirrored by similar increases in the death rate due to cirrhosis, the mortality associated with chronic HCV infection may potentially exceed the number of deaths due to AIDS.

Prospects for an HCV vaccine

Studies of the immune response in patients who have fully recovered following an acute HCV infection and who completely cleared the virus from the liver have unambiguously demonstrated the importance of a broadly directed cellular immune response involving both T helper (T_h) and cytotoxic T cells. Thus while the correlates of immunity are poorly understood, both humoral and cellular immune responses are likely to be essential to protection. Efforts to develop a vaccine against hepatitis C have been limited by a number of technical difficulties, compounded by intellectual property issues and probably also perceptions of a limited market for such a vaccine in developed countries. The compounding technical issues include the fact that HCV infects only humans and chimpanzees, resulting in the absence of small animal models in which the immunologic correlates of protection against the virus can be studied. Furthermore, the inability of HCV to be efficiently propagated in cell culture prevents the production of viral antigen in the abundance required for production of conventional inactivated vaccines, and limits prospects of creating an attenuated HCV vaccine. Attempts to

Table 1. Emerging features of hepatitis virus epidemiology

Hepatitis virus	Chronic hepatitis cirrhosis	Hepatocellular carcinoma	Effective vaccine	Trends in infection prevalence	Disease burden
A	No	No	Yes	Decreasing due to improved sanitation	Increasing disease incidence in some developing nations where infection is now encountered at older ages
B	Yes	Yes	Yes	Decreasing due to effective immunization programmes	Potential to decrease due to effective vaccines and antiviral therapy, but emergence of HBV variants is of uncertain long-term consequence
C	Yes	Yes	No	Recently increased in many societies due to illicit drug use	Lengthy infection period preceding development of clinically significant liver disease predicts future increases in incidence of hepatocellular carcinoma and cirrhosis
D	Yes	?	Yes*	Decreasing	Declining clinical significance in Europe, but remains a potential problem in S. America and possibly Asia
E	No	No	No	Unknown	Epidemiology not well established; fatal disease in 3rd trimester pregnancy

*HDV infection is prevented by hepatitis B immunization in HBV-naïve individuals.

protect chimpanzees with a candidate recombinant vaccine (Chiron) containing both HCV envelope proteins, E1 and E2, met with only limited success, as the vaccine was shown to be capable of inducing detectable antibody responses but only marginal protection against a homologous HCV challenge. Thus it appears unlikely that an effective vaccine will become available any time soon.

CONCLUSION

The most important features relating to disease burden and emergence characteristics of the five human hepatitis viruses are summarized in Table 1. Together, these viruses present a major challenge to those charged with maintaining the health of the public. The hepatitis B vaccine represents a remarkable success story, despite the appearance of virus variants with mutations in the envelope protein in rare patients who have failed immunization. However, the lack of affordable vaccines continues to stymie efforts to control new hepatitis A infections in the developing world, while the lack of any effective means of immunization limits control of hepatitis C to attempts to reduce transmission by modification of risky behaviours such as the illicit injection of drugs, or the screening of blood products for contaminating viruses. More effective and certainly more affordable therapies will be required to stem the almost certain increases in HCV-related liver-specific mortality that are to be expected in the next two decades.

REFERENCES

Abdel-Wahab, M. F., Zakaria, S., Kamel, M., Abdel-Khaliq, M. K., Mabrouk, M. A., Salama, H., Esmat, G., Thomas, D. L. & Strickland, G. T. (1994). High seroprevalence of hepatitis C infection among risk groups in Egypt. *Am J Trop Med Hyg* **51**, 563–567.

Abe, K., Edamoto, Y., Park, Y. N., Nomura, A. M., Taltavull, T. C., Tani, M. & Thung, S. N. (1998). In situ detection of hepatitis B, C, and G virus nucleic acids in human hepatocellular carcinoma tissues from different geographic regions. *Hepatology* **28**, 568–572.

Alter, M. J. (1997). Epidemiology of hepatitis C. *Hepatology* **26** (Suppl. 1), 62S–65S.

Alter, M. J., Hadler, S. C., Judson, F. N., Mares, A., Alexander, W. J., Hu, P. Y., Miller, J. K., Moyer, L. A., Fields, H. A., Bradley, D. W. & *et al.* (1990). Risk factors for acute non-A, non-B hepatitis in the United States and association with hepatitis C infection. *JAMA* **264**, 2231–2235.

Alter, M. J., Margolis, H. S., Krawczynski, K. & 9 other authors (1992). The natural history of community-acquired hepatitis C in the United States. *N Engl J Med* **327**, 1899–1905.

Alter, M. J., Gallagher, M., Morris, T. T., Moyer, L. A., Meeks, E. L., Krawczynski, K., Kim, J. P. & Margolis, H. S. (1997). Acute non-A-E hepatitis in the United States and the role of hepatitis G virus infection. Sentinel Counties Viral Hepatitis Study Team. *N Engl J Med* **336**, 741–746.

Alter, M. J., Kruszon-Moran, D., Nainan, O. V., McQuillan, G. M., Gao, F., Moyer, L. A., Kaslow, R. A. & Margolis, H. S. (1999). The prevalence of hepatitis C virus infection in the United States, 1988 through 1994. *N Engl J Med* **341**, 556–562.

Beasley, R. P., Hwang, L.-Y., Lin, C.-C. & Chien, C.-S. (1981). Hepatocellular carcinoma and hepatitis B virus: a prospective study of 22 707 men in Taiwan. *Lancet* **ii**, 1129–1133.

Bell, B. P., Shapiro, C. N., Alter, M. J., Moyer, L. A., Judson, F. N., Mottram, K., Fleenor, M., Ryder, P. L. & Margolis, H. S. (1998). The diverse patterns of hepatitis A epidemiology in the United States – implications for vaccination strategies. *J Infect Dis* **178**, 1579–1584.

Blum, H. E. (1997). Hepatitis viruses: genetic variants and clinical significance. *Int J Clin Lab Res* **27**, 213–224.

Blumberg, B. S., Alter, H. J. & Visnich, S. (1965). A "new" antigen in leukemia sera. *JAMA* **191**, 541–546.

Centers for Disease Control and Prevention (1995). *Hepatitis Surveillance Report No. 56.* (Generic).

Cerny, A. & Chisari, F. V. (1999). Pathogenesis of chronic hepatitis C: immunological features of hepatic injury and viral persistence. *Hepatology* **30**, 595–601.

Chiaramonte, M., Stroffolini, T., Lorenzoni, U., Minniti, F., Conti, S., Floreani, A., Ntakirutimana, E., Vian, A., Ngatchu, T. & Naccarato, R. (1996). Risk factors in community-acquired chronic hepatitis C virus infection: a case-control study in Italy. *J Hepatol* **24**, 129–134.

Chisari, F. V. (1992). Hepatitis B virus biology and pathogenesis. *Mol Genet Med* **2**, 67–104.

Choo, Q.-L., Kuo, G., Weiner, A. J., Overby, L. R., Bradley, D. W. & Houghton, M. (1989). Isolation of a cDNA clone derived from a blood-borne non-A, non-B viral hepatitis genome. *Science* **244**, 359–362.

Corey, L. & Holmes, K. K. (1980). Sexual transmission of hepatitis A in homosexual men: incidence and mechanism. *N Engl J Med* **302**, 435–438.

Darwish, M. A., Faris, R., Clemens, J. D., Rao, M. R. & Edelman, R. (1996). High seroprevalence of hepatitis A, B, C, and E viruses in residents in an Egyptian village in Nile Delta: a pilot study. *Am J Trop Med Hyg* **54**, 554–558.

De Cock, K. M., Govindarajan, S., Chin, K. P. & Redeker, A. G. (1986). Delta hepatitis in the Los Angeles area: a report of 126 cases. *Ann Intern Med* **105**, 108–114.

El-Serag, H. B. & Mason, A. C. (1999). Rising incidence of hepatocellular carcinoma in the United States. *N Engl J Med* **340**, 745–750.

Farci, P., Alter, H. J., Govindarajan, S. & 8 other authors (1992). Lack of protective immunity against reinfection with hepatitis C virus. *Science* **258**, 135–140.

Farci, P., Alter, H. J., Shimoda, A., Govindarajan, S., Cheung, L. C., Melpolder, J. C., Sacher, R. A., Shih, J. W. & Purcell, R. H. (1996). Hepatitis C virus-associated fulminant hepatic failure. *N Engl J Med* **335**, 631–634.

Feinstone, S. M., Kapikian, A. Z. & Purcell, R. H. (1973). Hepatitis A: detection by immune electron microscopy of a viruslike antigen associated with acute illness. *Science* **182**, 1026–1028.

Feinstone, S. M., Kapikian, A. Z., Purcell, R. H., Alter, H. J. & Holland, P. V. (1975). Transfusion-associated hepatitis not due to viral hepatitis type A or B. *N Engl J Med* **292**, 767–770.

Frank, C., Mohamed, M. K., Strickland, G. T. & 8 other authors (2000). The role of parenteral antischistosomal therapy in the spread of hepatitis C virus in Egypt. *Lancet* **355**, 887–891.

Frosner, G. G., Willers, H., Muller, R., Schenzle, D., Deinhardt, F. & Hopken, W. (1978). Decrease in incidence of hepatitis A infections in Germany. *Infection* **6**, 259–260.

Halliday, M. L., Kang, L.-Y., Zhou, T.-K., Hu, M.-D., Pan, Q.-C., Fu, T.-Y., Huang, Y. & Hu, S.-L. (1991). An epidemic of hepatitis A attributable to the ingestion of raw clams in Shanghai, China. *J Infect Dis* **164**, 852–859.

Hasegawa, K., Huang, J., Wands, J. R., Obata, H. & Liang, T. J. (1991). Association of hepatitis B viral precore mutations with fulminant hepatitis B in Japan. *Virology* **185**, 460–463.

Henning, K. J., Bell, E., Braun, J. & Barker, N. D. (1995). A community-wide outbreak of hepatitis A: risk factors for infection among homosexual and bisexual men. *Am J Med* **99**, 132–136.

Houghton, M., Weiner, A., Han, J., Kuo, G. & Choo, Q.-L. (1991). Molecular biology of the hepatitis C viruses: implications for diagnosis, development and control of viral disease. *Hepatology* **14**, 381–388.

Hutin, Y. J., Pool, V., Cramer, E. H. & 9 other authors (1999). A multistate, foodborne outbreak of hepatitis A. National Hepatitis A Investigation Team. *N Engl J Med* **340**, 595–602.

Kabrane-Lazizi, Y., Fine, J. B., Elm, J., Glass, G. E., Higa, H., Diwan, A., Gibbs, C. J. J., Meng, X. J., Emerson, S. U. & Purcell, R. H. (1999). Evidence for widespread infection of wild rats with hepatitis E virus in the United States. *Am J Trop Med Hyg* **61**, 331–335.

Khuroo, M. S. (1980). Study of an epidemic of non-A, non-B hepatitis: possibility of another human hepatitis virus distinct from post-transfusion non-A, non-B type. *Am J Med* **68**, 818–824.

Kiyosawa, K. & Furuta, S. (1994). Hepatitis C virus and hepatocellular carcinoma. In *Hepatitis C Virus*, pp. 98–120. Edited by H. W. Reesink. Basel: Karger.

Kiyosawa, K., Tanaka, E., Sodeyama, T., Yoshizawa, K., Yabu, K., Furuta, K., Imai, H., Nakano, Y., Usuda, S., Uemura, K., Furuta, S., Watanabe, Y., Watanabe, J., Fukuda, Y., Takayama, T. & South Kiso Hepatitis Study Group (1994). Transmission of hepatitis C in an isolated area in Japan: community-acquired infection. *Gastroenterology* **106**, 1596–1602.

Lau, D. T., Farooq, K. M., Doo, E. & 10 other authors (2000). Long-term therapy of chronic hepatitis B with lamivudine. *Hepatology* **32**, 828–834.

Lee, W. M. (1997). Hepatitis B virus infection. *N Engl J Med* **337**, 1733–1745.

Lemon, S. M. & Thomas, D. L. (1997). Vaccines to prevent viral hepatitis. *N Engl J Med* **336**, 196–204.

Lemon, S. M., Binn, L. N., Marchwicki, R., Murphy, P. C., Ping, L.-H., Jansen, R. W., Asher, L. V. S., Stapleton, J. T., Taylor, D. G. & LeDuc, J. W. (1990). In vivo replication and reversion to wild-type of a neutralization-resistant variant of hepatitis A virus. *J Infect Dis* **161**, 7–13.

Lemon, S. M., Murphy, P. C., Smith, A., Zou, J., Hammon, J., Robinson, S. & Horowitz, B. (1994). Removal/neutralization of hepatitis A virus during manufacture of high purity, solvent/detergent factor VIII concentrate. *J Med Virol* **43**, 44–49.

Lemon, S. M., Layden, T. J., Seeff, L., Suzuki, H., Nishioka, K., Mishiro, S. & Johnson, L. (2000). The 20th United States–Japan Joint Hepatitis Panel Meeting. *Hepatology* **31**, 800–806.

Liang, T. J., Hasegawa, K., Munoz, S. J., Shapiro, C. N., Yoffe, B., McMahon, B. J., Feng, C., Bei, H., Alter, M. J. & Dienstag, J. L. (1994). Hepatitis B virus precore mutation and fulminant hepatitis in the United States. A polymerase chain reaction-based assay for the detection of specific mutation. *J Clin Investig* **93**, 550–555.

Liaw, Y. F., Leung, N. W., Chang, T. T. & 8 other authors (2000). Effects of extended lamivudine therapy in Asian patients with chronic hepatitis B. Asia Hepatitis Lamivudine Study Group. *Gastroenterology* **119**, 172–180.

Linnen, J., Wages, J., Zhang-Keck, Z.-Y. & 27 other authors (1996). Molecular cloning and disease association of hepatitis G virus: a transfusion-transmissible agent. *Science* **271**, 505–508.

Major, M. E. & Feinstone, S. M. (1997). The molecular virology of hepatitis C. *Hepatology* **25**, 1527–1538.

Mannucci, P. M., Gdovin, S., Gringeri, A., Colombo, M., Mele, A., Schinaia, N., Ciavarella, N., Emerson, S. U., Purcell, R. H. & Italian Collaborative Group (1994). Transmission of hepatitis A to patients with hemophilia by factor VIII concentrates treated with organic solvent and detergent to inactivate viruses. *Ann Intern Med* **120**, 1–7.

Meng, X. J., Purcell, R. H., Halbur, P. G., Lehman, J. R., Webb, D. M., Tsareva, T. S., Haynes, J. S., Thacker, B. J. & Emerson, S. U. (1997). A novel virus in swine is closely related to the human hepatitis E virus. *Proc Natl Acad Sci U S A* **94**, 9860–9865.

Nakashima, K., Ikematsu, H., Hayashi, J., Kishihara, Y., Mitsutake, A. & Kashiwagi, S. (1995). Intrafamilial transmission of hepatitis C virus among the population of an endemic area of Japan. *JAMA* **274**, 1459–1461.

Nishiguchi, S., Kuroki, T., Nakatani, S., Morimoto, H., Takeda, T., Nakajima, S., Shiomi, S., Seki, S., Kobayashi, K. & Otani, S. (1995). Randomised trial of effects of interferon-α on incidence of hepatocellular carcinoma in chronic active hepatitis C with cirrhosis. *Lancet* **346**, 1051–1055.

Norman, J. E., Beebe, G. W., Hoofnagle, J. H. & Seeff, L. B. (1993). Mortality follow-up of the 1942 epidemic of hepatitis B in the U.S. Army. *Hepatology* **18**, 790–797.

Ogata, N., Cote, P. J., Zanetti, A. R., Miller, R. H., Shapiro, M., Gerin, J. & Purcell, R. H. (1999). Licensed recombinant hepatitis B vaccines protect chimpanzees against infection with the prototype surface gene mutant of hepatitis B virus. *Hepatology* **30**, 779–786.

Okamoto, H., Nishizawa, T. & Ukita, M. (1999). A novel unenveloped DNA virus (TT virus) associated with acute and chronic non-A to G hepatitis. *Intervirology* **42**, 196–204.

Prince, A. M. (1968). An antigen detected in the blood during the incubation period of serum hepatitis. *Proc Natl Acad Sci U S A* **60**, 814–821.

Prince, A., Grady, G., Hazzi, C., Brotman, B., Kuhns, W., Levine, R. & Millian, S. (1974). Long-incubation post-transfusion hepatitis without serological evidence of exposure to hepatitis-B virus. *Lancet* **2**, 241–246.

Reyes, G. R., Purdy, M. A., Kim, J. P., Luk, K.-C., Young, L. M., Fry, K. E. & Bradley, D. W. (1990). Isolation of a cDNA from the virus responsible for enterically transmitted non-A, non-B hepatitis. *Science* **247**, 1335–1339.

Rizzetto, M., Hoyer, B., Canese, M. G., Shih, J. W. K., Purcell, R. H. & Gerin, J. L. (1980). δ Agent: association of δ antigen with hepatitis B surface antigen and RNA in serum of δ-infected chimpanzees. *Proc Natl Acad Sci U S A* **77**, 6124–6128.

Sakugawa, H., Nakasone, H., Nakayoshi, T., Kawakami, Y., Miyazato, S., Kinjo, F., Saito, A., Ma, S. P., Hotta, H. & Kinoshita, M. (1999). Hepatitis delta virus genotype IIb predominates in an endemic area, Okinawa, Japan. *J Med Virol* **58**, 366–372.

Seeff, L. B. (1997). Natural history of hepatitis C. *Hepatology* **26**, 21S–28S.

Seeff, L. B., Beebe, G. W., Hoofnagle, J. H. & 9 other authors (1987). A serologic follow-up of the 1942 epidemic of post-vaccination hepatitis in the United States Army. *N Engl J Med* **316**, 965–970.

Shaw, D. D., Whiteman, D. C., Merritt, A. D., el-Saadi, D. M., Stafford, R. J., Heel, K. & Smith, G. A. (1999). Hepatitis A outbreaks among illicit drug users and their contacts in Queensland, 1997. *Med J Aust* **170**, 584–587.

Simons, J. N., Leary, T. P., Dawson, G. J., Pilot-Matias, T. J., Muerhoff, A. S., Schlauder, G. G., Desai, S. M. & Mushahwar, I. K. (1995). Isolation of novel virus-like sequences associated with human hepatitis. *Nat Med* **1**, 564–569.

Szmuness, W., Dienstag, J. L., Purcell, R. H., Harley, E. J., Stevens, C. E. & Wong, D. C. (1976). Distribution of antibody to hepatitis A antigen in urban adult populations. *N Engl J Med* **295**, 755–759.

Szmuness, W., Dienstag, J. L., Purcell, R. H., Stevens, C. E., Wong, D. C., Ikram, H., Bar-Shany, S., Beasley, R. P., Desmyter, J. & Gaon, J. A. (1977). The prevalence of antibody to hepatitis A antigen in various parts of the world: a pilot study. *Am J Epidemiol* **106**, 392–398.

Taylor, J. M. (1999). Hepatitis delta virus. *Intervirology* **42**, 173–178.

Waters, J. A., Kennedy, M., Voet, P., Hauser, P., Petre, J., Carman, W. & Thomas, H. C. (1992). Loss of the common "A" determinant of hepatitis B surface antigen by a vaccine-induced escape mutant. *J Clin Investig* **90**, 2543–2547.

Widell, A., Hansson, B. G., Moestrup, T. & Nordenfelt, E. (1983). Increased occurrence of hepatitis A with cyclic outbreaks among drug addicts in a Swedish community. *Infection* **11**, 198–200.

Wilson, J. N., Nokes, D. J. & Carman, W. F. (2000). Predictions of the emergence of vaccine-resistant hepatitis B in The Gambia using a mathematical model. *Epidemiol Infect* **124**, 295–307.

Wong, D. C., Purcell, R. H., Sreenivasan, M. A., Prasad, S. R. & Pavri, K. M. (1980). Epidemic and endemic hepatitis A in India: evidence for a non-A, non-B hepatitis virus etiology. *Lancet* **ii**, 876–879.

The emergence of human immunodeficiency viruses and AIDS

Robin A. Weiss[1] and Helen A. Weiss[2]

[1]Windeyer Institute of Medical Sciences, University College London, 46 Cleveland Street, London W1P 6DB, UK

[2]Infectious Disease Epidemiology Unit, London School of Hygiene and Tropical Medicine, Keppel Street, London WC1E 7HT, UK

INTRODUCTION

The emergence of human immunodeficiency virus type 1 (HIV-1) in the human population is one of nature's most recent and highly successful adaptive radiations. This microbe illustrates Darwinian natural selection at a fast-forward pace. In the West, HIV readily exploited various niches of our late 20th century lifestyles, including air travel, narcotics dependence and steamy, promiscuous bath houses (Shilts, 1987). Yet it is wreaking most havoc among the world's poorest and underprivileged communities. HIV/AIDS presents a fascinating but frightening *danse macabre* of sex, drugs and death, with no end in sight to the burgeoning pandemic.

Of course, one does not need to be a microbiologist to know about AIDS. The disease was first recognized when small clusters of young homosexual men in American cities were reported to suffer rare opportunistic infections (OIs) (*Pneumocystis carinii* pneumonia, cytomegalovirus retinitis) and Kaposi's sarcoma (Centers for Disease Control, 1981). The underlying immune deficiency was soon shown to involve a selective depletion of CD4-positive, T-helper lymphocytes (Gottlieb *et al.*, 1981). Initially, it was not clear whether 'gay compromise syndrome' as it was then named was transmissible or was related to non-infectious lifestyle factors. By early 1982, however, reports of AIDS in recipients of blood transfusions and pooled clotting factors, as well as among injecting drug users, indicated that an infectious agent was to blame. The appearance of AIDS in African countries and in Haiti suggested that the unknown pathogen was already widespread. In the past 20 years, the causative agent, HIV, has rampaged across the globe, with an estimated 53 million infected people among whom 19 million have died (Piot, 2000; UNAIDS, 2000b).

SGM symposium 60: New challenges to health: the threat of virus infection.
Editors G. L. Smith, W. L. Irving, J. W. McCauley, D. J. Rowlands. Cambridge University Press. ISBN 0 521 80614 3 ©SGM 2001.

Since HIV-1 (Barré-Sinoussi *et al.*, 1983) and HIV-2 (Clavel *et al.*, 1987) were first identified, we have gained exquisitely detailed knowledge of the molecular biology of these viruses as well as the related simian immunodeficiency viruses (SIV). Yet we still do not understand sufficiently how they cause disease. Neither do we know why chimpanzees and some African monkeys, which are the natural reservoir of the precursors of HIV-1 and HIV-2, can harbour similar levels of virus without progressing to AIDS. Our knowledge of the replication cycle of HIV as a retrovirus has been crucial for the development of anti-viral drugs which, where available, have led to $>80\,\%$ drop in mortality. The elucidation of the cellular tropism of HIV and the cell surface receptors it utilizes helps to explain the pattern of immune deficiency leading to OIs, the wasting disease and the dementia that comprise many of the clinical manifestations of AIDS. The genetic and phenotypic evolution of HIV proceeds at a prodigious rate, both within the infected individual and across the worldwide pandemic. This high rate of variation has impeded vaccine development, which to date has been unsuccessful in affording broad protection to infection.

Françoise Barré-Sinoussi and colleagues (1983) at the Institut Pasteur isolated their first virus from a patient with persistent lymphadenopathy and hence named it lymphadenopathy-associated virus. By April 1984, the French group had reported two further isolates, one from a person with AIDS (Vilmer *et al.*, 1984), and a month later Gallo's group at the US National Institutes of Health (NIH) reported retroviruses that they named human T-lymphotropic virus type III or HTLV-III (Gallo *et al.*, 1984). Levy *et al.* (1984) also independently isolated AIDS-related retroviruses. While the NIH group thought the virus was related to human T-cell leukaemia viruses (HTLV) types I and II, the French correctly identified it as a lentivirus (Montagnier *et al.*, 1984), an observation confirmed when the complete genome sequences of HIV and sheep maedi-visna virus were analysed (Wain-Hobson *et al.*, 1985a, b). The term HIV was adopted in 1986.

When the virus was adapted to grow to high titre in CD4-positive leukaemic cell lines (Cheingsong-Popov *et al.*, 1984; Levy *et al.*, 1984; Popovic *et al.*, 1984), sufficient viral antigen became available to develop the first reliable serological assays for epidemiological studies (Cheingsong-Popov *et al.*, 1984; Sarngadharan *et al.*, 1984). The British 'HIV test' proved to be robust in the African setting. Working with African colleagues it enabled us to show that the new, aggressive form of Kaposi's sarcoma was linked to HIV infection (Bayley *et al.*, 1985), and that 'Slim' disease was, as feared, indeed AIDS caused by HIV (Serwadda *et al.*, 1985).

A great deal is now known about the dynamics of HIV replication *in vivo* (Perelson *et al.*, 1996; Phillips, 1999) but there is little understanding of what eventually tips the balance of infection away from host immunity towards the development of AIDS

(Weiss, 1993; Cloyd *et al.*, 2000), or how to design a thoroughly safe yet efficacious vaccine (Gotch *et al.*, 2000). From the early days we found that, in comparison to HTLV-I and -II, HIV elicited only weakly protective humoral immunity (Weiss *et al.*, 1985) and that the neutralization antigens in the HIV envelope were highly variable (Weiss *et al.*, 1986). This antigenic variability is also evident for cell-mediated immunity (Goulder & Walker, 1999).

Here, we provide an outline of the origins, epidemiology, evolution, pathogenesis and control of HIV. For a more detailed account of the molecular biology and replication of HIV, the reader is referred to major texts (Coffin *et al.*, 1997; Levy, 1998). Further discussion of the epidemiology and public health impact of HIV and other sexually transmitted diseases (STDs) can be found in a comprehensive book (Holmes *et al.*, 1998).

THE VIRUSES

Human immunodeficiency viruses comprise two distinct viruses, HIV-1 and HIV-2, which differ in origin and gene sequence. Both viruses cause AIDS with a similar spectrum of symptoms, although central nervous system (CNS) disease may be more frequent in HIV-2 disease (Lucas *et al.*, 1993). It appears that HIV-2 is less virulent than HIV-1 as HIV-2 infection takes longer to progress to AIDS (Whittle *et al.*, 1994). However, this observation may obscure a bimodal pattern of disease, in which some HIV-2-infected people progress to AIDS at a similar pace to HIV-1 infection, while a higher proportion than is the case for HIV-1 remain long-term non-progressors.

The genomes of HIV-1 and HIV-2 are shown in Fig. 1. Both viruses belong to the lentivirus subfamily of retroviruses and have a similar structure and order of genes (Coffin *et al.*, 1997), except that HIV-1 has a *vpu* gene while HIV-2 and most SIVs have *vpx*. In common with all retroviruses, the *gag* gene encodes the structural proteins of the core (p24, p7, p6) and matrix (p17) proteins of the virus particle, and the *env* gene encodes the glycoproteins (gp120, gp41) that comprise the viral envelope antigens, which interact with cell surface receptors (Sommerfelt, 1999). The p24, gp120 and gp41 proteins are the ones most commonly incorporated into diagnostic tests for HIV antibodies.

The *pol* gene encodes the enzymes crucial for viral replication: reverse transcriptase (RT) to convert viral RNA into DNA, integrase (IN) to incorporate the viral DNA into host chromosomal DNA (the provirus) and protease (PR) to cleave the precursor Gag and Pol proteins into their component parts. Because these enzymes are peculiar to retroviruses, molecules that inhibit them or act as chain terminators for RT block HIV replication *in vivo* without causing too severe toxicity to the host. RT and PR inhibitors represent the current generation of anti-retroviral drugs given in combination to lower viral load.

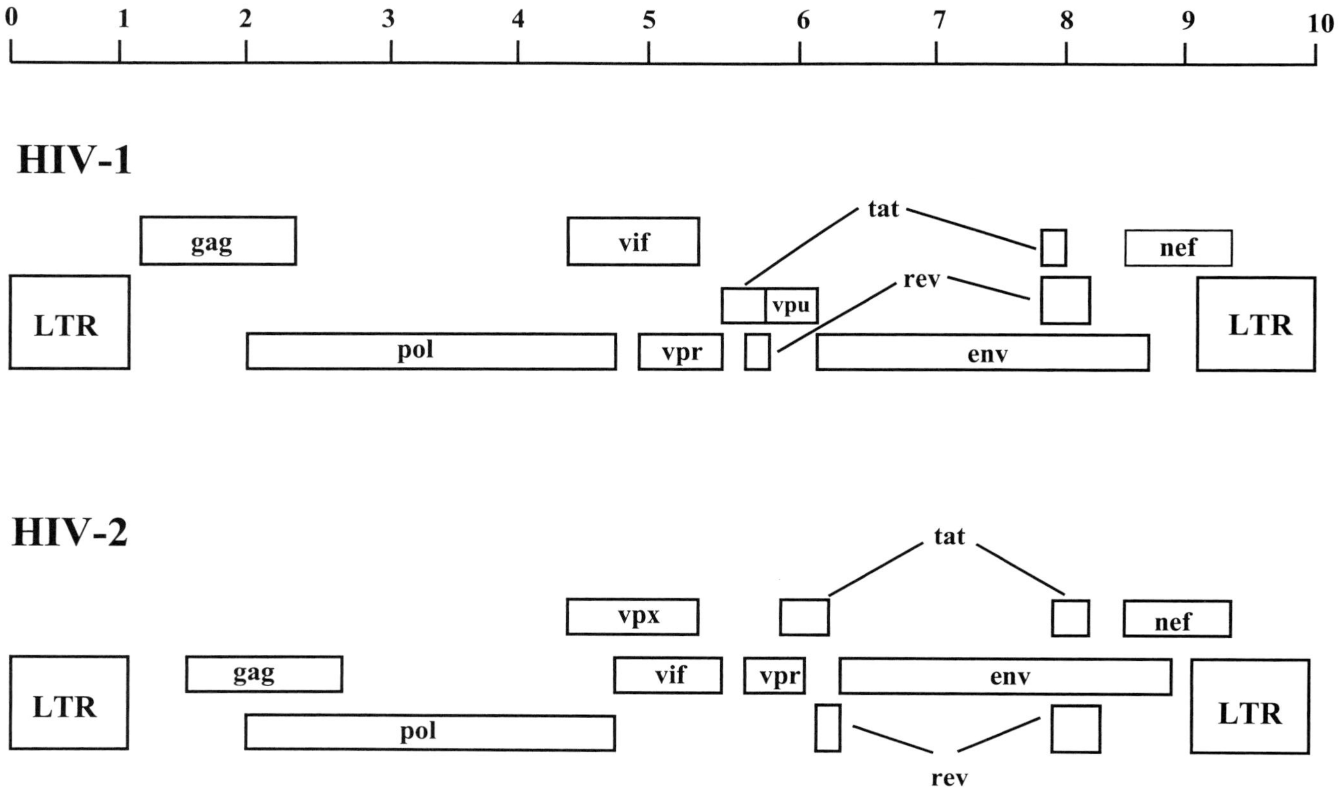

Fig. 1. Genomic organization of HIV-1 and HIV-2. The long terminal repeats (LTRs) control integration and gene expression. The genes are shown in their respective reading frames. The scale is in kilobases of proviral DNA.

The *tat* gene encodes a protein that promotes transcription or production of HIV RNA from the DNA provirus while *rev* ensures that the correctly processed mRNA and genomic RNA is exported from nucleus to cytoplasm by binding to a *rev*-response element (Alonso & Peterlin, 1999). The function of the other, accessory HIV genes is less well understood. *Vpr* helps to arrest the cell cycle. In HIV-1, *vpr* may also enable the reverse-transcribed DNA to gain access to the nucleus in non-dividing cells such as macrophages, a function performed by *vpx* in HIV-2. *Vpu* is important for virus particle release. *Vif* encodes a small protein that enhances the infectivity of progeny virions. *Nef* has multiple functions including signal transduction, down-regulating major histocompatibility antigens, and removing the CD4 receptor from the cell surface late in the cellular infection cycle to aid budding and release of virus particles.

Reverse transcription and integration of the proviral genome are the hallmarks of retroviruses which thus establish persistent infection in the host. The integrated DNA provirus can remain latent, and be passed to daughter cells during chromosomal replication and cell division. In activated cells, however, RNA molecules for mRNA and for new genomes are transcribed to produce the next generation of virus particles. Full replication of HIV in T-lymphocytes usually results in cell death, whereas in macrophages lower levels of virus production permit the host cell to survive for long periods. As with lentiviruses of animals, tissue macrophages represent a substantial virus reservoir in the infected host.

ORIGINS AND DIVERSITY

Recent evidence indicates that both HIV-1 and HIV-2 represent novel, zoonotic introductions into the human population within the past 100 years (Hahn *et al.*, 2000; Korber *et al.*, 2000). The animal lentivirus most related to HIV-1 is SIVcpz of chimpanzees. Among the SIVcpz genomes that have been sequenced, those of the subspecies *Pan troglodytes troglodytes* are closest to HIV-1 whereas one from *P.t. schweinfurthii* (the East African chimpanzee) is distinct (Gao *et al.*, 1999). Humans harbour three major groups of HIV-1, named M, N and O, with group M representing all the subtypes or 'clades' A–H that have spread to cause the worldwide pandemic. HIV-1 groups O and N, in contrast, are largely confined to Gabon, Cameroon and neighbouring countries (Peeters *et al.*, 1997; Simon *et al.*, 1998), close to the natural habitat of *P.t. troglodytes*. The gene sequences of groups M, N and O are as distinct from each other as they are from SIVcpz (Gao *et al.*, 1999). This finding implies that they are derived from three separate introductions from chimpanzees to humans, yet only one of them has become pandemic (Hahn *et al.*, 2000). However, the studies by Gao *et al.* (1999) are based on a small number of SIVcpz genomes. Further field studies are required to confirm whether chimpanzees are a natural reservoir in the wild (Weiss & Wrangham, 1999).

HIV-2 is endemic in West Africa, but has spread to Europe (especially Portugal) and to India. Like HIV-1, HIV-2 can be subdivided into a number of major groups which appear to represent separate zoonoses from a primate host (Gao *et al.*, 1992). In this case, the original primate reservoir is not a great ape but the West African monkey sooty mangabey (*Cercocebus atys*), which is infected with SIVsm. The infection of captive Asian macaque species by SIVmac also appears to be a cross-species transfer of SIVsm, probably when sooty mangabeys and macaques were housed together. While SIVsm is not pathogenic in mangabeys, its derivatives SIVmac and HIV-2 cause AIDS in their new hosts. Indeed, inoculation of SIVsm into macaques leads to disease.

Many species of African monkey harbour other types of SIV, but there is no evidence that any of these SIV strains have infected humans. African green monkeys are frequently infected by SIVagm, but this virus less readily adapts to propagate in human cells. This is fortunate as African green monkey kidneys have been used for over 35 years to propagate poliovirus for the live attenuated vaccine, although there is no evidence that SIVagm has crossed to humans. Thus a hypothesis that HIV emerged from polio vaccines contaminated with SIVagm appears untenable. However, it was recently proposed that if chimpanzee kidneys were used in experimental polio vaccine batches in early trials, the introduction of HIV-1 could have occurred by this route (Hooper, 1999). This notion has been hotly debated. There is no direct evidence that chimpanzee cell substrates were ever used for polio vaccine production, and the timing of the common origin of HIV group M dates well before the polio vaccine trials of the late 1950s (Korber *et al.*, 2000). Assuming group M's diversity began in humans, that would appear to exonerate polio vaccine.

Nevertheless, iatrogenic factors may have helped the cross-species transfer of HIV and its successful early take in humans. Experiments with cross-transfer of malaria parasites between simian and human by blood transfer could have carried over lentivirus contaminants (Gilks, 1991). In animal retrovirus infections, such as bovine leukosis virus, a major means of spread throughout a herd was the multiple use of veterinary hypodermic needles used for immunization against other pathogens. The smallpox vaccination campaigns in central Africa and other immunizations might have given early HIV strains not yet fully adapted to the human host that extra boost towards natural transmissibility. This line of argument could be dismissed as idle conjecture, when faced with the urgent and appalling problem of today's pandemic of HIV/AIDS. Yet we should be aware of the potential cross-species transmissibility of retroviruses lest we set off another new human disease via the medical use of animal tissue, e.g. in xenotransplantation (see the chapter by Stoye, this volume).

Although HIV-1 group M has probably not infected humans for more than about 70 years (Korber *et al.*, 2000), and did not emerge epidemically until 20 years ago, the virus has generated enormous variation (Leigh-Brown, 1999). There are several reasons for this great diversity. First, the process of reverse transcription does not include an editing device to correct mutations. However, lentiviruses appear to vary more than other retroviruses or indeed other RNA viruses such as polio or measles, although none of them scan and correct nascent replicons. Second, the RNA genomes of retrovirus particles are diploid and genetic recombination occurs during reverse transcription. In dually infected persons, therefore, recombinant forms can emerge and some of the HIV-1 clades show evidence of past recombinational origin, e.g. clade E in Thailand. However, recombinants between HIV-1 and HIV-2 have not yet been reported although dual infections occur in West Africa, perhaps because mixed assembly of genomes occurs rarely if at all (Kaye & Lever, 1999). Third, during the long asymptomatic incubation period before AIDS develops, the virus is not latent, but is actively replicating, producing as many as 10^9 virions and over 10^7 newly infected cells each day (Perelson *et al.*, 1996; Phillips, 1999).

Thus each infected individual possesses an immense pool of HIV variants, or quasispecies, allowing substantial genetic and antigenic drift to occur within each infected individual. It is this high rate of virus replication that provides the conditions for numerous immune escape and drug-resistant mutants to be generated. The evolution and selection of variants is complex, allowing both colonization of new cells and antigenic variation. For example, selection by point mutation in gp120 for escape from neutralization can also result in a change of cell tropism, and vice versa (McKnight *et al.*, 1995).

Despite the generation of large quasispecies or polymorphic virus populations in the infected human, there are certain constraints on passing infection from one person to another which help to reset the evolutionary clock of HIV with regard to phenotype and cell tropism each time transmission occurs. Virus fitness for infection to spread within an individual host does not necessarily match fitness for transmission.

EPIDEMIOLOGY

The prevalence and rate of spread of HIV in populations varies dramatically across continents, within continents, and within countries (Fig. 2). Of the 5.6 million new infections in 1999, two-thirds occurred in sub-Saharan Africa, and almost a quarter in South and Southeast Asia. Africa has been, and continues to be, hardest hit. The predominant mode of transmission in Africa is heterosexual contact, reflected in the fact that over half of those infected are women. In many countries in eastern and southern Africa, up to 40 % of young women are now infected. Given the high birth rates in many of these countries, and lack of provision of anti-retroviral therapy in pregnancy, the epidemic is being rapidly passed to the next generation.

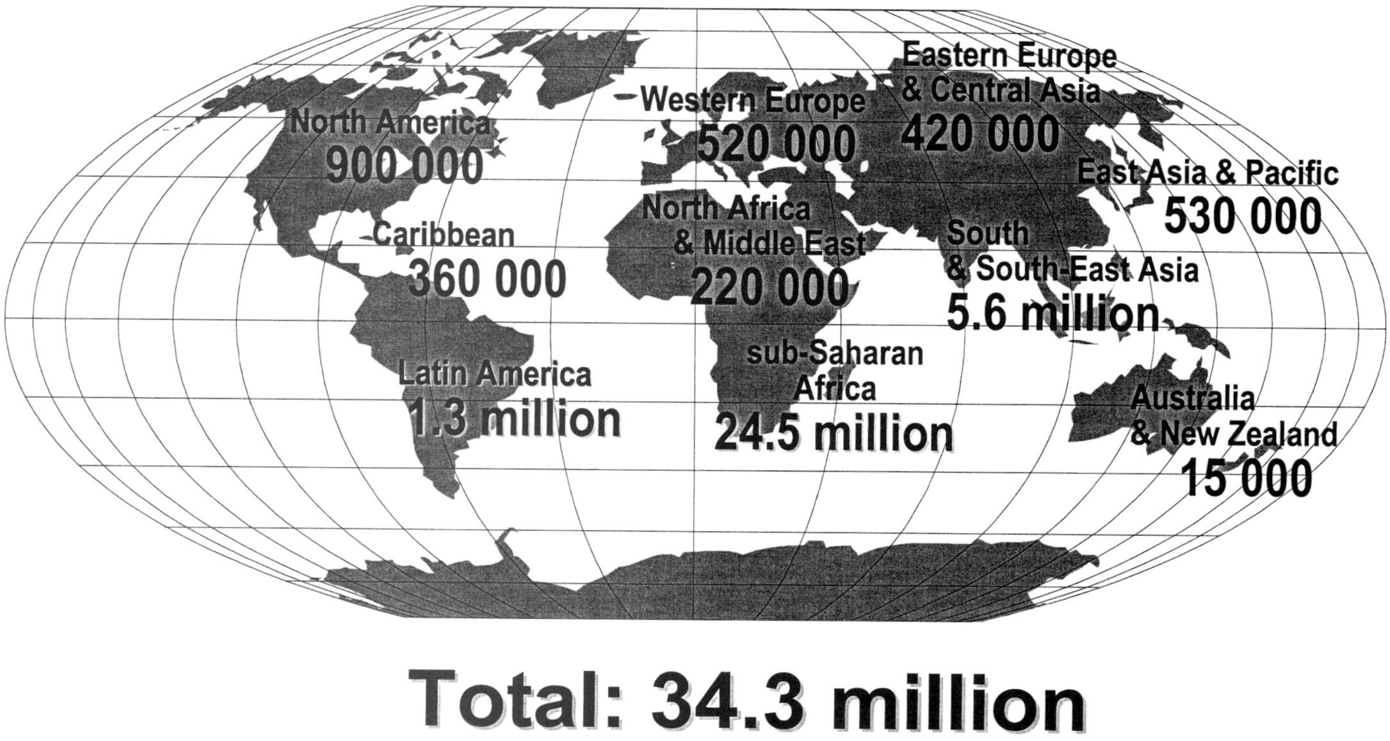

Fig. 2. Adults and children estimated to be living with HIV/AIDS at the turn of the millennium. Reproduced by kind permission of the Joint United Nations Programme on HIV/AIDS (UNAIDS) (www.UNAIDS.org/hivaidsinfo/documents.html).

Table 1. Geographical distribution of HIV types and subtypes

Region	Main modes of transmission for adults living with HIV/AIDS*	Main HIV type and subtype
Sub-Saharan Africa		
West Africa	Heterosexual	HIV-2
Southern Africa	Heterosexual	HIV-1: C
Central/East Africa	Heterosexual	HIV-1: A, D
Asia		
Thailand, Myanmar	Heterosexual, IDU	HIV-1: E, B
Cambodia, Vietnam	Heterosexual, IDU	HIV-1: E
India	Heterosexual	HIV-1: C
Latin America		
Brazil, Argentina	MSM, heterosexual, IDU	HIV-1: B, F
Chile	MSM, heterosexual, IDU	HIV-1: B
North America		
United States	MSM, IDU, heterosexual	HIV-1: B
Western Europe	MSM, IDU	HIV-1: B
Eastern Europe and central Asia	IDU	HIV-1: A, B
Australia and New Zealand	MSM, IDU	HIV-1: B

*MSM, Men who have sex with men; IDU, injecting drug users.

Within Africa, rates of HIV-1 infection vary widely, with relatively low and steady rates in most countries in West and central Africa (less than 10 % prevalence), including those where HIV-1 originated. For instance, the first documented case of HIV-1 infection is a blood sample taken in 1959 from a man in Kinshasa, Congo (Korber *et al.*, 2000; Zhu *et al.*, 1998). The estimated prevalence in Kinshasa was ~5 % in 1985 but rose to only ~7 % by 1998. In contrast, the epidemic spread of HIV southwards to Zimbabwe, Botswana and South Africa took off only in the last 12 years, yet these countries now have the highest prevalence rates. The reasons for the discrepancy are not clear.

The prevalence of HIV-1 subtypes varies throughout the world, and also according to mode of transmission (Table 1). There is little evidence to date that the genetic subtypes differ in virulence for transmission or pathogenesis (Hu *et al.*, 1999). However, this point requires further study. In Uganda, persons infected with subtype D may progress to AIDS faster than those with subtype A (P. Kaleebu, personal communication). Subtype C now accounts for nearly 50 % of all new infections. Although subtype C does not generate co-receptor variants (Abebe *et al.*, 1999; Ping *et al.*, 1999), it may prove to be a rather virulent strain (Cohen, 2000).

The rate of spread of HIV depends on both behavioural factors associated with risk of exposure to the virus, and 'co-factors' which modulate the risk of becoming infected given exposure. The main behavioural variables are rates of change of sexual partners, and lack of condom use. Patterns of sexual networking are probably one of the main determinants of the spread of the virus. In some populations, for example, commercial sex ('core group' sex in which a small number of women have sexual contact with a large number of men) appears to play a crucial role in the HIV epidemic, as it does with regard to the classical STDs. Patterns of sexual behaviour are themselves influenced by a wide range of social, economic and cultural factors, including patterns of migration and mobility, urbanization, marriage patterns, major routes of commerce such as truck routes, military conflict and civil unrest.

It is now clear from both epidemiological and biological studies that sexual transmission of HIV is enhanced in the presence of other STDs (Fleming & Wasserheit, 1999), and this may help to explain the rapid dissemination of HIV in countries with high population rates of STDs. A recent multicentre analysis in four African cities concluded that factors such as herpes simplex virus 2 (HSV-2) infection and lack of male circumcision play a more important role than population-level differences in sexual behaviour or HIV-1 subtype (Buvé, 2000; Hu *et al.*, 1999). HSV-2 infection is highly prevalent in many developing countries, and could increase the risk of HIV transmission by causing genital ulcers that provide a portal of entry for the virus. A protective effect of male circumcision on acquiring HIV was first postulated by Fink (1987), and the epidemiological evidence for a strong relative protection in female to male transmission is now compelling, especially in populations at high risk for STDs and HIV (H. A. Weiss *et al.*, 2000). Such a protective effect is biologically plausible owing to the high concentration of Langerhans cells on the inner, mucosal surface of the prepuce (Szabo & Short, 2000), yet any translation of this finding into public health practice would be fraught with difficulties (Halperin & Bailey, 1999).

In the United States and Europe, the pattern of spread of HIV has been different, partly because of the much lower levels of other STDs in the population, and perhaps less dependence on commercial sex workers as a sexual outlet. The risk of HIV infection remains highest among men who have sex with men, and among injecting drug users. However, the pattern is changing, with heterosexual activity now the main mode of HIV transmission in the UK (Communicable Disease Surveillance Centre, 2000) and other countries in northern Europe. Similarly, in the US, the proportion of infections due to heterosexual transmission is increasing (Fig. 3).

Over the last 5 years, the availability of anti-retroviral drugs has made a dramatic impact on mortality due to HIV/AIDS in developed countries. Fig. 4 shows the rise and fall in mortality due to AIDS in the USA. It grew from insignificance in 1982 to become the leading cause of death in men and women below the age of 45 years. The sharp drop in AIDS mortality since 1996 is almost entirely due to combination anti-retroviral therapy. However, it is important to note that the number of new infections in the USA has remained steady at about 40 000 per year, emphasizing the need for sustained public health education campaigns – there is no room for complacency in primary prevention.

HIV PATHOGENESIS

Disease parameters

The clinical manifestations of AIDS include OIs by micro-organisms that thrive in the immunosuppressed host (R. A. Weiss *et al.*, 2000). Some of these are common parasitic or commensal microbes that are seldom pathogenic in immunocompetent individuals. *Pneumocystis carinii*, for example, as well as *Candida albicans* and *Aspergillus* are fungal forms that do not cause more than superficial colonization or mild thrush in healthy subjects. Others probably have animal reservoirs, such as *Toxoplasma gondii* and *Mycobacterium avium–intracellulare*. However, known human pathogens may cause more severe illness in AIDS, e.g. *M. tuberculosis*, which often extends to extrapulmonary sites.

Among virus infections, various latent herpesvirus infections frequently become reactivated to cause severe illness in AIDS. The alphaherpesviruses herpes simplex types 1 and 2 and varicella-zoster virus can develop as life-threatening OIs in AIDS, and likewise the betaherpesvirus cytomegalovirus (CMV). Another betaherpesvirus, human herpesvirus 6, was first isolated from AIDS patients. The gammaherpesviruses Epstein–Barr virus and Kaposi's sarcoma herpesvirus (human herpesvirus 8) allow the tumours to which they are aetiologically linked to occur at much higher frequency in AIDS (see the chapter by Boshoff, this volume).

Thus certain selective OIs which are normally well controlled by cell-mediated immunity present as serious disease conditions in AIDS. It was on the account of the extraordinary rarity of *Pneumocystis* pneumonia and Kaposi's sarcoma in young adults that led surveillance officers at the Centers for Disease Control in USA to alert the world to AIDS (Centers for Disease Control, 1981). Other chronic infections (and indeed other

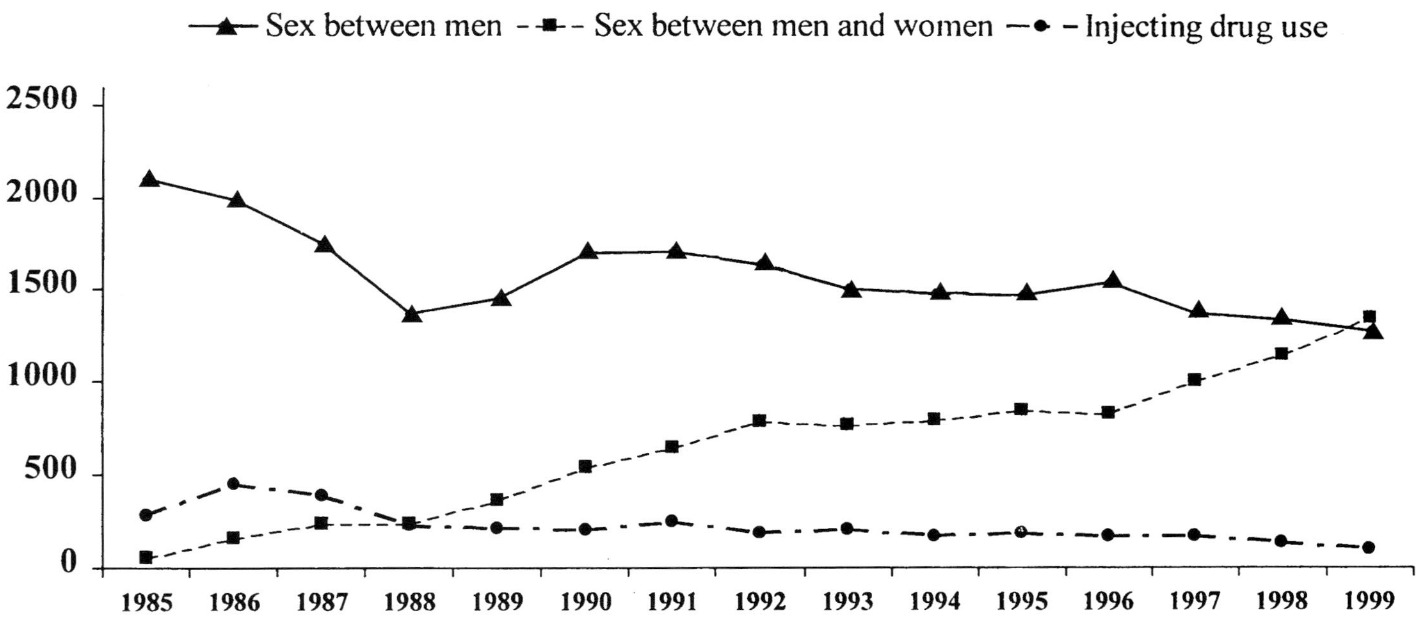

Fig. 3. Changing proportion of major risk groups for HIV infection in the UK, 1985–1999. Adapted with permission from the Communicable Disease Surveillance Centre, Public Health Laboratory Service.

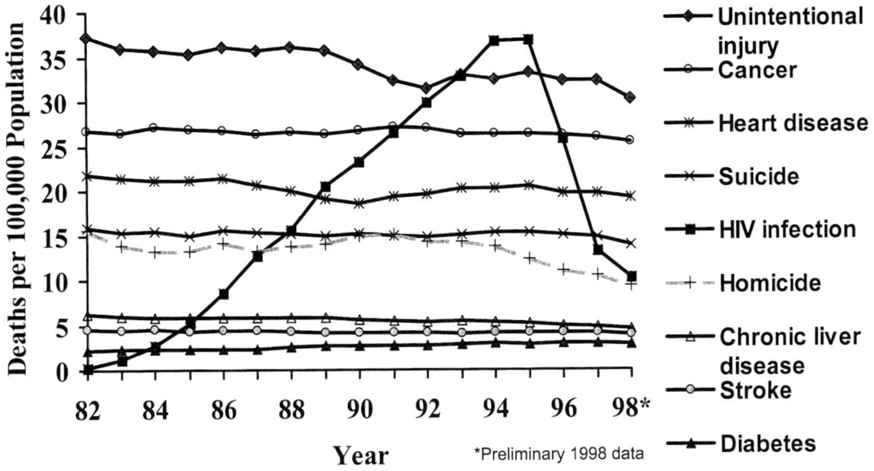

Fig. 4. Most frequent causes of death among men and women in the USA, aged 25–44. Adapted with permission from the US Centers for Disease Control and Prevention, Division of HIV/AIDS Prevention.

cancers save for non-Hodgkin's lymphoma) do not appear to be significantly exacerbated in AIDS, e.g. malaria, leprosy and viral hepatitis B and C. This leads one to question to what extent cell-mediated immunity controls these conditions in the immunocompetent individual.

HIV dynamics and progression

The typical course of HIV infection is depicted in Fig. 5. Primary infection by HIV may be unnoticed or, occasionally, cause a transient fever and lymphadenopathy (HIV seroconversion illness). HIV titres in the peripheral blood peak 3–4 weeks after exposure and fall concomitantly with the appearance of specific cytotoxic T-lymphocytes (CTLs). Seroconversion, with the appearance of antibodies to Gag, Env and other viral proteins, also occurs at this time or shortly afterwards. Thus the host's short-term response to HIV is one of cellular and humoral immunity as one would expect to other viral infections.

Following primary infection, however, HIV is neither eliminated nor does it become truly latent, like alphaherpesvirus infections. Rather it remains active in the asymptomatic individual though at a lower level than in primary infection. That level is sometimes called the 'set point' and the higher it is, the sooner progression to AIDS is likely to occur (Mellors *et al.*, 1995). It is not known what determines the set point. It may be compounded by the virulence of the virus strain and the number of T-lymphocytes and

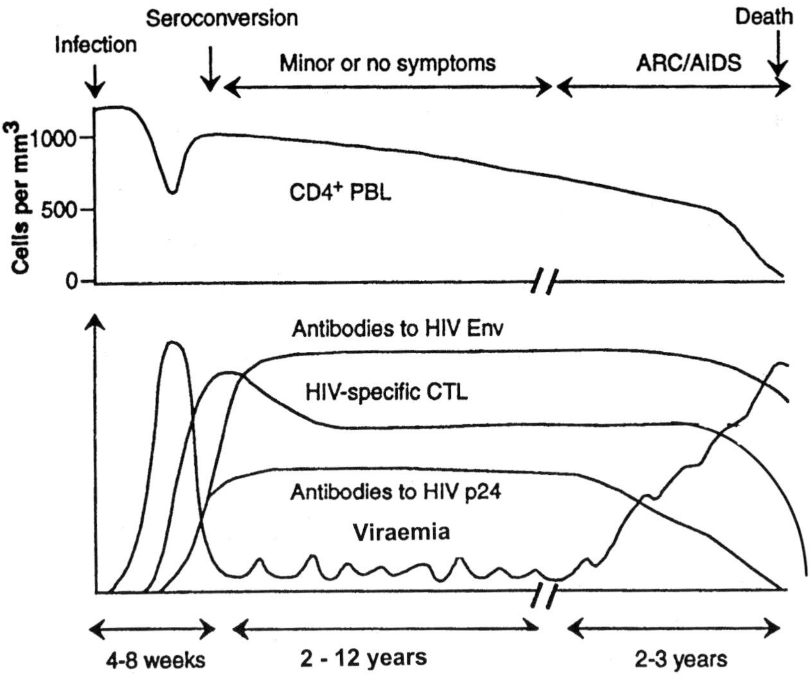

Fig. 5. Typical time course of HIV infection. The vertical axis shows the relative amount of HIV CD4 cells and immune responses from primary infection to AIDS and death. ARC, AIDS-related complex; CTL, cytotoxic T-lymphocytes; PBL, peripheral blood lymphocytes; p24, major core (capsid) antigen of HIV Gag. Reprinted with permission from Weiss (1993). Copyright 1993 American Association for the Advancement of Science.

macrophages in a susceptible state for infection. Immune activation by other infections such as influenza or by immunization such as with tetanus toxoid can transiently elevate HIV load.

The stark reality is that the vast majority of untreated people infected with HIV eventually succumb to AIDS-related death. The asymptomatic period varies greatly among HIV-infected people. In the West, about 50 % progress to AIDS within 9–10 years, with slowly declining levels of CD4 lymphocytes (Fig. 5). Some individuals, called long-term non-progressors, maintain healthy levels of CD4 cells and low viral loads for much longer periods, whereas others progress to AIDS within 3–5 years. Host genetic markers may contribute to the rate of progression, with good or poor prognosis alleles of major histocompatibility and co-receptor genes (see below). More importantly, increasing age among adults infected by HIV is strongly correlated with more rapid progression (Darby *et al.*, 1996). Because CD4 lymphocytes are being destroyed throughout the asymptomatic period (Phillips, 1999), the rate of immune cell replacement or restoration may be linked to the age risk. CD4 cell replenishment may come from the periphery and also from new T-lymphocytes processed in the thymus.

The overall dynamics *in vivo* of HIV replication and T-lymphocyte turnover has been estimated to be remarkably high, with approximately 10^9 new virions and 3.5×10^7 new cell infections per day (Perelson *et al.*, 1996; Phillips, 1999). CD4-lymphocyte destruction may be caused by the cytopathic effect of HIV itself, by cytotoxic CD8-positive T-lymphocytes, and by indirect apoptosis, especially after homing to lymph nodes (Cloyd *et al.*, 2000; Weiss, 1993).

Rising HIV load and falling CD4 cell counts independently presage the onset of AIDS (Fig. 5). As discussed below, the phenotype of the virus may also change with disease progression. Certain HIV variants tend to appear late in the course of infection which grow to higher titre *in vitro* (Asjo *et al.*, 1986), often induce cell fusion as syncytium-inducing variants (Tersmette *et al.*, 1988), have positively charged amino acids in the V3 loop of the viral envelope glycoprotein gp120 (Fouchier *et al.*, 1992), and have an increased ability to switch co-receptor usage (Berger *et al.*, 1999). These variants of HIV appear to be at a selective disadvantage when transmitted from a late stage person to a newly infected one. In a sense, late stage variants of HIV may be viewed as OIs in their own right, exacerbating progression to AIDS once the host immune system is sufficiently damaged to allow their emergence.

Cellular tropism and disease

From the beginning, AIDS was recognized to be an immunodeficiency of CD4-positive helper T-lymphocytes (Gottlieb *et al.*, 1981). Since CD4 cells prime CD8-positive CTLs (or killer cells) to destroy other cells expressing foreign antigens, and also enhance antibody production by B-lymphocytes, they represent a key component of the immune system. In the healthy individual, about 1200 CD4 cells circulate (μl blood)$^{-1}$ but when CD4 counts drop below 400 μl^{-1} OIs can occur. Soon after HIV-1 was first isolated, Klatzmann *et al.* (1984) showed that the virus selectively infected and destroyed CD4 cells in culture, and we then demonstrated that it does so by binding to the CD4 antigen itself (Dalgleish *et al.*, 1984). Thus the cellular tropism of the virus already explained much about the pattern of disease.

Another cell type that expresses low levels of CD4 antigen are the tissue macrophages, derived from monocytes circulating in the blood, which function as scavengers and as antigen-presenting cells. Macrophages become infected by HIV, and act as an important reservoir of the virus in the body (Levy, 1998). It is macrophage infection that probably accounts for the wasting syndrome and CNS disease in AIDS. Ruminant lentiviruses, such as maedi-visna virus of sheep, cause similar wasting and CNS diseases, and infect macrophages but not lymphocytes. Microglia in the brain are a type of macrophage. Their infection by HIV leads to aberrant signalling of cytokines and chemokines via astrocytes, leading to loss of neurons and the dementia that sometimes occurs in AIDS. Dendritic cells, another important type of antigen-presenting cell derived from monocytes, are also affected by HIV. Dendritic cells include the Langerhans cells of the mucous membranes and these may be involved as an early target of infection during sexual transmission. Dendritic cells carry HIV to the lymph nodes, where CD4-positive lymphocytes become infected (Pope *et al.*, 1994; Geijtenbeek *et al.*, 2000).

It became apparent that while CD4 is necessary for attachment, it is not sufficient for HIV entry into host cells (Maddon *et al.*, 1986). In fact, some HIV-2 strains do not depend on CD4 at all (Clapham *et al.*, 1992). After much research by many groups, two chemokine receptors, known as CCR5 and CXCR4, were identified as co-receptors to CD4 that permitted virus entry (Berger *et al.*, 1999). Several other chemokine receptors can be utilized by HIV-1, and especially HIV-2 and SIV in culture, but only CCR5 or CXCR4 have been shown to be crucial *in vivo*. Most HIV-1 strains of each subtype utilize CCR5 as a co-receptor, which is expressed on both lymphocytes and macrophages. However, in about 50 % of HIV-infected persons progressing to AIDS, CXCR4-utilizing viruses emerge late in the course of disease (Tersmette *et al.*, 1988). These 'X4' strains (Berger *et al.*, 1998) are more virulent than the initial 'R5' strains and probably hasten the depletion of CD4 cells and the onset of disease. It is remarkable that, while X4 viruses are much less efficient for person-to-person transmission than R5 viruses, they emerge as 'opportunistic' new variants in so many infected individuals as disease progresses. However, X4 viruses seldom arise in infection by HIV-1 subtype C (Abebe *et al.*, 1999; Ping *et al.*, 1999).

Certain individuals carry genetically defective CCR5 receptors (Carrington *et al.*, 1999; Michael, 1999). A 32 bp deletion in the CCR5 gene is a frequent polymorphism in the Caucasian population, with approximately 1 in 400 people homozygous for the deleted allele. Such homozygotes are apparently in good health and they are highly resistant to HIV infection. They are significantly overrepresented among the uninfected partners of discordant couples who are frequently exposed to HIV. Moreover, it is evident that the few exceptions among the homozygotes, who have become infected by HIV, exclusively harbour X4 strains of virus which do not require the CCR5 co-receptor. Deletions in the CCR5 gene are rare among Africans. Mutations in the promoter region of this gene, however, occur frequently and low expression of CCR5 correlates with a long incubation period from infection to AIDS (Carrington *et al.*, 1999; Michael, 1999).

While the foregoing story of the cellular tropism and pathogenesis of HIV explains why the virus causes immune deficiency and its associated syndromes, it does not explain why the virus eventually wins the balance of power over the immune system. Other viruses that infect CD4 lymphocytes (e.g. human herpesvirus 7) and macrophages (e.g. cytomegalovirus) typically do not cause disease unless the immune system is already impaired. Furthermore, the non-human primates from whence HIV-1 and HIV-2 came, chimpanzees and sooty mangabey monkeys, carry SIV without evidently developing disease. Indeed, the sooty mangabeys carry as high viral loads of SIVsm as many humans do with HIV, yet only the humans and SIV-infected Asian macaques succumb to AIDS. A partial answer to this conundrum is that HIV infection in humans elicits a general activation and proliferation of lymphocytes, including B-cells (Levy, 1998), which in turn assists virus replication, whereas in infected chimpanzees and sooty mangabeys, perturbation of lymphocyte regulation is less evident. If a means could be found of converting humans to a simian-type, commensal co-existence with the virus, people might not develop AIDS, although this would not impede HIV transmission.

CONTROLLING HIV INFECTION

The most pressing task in controlling HIV/AIDS is to devise better, cheaper treatment for those already infected, and to develop strategies to prevent infection, especially an efficacious HIV vaccine. When AIDS first emerged, the only treatments were those targeted to the OIs rather than the underlying cause. The antibiotics, antifungal agents, and antivirals such as ganciclovir, inhibiting herpesvirus infections, still have a role today.

Anti-retroviral therapy

The first anti-retroviral drug to be tested clinically was 3′-azidothymidine (AZT, zidovudine), an analogue of thymidine which acts as a chain-terminator to the elongation of DNA during reverse transcription. AZT is less efficiently incorporated in DNA synthesis catalysed by cellular DNA polymerases. It therefore provided a good therapeutic ratio with relatively few toxic side effects at doses achievable *in vivo*. A number of other chain-terminating nucleoside analogues have also come into clinical use.

Non-nucleoside RT inhibitors such as nevirapine strongly inhibit HIV-1 by binding to the RT enzyme at a site other than the active site. HIV-2, however, does not have this binding site, and a point mutation in RT of HIV-1 can lead to complete resistance to nevirapine with cross-resistance to related compounds. The other enzyme target that has yielded clinically useful drugs is the viral protease needed for cleavage of Gag and Pol precursors during maturation of virions after budding. The protease inhibitors are based on uncleavable peptides that bind to the active site of the enzyme.

Monotherapy with each of these anti-HIV drugs sooner or later results in resistance in the treated patient (Hirsch *et al.*, 2000; R. A. Weiss *et al.*, 2000). Given the high turnover and large reservoir of viral genomes in the infected individual, drug treatment exerts a strong selective pressure for resistant, mutant forms. Resistance to the non-nucleosides is achieved by a single amino-acid substitution, as is resistance to 3TC (lamivudine), often by a methionine to valine change at codon 184 in the active site of RT. Full resistance to AZT, however, requires four of six distinct mutations. Two of these (codons 70 and 215) confer resistance, but others may be compensating mutations (codons 41, 67, 210 and 219) to re-establish relative fitness to the RT and to virus replication. There does seem to be a cost in fitness to HIV in selecting resistance because after cessation of drug treatment, wild-type viruses come to predominate again.

The almost inevitable evolution of HIV-resistant variants in patients undergoing anti-retroviral treatment meant that the clinical benefit of monotherapy was short-term. Since 1996, however, dramatic improvement in survival has been achieved by multiple therapy with drugs that do not exhibit cross-resistance. It is more difficult for HIV to develop resistance to combination anti-retroviral therapy for at least three reasons: (a) multiple resistance takes much longer to develop; (b) multiply resistant HIV strains are less fit and therefore achieve lower viral loads and virulence in the patient; (c) resistance to one drug can sometimes confer new sensitivity to another drug to which HIV had previously become resistant. Thus AZT-resistant strains of HIV-1 bearing mutations in RT at nucleotides 41, 70, 210 and 215 become phenotypically sensitive to AZT again if position 184 mutates to afford resistance to 3TC, without reversion at any of the AZT mutation sites. Successful treatment leads to a rapid and sustained fall in viral load, often to undetectable levels, even by sensitive PCR-based assays (Hirsch *et al.*, 2000; R. A. Weiss *et al.*, 2000).

There have been problems in compliance with the complex drug regimes to which HIV-positive patients are expected to adhere as well as their unpleasant side effects. Better formulations of anti-retroviral drugs may help in the future. The financial burden of these successful drug treatments is considerable. Despite a recent announcement of drastic reduction in the price of certain drugs to developing countries, these drugs may still be beyond the reach of many countries and most infected individuals. Annual treatment with combination therapy can cost $10 000 per person, yet the annual government budget for health care in many African countries is less than $10 per person per year. Further problems include competing health care needs, and the many logistical problems in terms of timing of HIV diagnosis (often at advanced stages of disease), laboratory facilities and monitoring of drug therapy. Even the wealthiest nations would not be able to afford the pharmaceutical cost if the prevalence of infection was as high as in southern Africa. Nevertheless, economic and social strategies to treat HIV infection can be designed (Bloom & River Path Associates, 2000).

The dramatic impact of HIV drugs on HIV/AIDS in the West is remarkable. As mentioned, there has been a sharp drop in mortality due to AIDS in the USA over the past 4 years (Fig. 4). This fall is almost entirely due to combination anti-retroviral therapy. This effective drug treatment, however, does not eliminate HIV from the infected person, and drug 'holidays' result in rapid re-appearance of virus in the peripheral blood. It remains an open question whether combination anti-retroviral therapy will keep the viral load permanently low in the majority of patients who respond to treatment, or whether resistant virus will eventually creep up and cause disease. For the time being, it is good news indeed, but there is a need to develop new drugs blocking different viral targets.

For future drug development, several of the steps in the virus replication cycle (Fig. 6) offer promising means of pharmaceutical intervention. To date, all the drugs in clinical use target either reverse transcription (Fig. 6, step 4) or protease cleavage (step 13). To block steps 1 and 2, for example, soluble mimics to the CD4 binding receptor can inhibit initial attachment, and compounds have been found that potently block HIV interaction with CXCR4 (Simmons *et al.*, 1997) and CCR5 (Dragic *et al.*, 2000). Because these drugs bind to the co-receptor rather than to viral proteins, the rapid emergence of resistance is less likely to occur. CCR5 is a particularly attractive host target because the healthy status of deletion homozygotes indicates that this receptor is not needed physiologically. Another promising lead which may soon go into clinical trials is that of inhibitors of the third viral enzyme, integrase (Hazuda *et al.*, 2000), to block step 6, the insertion of the provirus into host chromosomal DNA.

HIV vaccine development

The best means of putting a brake on the HIV pandemic would be a safe, efficacious prophylactic vaccine that is easy to administer. Alas, this is not yet a success story (Gotch *et al.*, 2000). Almost every conceivable vaccine approach that has been successful for other viruses has been examined for HIV and SIV, and found wanting. SIV constructs bearing HIV genes, called SHIV hybrid viruses, have been tested in macaques. Whole killed virions and subunit recombinant envelope antigens can give protection against the particular HIV strain used as an immunogen. However, little protection is afforded even against other virus isolates of the same envelope subtype, and what immunity is elicited soon wanes.

Live attenuated vaccines have been tested with SIVmac strains. Macaques can be protected from challenge with a virulent strain by prior infection with vaccine strains

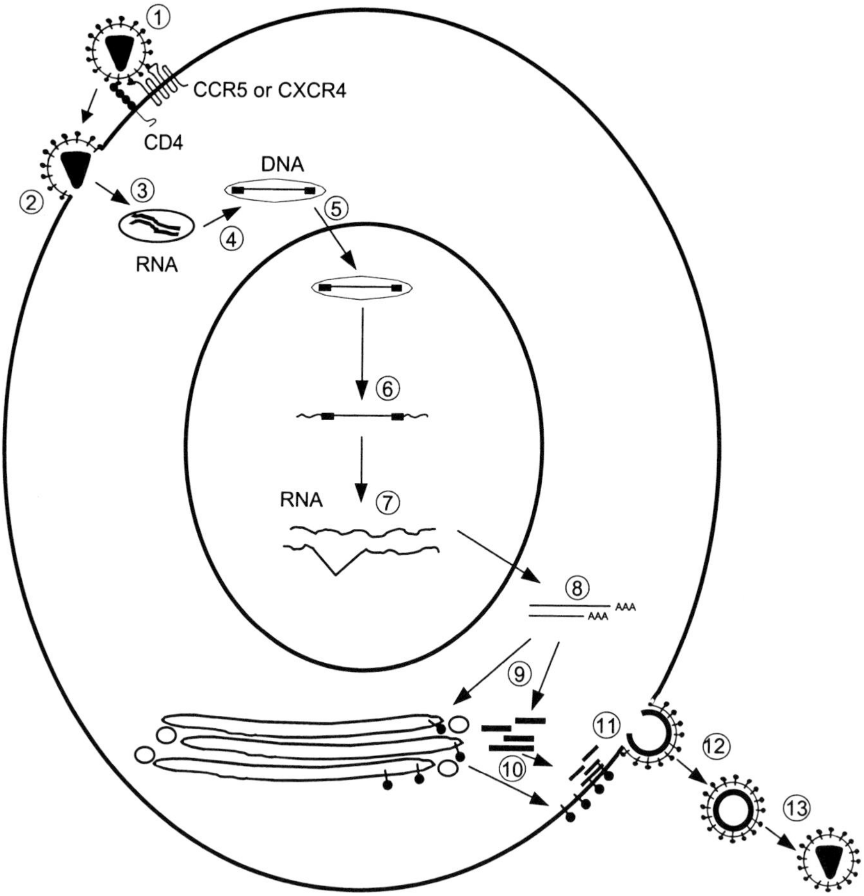

Fig. 6. Stages in the replication cycle of HIV. (1) Attachment; (2) fusion; (3) entry; (4) reverse transcription; (5) nuclear transport; (6) chromosomal integration of DNA provirus; (7) transcription of RNA; (8) nuclear export of RNA; (9) translation and processing; (10) membrane transport; (11) assembly; (12) budding; (13) maturation.

carrying deletions in one to three accessory genes such as *nef* and *vif* (Wyand *et al.*, 1996). The mechanism of immunity is not known. While live attenuated viruses elicit both humoral and cell-mediated immune responses, it is not clear that these mediate all of the protective effect; they might also act as defective, interfering virus populations. However, such 'attenuated' viruses have caused disease when administered to infant macaques (Baba *et al.*, 1995). At this stage, it would be foolhardy to engage in human clinical trials of live attenuated HIV vaccines. Such viruses would persist in the host, and no-one knows if the vaccinees might develop some aspect of HIV disease, e.g. the wasting syndrome, perhaps 30 years or more after exposure. With the virulent strains having a mean incubation period from infection to disease of approximately 9 years, how long would regulatory bodies need to wait before pronouncing phase I clinical trials of an attenuated strain safe?

A wide variety of viral and bacterial gene vectors have been examined to test the immunogenicity of HIV antigens expressed in this way. These include *Salmonella* and polio vaccine constructs, affording mucosal immunity, vaccinia and avipox vectors that elicit good cytotoxic T-cell responses, and RNA virus vectors such as alphaviruses and Venezuelan equine encephalitis virus. The latter virus targets dendritic cells and hence expresses recombinant HIV proteins in antigen-presenting cells.

Immune escape has been recorded from neutralizing antibodies and from epitopes recognized by CTLs (Goulder & Walker, 1999). The problem of both natural and selected antigenic variation is severe. One of the more promising immunization approaches is to combine a modified vaccinia Ankara vector with DNA vaccination (Gotch *et al.*, 2000). If a successful vaccine were to be developed, it would need to be multivalent, given the considerable antigenic variability of the HIV subtypes and strains. DNA vaccines could readily include multiple strains. However, DNA vaccines are novel and have not yet been licensed for any veterinary or human vaccine; proof of principle, safety and efficacy will be required for a virus for which another successful vaccine already exists. Protective vaccines have not been developed for any of the animal lentiviruses, and the obstacles to development of an HIV vaccine remain daunting.

A small number of sex workers in East and West Africa have been identified who appear to be uninfected by HIV, i.e. they remain persistently seronegative and lack HIV genomes in the blood. These women are continually exposed to infected men and appear to have specific cellular immunity to HIV in the genital tract (Kaul *et al.*, 2000). If the precise mechanism by which they have become immune can be elucidated, it might be possible to generate a vaccine that elicits similar protection. However, several of the women in a cohort study have become infected by HIV a few months after stopping commercial sex practices, suggesting that continual exposure to antigen may be necessary to maintain protective immunity.

Therapeutic vaccines have also raised considerable interest. With drug therapy blocking HIV replication and permitting restoration of vigorous immune function, specific immune stimulation by HIV antigens might act synergistically with anti-retroviral therapy and, possibly, cytokine therapy by interleukin-2 (IL-2) or IL-12 to boost T-helper type 1 immune responses. These are experimental approaches worth exploring in the window of opportunity while drugs keep the viral load low. In a sense, brief drug 'holidays' allowing a resurgence of virus replication might itself act as a useful booster immunization provided that reintroduction of anti-viral therapy was as effective as before in reducing viral load.

Public health control and prevention

The difficulty in providing effective treatment in resource-poor settings means that prevention of new infections continues to be an urgent priority. The few countries which have demonstrated success in controlling the HIV/AIDS epidemic are those that have implemented a multi-sectoral approach, with rapid responses from the Government, non-governmental organizations, and the community. In the absence of a vaccine, the main prevention strategies include the following.

Blood safety. Western countries introduced HIV antibody screening for all blood donations during 1985. HIV genome screening can detect primary infection before seroconversion (Fig. 5) but is much more expensive. The probability of becoming infected through a transfusion of HIV-contaminated blood is estimated to be over 90 % (UNAIDS/WHO, 1997), and development of a safe blood supply through an accountable national blood transfusion service is a high priority in those countries which have yet to develop one.

Needle-exchange programmes. Injecting drug users are also at high risk for HIV infection, again due to the high risk associated with parenteral transmission. Many countries have now implemented comprehensive packages to prevent transmission, including needle-exchange programmes which provide sterile injecting equipment. This also raises awareness among injectors and their sex partners about HIV risks and safe practices, and increases provision for drug treatment.

Sexual counselling. Heterosexual intercourse causes the vast majority of HIV infection globally, yet changes in sexual behaviour are particularly difficult to engineer through health education. Nonetheless, condom use and other 'safe-sex' counselling has also made a substantial impact in reducing homosexual HIV transmission. STD care and prompt treatment is not only beneficial in itself but may help to reduce HIV infection. Uganda is one of the very few countries where a population-level reduction of incident HIV infection has occurred, including delayed onset of sexual activity and marriage, and substantial increased reported condom use (Asiimwe-Okiror *et al.*, 1997). These changes in behaviour have been reflected in a 40 % decrease in HIV prevalence among women attending antenatal clinics between 1989 and 1995. Large-scale change in behaviour has also been documented in Thailand, where the Government acted swiftly with a '100 % condom programme' to enforce condom use within the many commercial sex establishments. Again, prevalence of HIV has stabilized in many groups. However, these success stories are sporadic, and rates of heterosexually transmitted HIV are increasing among young people, especially among young women, in many countries. The importance of comprehensive, sustained, safer sex education programmes in all countries cannot be overemphasized.

Mother-to-child transmission. One major recent development in HIV control has been the use of zidovudine or nevirapine to reduce maternal–child transmission. The initial randomized trial, conducted in the USA and France, showed that an intensive regimen of zidovudine therapy given to pregnant women and their newborn infants reduced the rate of vertical transmission of HIV by nearly 70 % (Connor *et al.*, 1994). This regimen is not feasible for implementation in the developing countries in which vertical transmission is most common. However, a trial in Thailand has now shown that short-term ziduvudine treatment starting at 36 weeks gestation and given orally in labour reduced maternal–child transmission of HIV by 50 %, from 18.9 % to 9.4 % (Shaffer *et al.*, 1999). Further, a trial in Uganda showed that a single dose of 200 mg nevirapine to the mother at onset of labour and a single dose of 2 mg kg^{-1} to the baby within 72 h of birth lowered the risk of HIV-1 transmission during the first 14–16 weeks of life by nearly 50 % compared with zidovudine in a breastfeeding population (Guay *et al.*, 1999). Work on translating these promising results into public health policy is under way.

Virucides. Research is in progress to develop and evaluate vaginal products that can be used by women to prevent HIV transmission from their male partners. Unfortunately, recent trials of the microbicide nonoxynol-9, used by female sex workers in Kenya, Cameroon, Benin, Côte d'Ivoire, South Africa and Thailand (UNAIDS, 2000a), have failed to show any impact on HIV incidence. Trials of alternative products are now in progress.

HIV denial. A handful of scientists and commentators, inspired by the retrovirologist Peter Duesberg, have argued that HIV is harmless, has never been properly isolated and characterized, or even that AIDS does not exist as a distinct entity (Duesberg, 1991). While there was some value to challenging the role of HIV in AIDS when less was known about the virus and the syndrome in the mid-1980s, these arguments have been thoroughly examined and found untenable (Weiss & Jaffe, 1990; O'Brien & Goedert, 1996).

The evidence linking HIV aetiologically to AIDS is as soundly based on the same criteria as for any other viral disease, e.g. poliomyelitis or hepatitis. Nonetheless, a small but vocal group continue to spread propaganda by the internet and the media that HIV is irrelevant, that AIDS is not an infectious disease, and that anti-retroviral drugs do more harm than good. In southern Africa, faced with 25 % or more young adults positive in

HIV tests with all that implies for the future health, demography and economy of the region, it is perhaps not surprising that some politicians are tempted to listen to these siren voices singing HIV denial, and to seek their advice. The danger of heeding these voices has recently been addressed by the president of the South African Medical Research Council (Makgoba, 2000) and by 5000 members of the international scientific and medical community (Commentary, 2000).

CONCLUSIONS AND PROSPECTS

As progress in medical research goes, the study of HIV/AIDS has been remarkably rapid. AIDS was first recognized as a new disease in mid-1981, and to be transmissible by early 1982. The first HIV isolate was made in 1983. This led not only to our understanding the nature of AIDS, but also to prevention, diagnosis and treatment. By the end of 1985, within a year or so of learning to propagate the virus to high titres in culture, screening of blood donations had been introduced throughout developed countries. A year later, zidovudine therapy was introduced to clinical trial, though it took a further 10 years to achieve the efficacy of combination anti-retroviral therapy that has so dramatically reduced the mortality from AIDS in developed countries. Delivering such therapy where it is most needed is prohibitively expensive and adherence for the patient also has its problems. We do not yet know whether the therapeutic success experienced in the West is a temporary pause before multi-resistant strains of HIV appear and spread, or whether virus load can be permanently controlled, thus abrogating progression to AIDS. Behavioural control of HIV transmission is notoriously difficult to sustain, although some health education campaigns have been successful, e.g. in Uganda. The most important, yet daunting, task facing virologists and immunologists is to provide an efficacious vaccine for mass immunization. Meanwhile, HIV continues to spread, having already killed 19 million people and infected 34 million more. In 2001, HIV is estimated to cause over 3.5 million deaths, surpassing tuberculosis and malaria. This small virus with just nine genes has become mankind's most formidable foe.

ACKNOWLEDGEMENTS

The authors' research is supported by the Medical Research Council.

REFERENCES

Abebe, A., Demissie, D., Goudsmit, J., Brouwer, M., Kuiken, C. L., Pollakis, G., Schuitemaker, H., Fontanet, A. L. & Rinke de Wit, T. F. (1999). HIV-1 subtype C syncytium- and non-syncytium-inducing phenotypes and coreceptor usage among Ethiopian patients with AIDS. *AIDS* **13**, 1305–1311.

Alonso, A. & Peterlin, B. M. (1999). HIV-1: control of gene expression by the viral regulatory proteins Tat and Rev. In *HIV and the New Viruses*, 2nd edn, pp. 1–28. Edited by A. G. Dalgleish & R. A. Weiss. London: Academic Press.

Asiimwe-Okiror, G., Opio, A. A., Musinguzi, J., Madraa, E., Tembo, G. & Carael, M. (1997). Change in sexual behaviour and decline in HIV infection among young pregnant women in urban Uganda. *AIDS* **11**, 1757–1763.

Asjo, B., Morfeldt Manson, L., Albert, J., Biberfeld, G., Karlsson, A., Lidman, K. & Fenyo, E. M. (1986). Replicative capacity of human immunodeficiency virus from patients with varying severity of HIV infection. *Lancet* **ii**, 660–662.

Baba, T. W., Jeong, Y. S., Pennick, D., Bronson, R., Greene, M. F. & Ruprecht, R. M. (1995). Pathogenicity of live, attenuated SIV after mucosal infection of neonatal macaques. *Science* **267**, 1820–1825.

Barré-Sinoussi, F., Chermann, J. C., Rey, F. & 9 other authors (1983). Isolation of a T-lymphotropic retrovirus from a patient at risk for acquired immune deficiency syndrome (AIDS). *Science* **220**, 868–871.

Bayley, A. C., Downing, R. G., Cheingsong-Popov, R., Tedder, R. S., Dalgleish, A. G. & Weiss, R. A. (1985). HTLV-III serology distinguishes atypical and endemic Kaposi's sarcoma in Africa. *Lancet* **i**, 359–361.

Berger, E. A., Doms, R. W., Fenyo, E.-M., Korber, B. T. M., Littman, D. R., Moore, J. P., Sattentau, Q. J., Schuitemaker, H., Sodroski, J. & Weiss, R. A. (1998). A new classification for HIV-1. *Nature* **391**, 240.

Berger, E. A., Murphy, P. M. & Farber, J. M. (1999). Chemokine receptors as HIV-1 coreceptors: roles in viral entry, tropism, and disease. *Annu Rev Immunol* **17**, 657–700.

Bloom, D. E. & River Path Associates (2000). Something to be done: treating HIV/AIDS. *Science* **288**, 2171–2173.

Buvé, A., for the Study Group on Heterogeneity of HIV Epidemics in African Cities (2000). HIV/AIDS in Africa: why so severe, why so heterogeneous? *Seventh National Conference on Retroviruses and Opportunistic Infections*, abstract S28. San Francisco, USA.

Carrington, M., Dean, M., Martin, M. P. & O'Brien, S. J. (1999). Genetics of HIV-1 infection: chemokine receptor CCR5 polymorphism and its consequences. *Hum Mol Genet* **8**, 1939–1945.

Centers for Disease Control (1981). Kaposi's sarcoma and *Pneumocystis* pneumonia among homosexual men – New York City and California. *Morb Mortal Wkly Rep* **30**, 205–208.

Cheingsong-Popov, R., Weiss, R. A., Dalgleish, A. & 16 other authors (1984). Prevalence of antibody to human T-lymphotropic virus type III in AIDS and AIDS-risk patients in Britain. *Lancet* **ii**, 477–480.

Clapham, P. R., McKnight, A. & Weiss, R. A. (1992). Human immunodeficiency virus type 2 infection and fusion of CD4-negative human cell lines: induction and enhancement by soluble CD4. *J Virol* **66**, 3531–3537.

Clavel, F., Mansinho, K., Chamaret, S., Guetard, D., Favier, V., Nina, J., Santos Ferreira, M. O., Champalimaud, J. L. & Montagnier, L. (1987). Human immunodeficiency virus type 2 infection associated with AIDS in West Africa. *N Engl J Med* **316**, 1180–1185.

Cloyd, M. W., Chen, J. J. & Wang, I. (2000). How does HIV cause AIDS? The homing theory. *Mol Med Today* **6**, 108–111.

Coffin, J., Hughes, S. H. & Varmus, H. E. (1997). *Retroviruses*. Cold Spring Harbor, NY: Cold Spring Harbor Laboratory.

Cohen, J. (2000). Is AIDS in Africa a distinct disease? *Science* **288**, 2153–2155.

Commentary (2000). The Durban Declaration. *Nature* **406**, 15–16.

Communicable Disease Surveillance Centre (2000). AIDS and HIV infection in the United Kingdom. *Communicable Dis Rep* **10**, 157–160.

Connor, E. M., Sperling, R. S., Gelber, R. & 17 other authors (1994). Reduction of maternal-infant transmission of human immunodeficiency virus type 1 with zidovudine treatment. Pediatric AIDS Clinical Trials Group Protocol 076 Study Group. *N Engl J Med* **331**, 1173–1180.

Dalgleish, A. G., Beverley, P. C., Clapham, P. R., Crawford, D. H., Greaves, M. F. & Weiss, R. A. (1984). The CD4 (T4) antigen is an essential component of the receptor for the AIDS retrovirus. *Nature* **312**, 763–767.

Darby, S. C., Ewart, D. W., Giangrande, P. L., Spooner, R. J. & Rizza, C. R. (1996). Importance of age at infection with HIV-1 for survival and development of AIDS in UK haemophilia population. *Lancet* **347**, 1573–1579.

Dragic, T., Trkola, A., Thompson, D. A. & 8 other authors (2000). A binding pocket for a small molecule inhibitor of HIV-1 entry within the transmembrane helices of CCR5. *Proc Natl Acad Sci U S A* **97**, 5639–5644.

Duesberg, P. H. (1991). AIDS epidemiology: inconsistencies with human immunodeficiency virus and with infectious disease. *Proc Natl Acad Sci U S A* **88**, 1575–1579.

Fink, A. J. (1987). Circumcision and heterosexual transmission of HIV to men. *N Engl J Med* **316**, 1546–1547.

Fleming, D. T. & Wasserheit, J. N. (1999). From epidemiological synergy to public health policy and practice: the contribution of other sexually transmitted diseases to sexual transmission of HIV infection. *Sex Transm Infect* **75**, 3–17.

Fouchier, R. A., Groenink, M., Kootstra, N. A., Tersmette, M., Huisman, H. G., Miedema, F. & Schuitemaker, H. (1992). Phenotype-associated sequence variation in the third variable domain of the human immunodeficiency virus type 1 gp120 molecule. *J Virol* **66**, 3183–3187.

Gallo, R. C., Salahuddin, S. Z., Popovic, M. & 10 other authors (1984). Frequent detection and isolation of cytopathic retroviruses (HTLV-III) from patients with AIDS and at risk for AIDS. *Science* **224**, 500–503.

Gao, F., Yue, L., White, A. T., Pappas, P. G., Barchue, J., Hanson, A. P., Greene, B. M., Sharp, P. M., Shaw, G. M. & Hahn, B. H. (1992). Human infection by genetically diverse SIVsm-related HIV-2 in west Africa. *Nature* **358**, 495–499.

Gao, F., Bailes, E., Robertson, D. L. & 9 other authors (1999). Origin of HIV-1 in the chimpanzee *Pan troglodytes troglodytes*. *Nature* **397**, 436–441.

Geijtenbeek, T. B., Kwon, D. S., Torensma, R. & 9 other authors (2000). DC-SIGN, a dendritic cell-specific HIV-1-binding protein that enhances trans-infection of T cells. *Cell* **100**, 587–597.

Gilks, C. (1991). AIDS, monkeys and malaria. *Nature* **354**, 262.

Gotch, F., Rutebemberwa, A., Jones, G., Imami, N., Gilmour, J., Kaleebu, P. & Whitworth, J. (2000). Vaccines for the control of HIV/AIDS. *Trop Med Int Health* **5**, A16–21.

Gottlieb, M. S., Schroff, R., Schanker, H. M., Weisman, J. D., Fan, P. T., Wolf, R. A. & Saxon, A. (1981). *Pneumocystis carinii* pneumonia and mucosal candidiasis in previously healthy homosexual men: evidence of a new acquired cellular immunodeficiency. *N Engl J Med* **305**, 1425–1431.

Goulder, P. J. & Walker, B. D. (1999). The great escape – AIDS viruses and immune control. *Nat Med* **5**, 1233–1235.

Guay, L. A., Musoke, P., Fleming, T. & 16 other authors (1999). Intrapartum and neonatal single-dose nevirapine compared with zidovudine for prevention of mother-to-child transmission of HIV-1 in Kampala, Uganda: HIVNET 012 randomised trial. *Lancet* **354**, 795–802.

Hahn, B. H., Shaw, G. M., De Cock, K. M. & Sharp, P. M. (2000). AIDS as a zoonosis: scientific and public health implications. *Science* **287**, 607–614.

Halperin, D. T. & Bailey, R. C. (1999). Male circumcision and HIV infection: 10 years and counting. *Lancet* **354**, 1813–1815.

Hazuda, D. J., Felock, P., Witmer, M. & 8 other authors (2000). Inhibitors of strand transfer that prevent integration and inhibit HIV-1 replication in cells. *Science* **287**, 646–650.

Hirsch, M. S., Brun-Vezinet, F., D'Aquila, R. T. & 11 other authors (2000). Antiretroviral drug resistance testing in adult HIV-1 infection: recommendations of an International AIDS Society – USA Panel. *J Am Med Assoc* **283**, 2417–2426.

Holmes, K. K., Sparling, P. F., Mardh, P.-A., Lemon, S. M., Stamm, W. E., Piot, P. & Wasserheit, J. M. (1998). *Sexually Transmitted Diseases*, 3rd edn. New York: McGraw-Hill.

Hooper, E. (1999). *The River*. New York: Little Brown.

Hu, D. J., Buvé, A., Baggs, J., van der Groen, G. & Dondero, T. J. (1999). What role does HIV-1 subtype play in transmission and pathogenesis? An epidemiological perspective. *AIDS* **13**, 873–881.

Kaul, R., Plummer, F. A., Kimani, J. & 8 other authors (2000). HIV-1 specific mucosal cytotoxic T-lymphocyte (CTL) responses in the cervix of HIV-1 resistant prostitutes in Nairobi. *J Immunol* **164**, 1602–1611.

Kaye, J. F. & Lever, A. M. (1999). Human immunodeficiency virus types 1 and 2 differ in the predominant mechanism used for selection of genomic RNA for encapsidation. *J Virol* **73**, 3023–3031.

Klatzmann, D., Barré-Sinoussi, F., Nugeyre, M. T. & 8 other authors (1984). Selective tropism of lymphadenopathy associated virus (LAV) for helper-inducer T lymphocytes. *Science* **225**, 59–63.

Korber, B., Muldoon, M., Theiler, J., Gao, F., Gupta, R., Lapedes, A., Hahn, B. H., Wolinsky, S. & Bhattacharya, T. (2000). Timing the ancestor of the HIV-1 pandemic strains. *Science* **288**, 1789–1796.

Leigh-Brown, A. J. (1999). Viral evolution and variation in the HIV pandemic. In *HIV and the New Viruses*, 2nd edn, pp. 29–42. Edited by A. G. Dalgleish & R. A. Weiss. London: Academic Press.

Levy, J. A. (1998). *HIV and the Pathogenesis of AIDS*. Washington, DC: American Society for Microbiology.

Levy, J. A., Hoffman, A. D., Kramer, S. M., Landis, J. A., Shimabukuro, J. M. & Oshiro, L. S. (1984). Isolation of lymphocytopathic retroviruses from San Francisco patients with AIDS. *Science* **225**, 840–842.

Lucas, S. B., Hounnou, A., Peacock, C. & 15 other authors (1993). The mortality and pathology of HIV infection in a west African city. *AIDS* **7**, 1569–1579.

McKnight, A., Weiss, R. A., Shotton, C., Takeuchi, Y., Hoshino, H. & Clapham, P. R. (1995). Change in tropism upon immune escape by human immunodeficiency virus. *J Virol* **69**, 3167–3170.

Maddon, P. J., Dalgleish, A. G., McDougal, J. S., Clapham, P. R., Weiss, R. A. & Axel, R. (1986). The T4 gene encodes the AIDS virus receptor and is expressed in the immune system and the brain. *Cell* **47**, 333–348.

Makgoba, M. W. (2000). HIV/AIDS: the peril of pseudoscience. *Science* **288**, 1171.

Mellors, J. W., Kingsley, L. A., Rinaldo, C. R., Jr, Todd, J. A., Hoo, B. S., Kokka, R. P. &

Gupta, P. (1995). Quantitation of HIV-1 RNA in plasma predicts outcome after seroconversion. *Ann Intern Med* **122**, 573–579.

Michael, N. L. (1999). Host genetic influences on HIV-1 pathogenesis. *Curr Opin Immunol* **11**, 466–474.

Montagnier, L., Dauguet, C., Axler, C., Chamanet, S., Gruest, J., Nugere, M., Rey, F., Barré-Sinnoussi, F. & Chermann, J.-C. (1984). A new type of retrovirus isolated from patients presenting with lymphadenopathy and acquired immune deficiency syndrome: structural and antigenic relatedness with equine infectious anaemia virus. *Ann Virol* **135E**, 119–134.

O'Brien, S. J. & Goedert, J. J. (1996). HIV causes AIDS: Koch's postulates fulfilled. *Curr Opin Immunol* **8**, 613–618.

Peeters, M., Gueye, A., Mboup, S. & 15 other authors (1997). Geographical distribution of HIV-1 group O viruses in Africa. *AIDS* **11**, 493–498.

Perelson, A. S., Neumann, A. U., Markowitz, M., Leonard, J. M. & Ho, D. D. (1996). HIV-1 dynamics *in vivo*: virion clearance rate, infected cell life-span, and viral generation time. *Science* **271**, 1582–1586.

Phillips, A. (1999). HIV dynamics: lessons from the use of antiretrovirals. In *HIV and the New Viruses*, 2nd edn, pp. 59–74. Edited by A. G. Dalgleish & R. A. Weiss. London: Academic Press.

Ping, L. H., Nelson, J. A., Hoffman, I. F. & 12 other authors (1999). Characterization of V3 sequence heterogeneity in subtype C human immunodeficiency virus type 1 isolates from Malawi: underrepresentation of X4 variants. *J Virol* **73**, 6271–6281.

Piot, P. (2000). Global AIDS epidemic: time to turn the tide. *Science* **288**, 2176–2178.

Pope, M., Betjes, M. G., Romani, N., Hirmand, H., Cameron, P. U., Hoffman, L., Gezelter, S., Schuler, G. & Steinman, R. M. (1994). Conjugates of dendritic cells and memory T lymphocytes from skin facilitate productive infection with HIV-1. *Cell* **78**, 389–398.

Popovic, M., Read Connole, E. & Gallo, R. C. (1984). T4 positive human neoplastic cell lines susceptible to and permissive for HTLV-III. *Lancet* **ii**, 1472–1473.

Sarngadharan, M. G., Popovic, M., Bruch, L., Schupbach, J. & Gallo, R. C. (1984). Antibodies reactive with human T-lymphotropic retroviruses (HTLV-III) in the serum of patients with AIDS. *Science* **224**, 506–508.

Serwadda, D., Mugerwa, R. D., Sewankambo, N. K. & 9 other authors (1985). Slim disease: a new disease in Uganda and its association with HTLV-III infection. *Lancet* **ii**, 849–852.

Shaffer, N., Chuachoowong, R., Mock, P. A. & 10 other authors (1999). Short-course zidovudine for perinatal HIV-1 transmission in Bangkok, Thailand: a randomised controlled trial. Bangkok Collaborative Perinatal HIV Transmission Study Group. *Lancet* **353**, 773–780.

Shilts, R. (1987). *And the Band Played On – Politics, People and the AIDS Epidemic*. New York: St Martin's Press.

Simmons, G., Clapham, P. R., Picard, L., Offord, R. E., Rosenkilde, M. M., Schwartz, T. W., Buser, R., Wells, T. N. C. & Proudfoot, A. E. (1997). Potent inhibition of HIV-1 infectivity in macrophages and lymphocytes by a novel CCR5 antagonist. *Science* **276**, 276–279.

Simon, F., Mauclere, P., Roques, P., Loussert-Ajaka, I., Muller-Trutwin, M. C., Saragosti, S., Georges-Courbot, M. C., Barré-Sinoussi, F. & Brun-Vezinet, F. (1998). Identification of a new human immunodeficiency virus type 1 distinct from group M and group O. *Nat Med* **4**, 1032–1037.

Sommerfelt, M. A. (1999). Retrovirus receptors. *J Gen Virol* **80**, 3049–3064.

Szabo, R. & Short, R. V. (2000). How does male circumcision protect against HIV infection? *Br Med J* **320**, 1592–1594.

Tersmette, M., de Goede, R. E., Al, B. J., Winkel, I. N., Gruters, R. A., Cuypers, H. T., Huisman, H. G. & Miedema, F. (1988). Differential syncytium-inducing capacity of human immunodeficiency virus isolates: frequent detection of syncytium-inducing isolates in patients with acquired immunodeficiency syndrome (AIDS) and AIDS-related complex. *J Virol* **62**, 2026–2032.

UNAIDS (2000a). Press Release (http://www.unaids.org/whatsnew/press/eng/geneva130600.html).

UNAIDS (2000b). Report on the global HIV/AIDS epidemic, June 2000 (http://www.UNAIDS.org/hivaidsinfo/).

UNAIDS/WHO (1997). Blood safety and AIDS. *UNAIDS Best Practice Collection: Point of View*. Geneva: UNAIDS.

Vilmer, E., Barré-Sinoussi, F., Rouzioux, C. & 8 other authors (1984). Isolation of new lymphotropic retrovirus from two siblings with haemophilia B, one with AIDS. *Lancet* **i**, 753–757.

Wain-Hobson, S., Sonigo, P., Alizon, M., Staskus, K., Klatzmann, D., Cole, S., Danos, O., Retzel, E., Tiollais, P. & Haase, A. (1985a). Nucleotide sequence of the visna lentivirus: relationship to the AIDS virus. *Cell* **42**, 369–382.

Wain-Hobson, S., Sonigo, P., Danos, O., Cole, S. & Alizon, M. (1985b). Nucleotide sequence of the AIDS virus, LAV. *Cell* **40**, 9–17.

Weiss, H. A., Quigley, M. A. & Hayes, R. J. (2000). Male circumcision and risk of HIV infection in sub-Saharan Africa: a systematic review and meta-analysis. *AIDS* **14**, 2361–2370.

Weiss, R. A. (1993). How does HIV cause AIDS? *Science* **260**, 1273–1279.

Weiss, R. A. & Jaffe, H. W. (1990). Duesberg, HIV and AIDS. *Nature* **345**, 659–660.

Weiss, R. A. & Wrangham, R. W. (1999). The origin of HIV-1: from *Pan* to pandemic. *Nature* **397**, 385–386.

Weiss, R. A., Clapham, P. R., Cheingsong Popov, R., Dalgleish, A. G., Carne, C. A., Weller, I. V. & Tedder, R. S. (1985). Neutralization of human T-lymphotropic virus type III by sera of AIDS and AIDS-risk patients. *Nature* **316**, 69–72.

Weiss, R. A., Clapham, P. R., Weber, J. N., Dalgleish, A. G., Lasky, L. A. & Berman, P. W. (1986). Variable and conserved neutralization antigens of human immunodeficiency virus. *Nature* **324**, 572–575.

Weiss, R. A., Dalgleish, A. G. & Loveday, C. (2000). Human immunodeficiency viruses. In *Principles and Practice of Clinical Virology*, 4th edn, pp. 659–694. Edited by A. J. Zuckerman, J. E. Banatvala & J. R. Pattison. Chichester: Wiley.

Whittle, H., Morris, J., Todd, J., Corrah, T., Sabally, S., Bangali, J., Ngom, P. T., Rolfe, M. & Wilkins, A. (1994). HIV-2-infected patients survive longer than HIV-1-infected patients. *AIDS* **8**, 1617–1620.

Wyand, M. S., Manson, K. H., Garcia-Moll, M., Montefiori, D. & Desrosiers, R. C. (1996). Vaccine protection by a triple deletion mutant of simian immunodeficiency virus. *J Virol* **70**, 3724–3733.

Zhu, T., Korber, B. T., Nahmias, A. J., Hooper, E., Sharp, P. M. & Ho, D. D. (1998). An African HIV-1 sequence from 1959 and implications for the origin of the epidemic. *Nature* **391**, 594–597.

Morbilliviruses: dangers old and new

Tom Barrett

Institute for Animal Health, Pirbright Laboratory, Ash Road, Pirbright, Surrey GU24 0NF, UK

INTRODUCTION

The emergence, or more often the re-emergence, of disease-causing organisms is a continuing threat to mankind either directly by impacting on human health, or indirectly through the effects they may have on wild and domesticated animals. Virus infections can cause disease outbreaks of plague dimensions in humans and animals, with serious social, economic and environmental consequences. In the past, diseases such as smallpox, measles, polio and influenza have caused devastation in the human population. Smallpox has the distinction of being the only virus disease to have been eradicated globally while the others mentioned above are, or can be, effectively controlled by vaccination.

Almost every year, new viruses emerge which threaten the lives of humans and animals. In particular, members of the family *Paramyxoviridae*, a group that includes viruses causing some of the most virulent diseases of animals, pose significant medical, economic and ecological threats (Chant *et al.*, 1998; Renshaw *et al.*, 2000). Of the many new paramyxoviruses that have emerged in the past decade, the so-called 'equine morbillivirus' received a high profile in the mid-1990s because it caused the deaths of several racehorses and two human fatalities in Australia (Murray *et al.*, 1995). The virus was initially grouped with the morbilliviruses because of a distant sequence relationship with members of the group, but it is now considered to belong to a new, as yet unnamed, genus in the *Paramyxoviridae*. It has been renamed Hendra virus, after the location of the first outbreak. In 1999, another newly discovered paramyxovirus, Nipah virus, caused a calamitous outbreak of disease in pigs in Malaysia. Within the

SGM symposium 60: New challenges to health: the threat of virus infection.
Editors G. L. Smith, W. L. Irving, J. W. McCauley, D. J. Rowlands. Cambridge University Press. ISBN 0 521 80614 3 ©SGM 2001.

space of 1 month, more than 100 fatal human cases occurred before the disease was brought under control by a draconian pig slaughter policy (Chua *et al.*, 2000).

Within the family *Paramyxoviridae*, members of the *Morbillivirus* genus cause some of the most severe diseases and this chapter concentrates on the effects of this group of viruses on animals. The type virus of the genus is measles virus (MV) and the name *Morbillivirus* is derived from *morbilli*, the dimunitive form of the Latin word for plague (*morbus*). This name was used to distinguish measles from smallpox and scarlet fever, which in former times were considered more serious diseases (Carmichael, 1997). The most feared of cattle diseases, rinderpest or cattle plague, is caused by a virus (RPV) that is antigenically and genetically most closely related to MV. Another old established member of the group, canine distemper virus (CDV), infects many carnivore species, including domestic dogs, mink and ferrets, and can have serious consequences when susceptible endangered wildlife species are threatened (Laurenson *et al.*, 1998). A virus disease in small ruminants which clinically resembles rinderpest was discovered in West Africa in the 1940s and named peste des petits ruminants virus (PPRV) (Gargadennec & Lalanne, 1942). It was originally thought to be a variant of RPV adapted to small ruminants but subsequently it was shown to be an antigenically and genetically distinct virus (Diallo *et al.*, 1989; Gibbs *et al.*, 1979). More recently, new morbilliviruses have been discovered in marine mammals. In 1988, a seal morbillivirus, now known as phocid distemper virus (PDV), was identified as the cause of death in thousands of seals around the coasts of northern Europe (Osterhaus & Vedder, 1988). Similar viruses were then isolated from porpoises (Kennedy *et al.*, 1988) and dolphins (Domingo *et al.*, 1990). These 'cetacean' viruses are genetically very closely related to each other (Barrett *et al.*, 1993) and are often referred to as the cetacean morbillivirus (CeMV).

In this chapter, diseases caused by old and new members of the genus are described and their impact on man and animals discussed. The focus will mainly be on the veterinary aspects of morbillivirus infections.

HISTORY

The earliest descriptions of human MV date back to Persian writings of the 10th century AD. Before that, measles was often confused with other great killer diseases, such as smallpox. Measles and scarlet fever, along with smallpox virus, decimated indigenous populations when taken from Europe to the Americas and the Pacific islands (Carmichael, 1997). The virus remains a major cause of childhood mortality in the developing world, where it is estimated that about 1 million children die each year as a result of MV infection (Sabin, 1992). Vaccination has reduced the number of measles cases by about 75 % worldwide and before effective vaccines were available perhaps as many as 7–8 million childhood deaths occurred annually as a result of infection (Wild *et al.*, 2000).

Recognizable descriptions of rinderpest date back to late Roman times (Scott, 1996; Barton, 1956) and throughout subsequent history the virus has been responsible for devastating cattle plagues and consequent famines. In Europe, rinderpest often entered with invading armies from the east and later as a result of cattle trading, mainly from Russia. Because of this it was also known as the 'Steppe Murrain'. It remained endemic in the Netherlands for many years but by the end of the 19th century the virus had been eradicated from western Europe by measures which included import controls and the adoption of effective quarantine and zoosanitary procedures (Scott, 1997). Until the middle of the 20th century, when safe and effective vaccines were first developed, the virus was endemic in vast areas of Africa and Asia, including Russia, China and Korea (Fig. 1). Even today in war situations, where there is uncontrolled movement of soldiers and refugees, often bringing with them live animals as a source of food, the virus is carried to new regions. Asian, but not African, strains of rinderpest can productively infect sheep and goats, which may or may not show mild clinical signs of disease, and which are then capable of passing the infection to cattle. Small ruminants taken for food with Indian peacekeeping troops and then traded locally were the source of the virus re-introduced into Sri Lanka in 1987 after an absence of 40 years (Anderson *et al.*, 1990). Following the Gulf War, several outbreaks occurred in Turkey in the early 1990s as a result of the displacement of the Kurdish refugees.

Rinderpest was introduced into sub-Saharan Africa towards the end of the 19th century, most probably with infected Indian or Arabian cattle imported into Eritrea by the Italian army sometime between 1887 and 1889 (Mack, 1970; Rossiter, 1994). The resulting pandemic laid waste the cattle and buffalo populations as well as many susceptible wildlife species. Over a period of about 10 years, the virus spread west to the Atlantic coast and south to Cape Town. Effective veterinary control measures instigated in southern Africa, along with primitive vaccination protocols, eventually eliminated the virus from this part of Africa but it continued to circulate in vast areas of sub-Saharan Africa for most of the past century (Mack, 1970). Extensive vaccination and control programmes carried out over the past 50 years have successfully reduced its global incidence, but it remains in areas of the world where conflict and civil wars hamper effective control measures being introduced. These areas include the Horn of Africa, southern Sudan, Yemen, Afghanistan and the 'Kurdish Triangle', the area of Iran, Iraq and Turkey occupied by Kurdish people (see Fig. 1c). The virus now appears to have been eradicated from India, a major achievement given the numbers of cattle and buffalo involved and the difficulties of restricting cattle movements in such a vast country. An eradication programme has been established with the

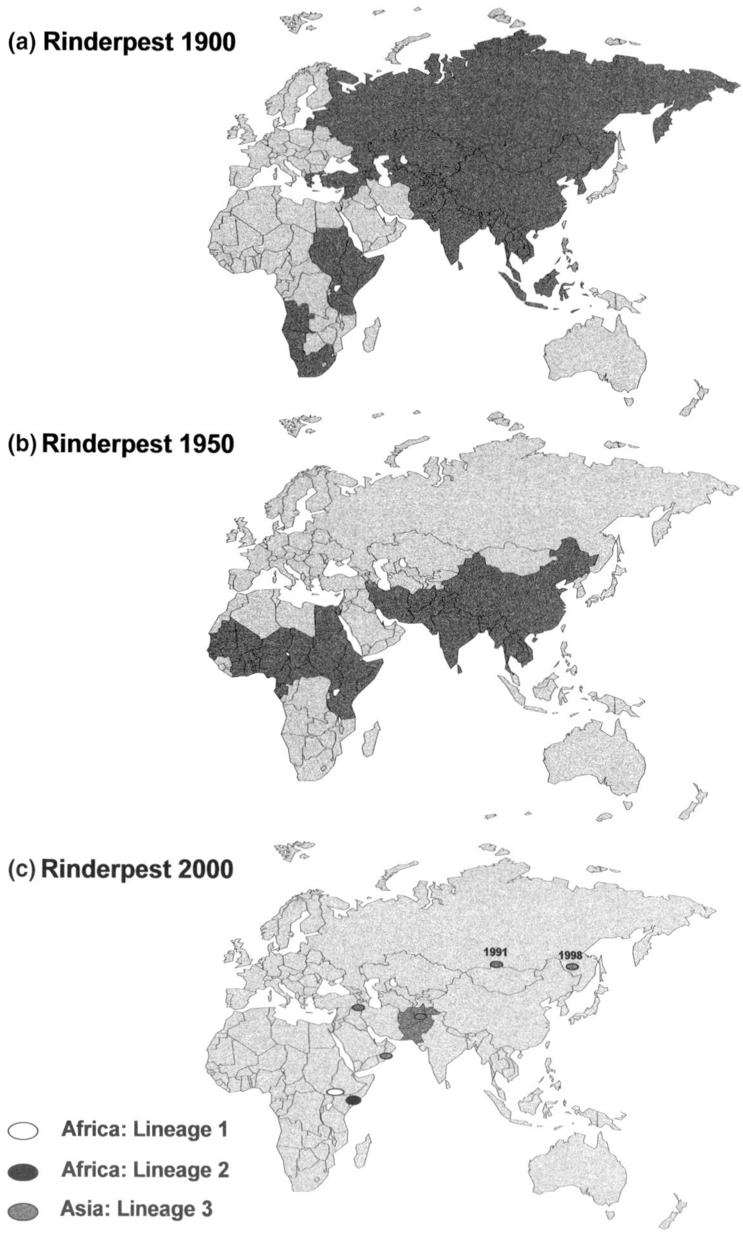

Fig. 1. Map showing the progress of rinderpest eradication over the past century. Dark shading shows affected countries. Parts (a) and (b) were taken from Scott (1981) and are reproduced with permission. In (c), the locations of the three extant lineages of RPV are indicated along with the locations of the two unusual Russian outbreaks in 1991 and 1998.

target of global eradication of the virus by the year 2010 (Rweyemamu & Cheneau, 1995).

CONTROL

Morbillivirus transmission requires close contact between an acutely infected and a susceptible host since, due to its fragile nature, the virus remains viable for only a very short time in the environment. Also, these viruses do not generally persist in the host following recovery from the acute infection, although MV and CDV can persist in a non-infectious form in rare cases, a situation that is fatal after a long incubation period (Schneider-Schaulies *et al.*, 1999). In addition, a single infection results in life-long immunity with the consequence that all morbilliviruses depend on a continuous supply of susceptible hosts for their maintenance; stopping the supply of new hosts (by vaccination) is an effective way of eliminating these viruses (Anderson & May, 1991; Black, 1991). The development of live attenuated vaccines was the key to achieving success in controlling diseases such as measles, distemper and rinderpest since, as with natural infection, the immunity they induce is long-lived and involves a cell-mediated response. In contrast, the original inactivated vaccines did not always protect from severe disease on subsequent natural challenge, although in most cases they generated a strong virus-neutralizing antibody response in the vaccinees (Appel *et al.*, 1984).

Vaccination has been used to control MV in the developed world but re-introduction from endemic regions in Asia and Africa remains a threat because of rapid international air travel. Until global eradication can be achieved, there is no prospect that vaccination can cease and this places a huge financial burden on health resources, particularly in developing countries that lack resources to provide universal vaccination and where poor hygiene and malnutrition compound its seriousness. The World Health Organization has a programme to eradicate measles by vaccination, a strategy that was successful in the case of smallpox and has almost been achieved for polio with the use of the oral vaccine. However, there are many logistical and financial problems associated with delivering a fragile, heat-labile vaccine in the poorer countries of the world. Another obstacle facing the measles eradication campaign is the inability to vaccinate children early in life while maternal antibodies against the virus are still present in the blood. New recombinant vaccines are being investigated as a possible approach to solving some of the problems (Wild *et al.*, 2000).

Similar difficulties are encountered when delivering rinderpest vaccine; nevertheless, it has been used with great success to eradicate the disease from many parts of Africa and Asia. However, unlike the case with MV, movement and import restrictions can be enforced on infected herds and countries to control the spread of RPV infection. When a safe and effective attenuated live vaccine was developed for rinderpest in the late

1950s, a determined effort was made to eliminate the disease from Africa. An international campaign, known as Joint Programme 15 (JP 15), was set up under the auspices of OAU, FAO and with funding from the EC with the aim of vaccinating rinderpest out of existence on that continent. The programme was highly successful and within 10 years had reduced the incidence of disease to a level where, in 1976, only three African countries, Mali, Mauritania and Senegal, reported disease. The false sense of security which this success encouraged, along with the political and economic crises of the 1970s and the resulting lack of national and international funds to continue the vaccination, eventually led in 1979 to the resurgence of the disease and the second African pandemic. In 1984, another international campaign to eradicate rinderpest from Africa was initiated with funding from the EC. Again this campaign, known as the Pan African Rinderpest Campaign (PARC), was highly successful in controlling and eliminating the disease from most of the continent. Within a period of less than 10 years, the virus was eradicated from all of West Africa and from most of East Africa (Barrett & Rossiter, 1999).

The ability to prevent and effectively control rinderpest incursions is considered to be a measure of the effectiveness of a country's veterinary service and the major factor blocking global eradication is the continuation of ethnic conflicts in the endemic regions. Effective government control is absent in the remaining endemic foci in Africa. For countries wishing to declare freedom from rinderpest, an internationally agreed protocol has been established, known as the O.I.E. pathway. To enter this pathway, a country must be free of overt disease, stop vaccination and embark on a prescribed process of disease monitoring (Taylor *et al.*, 1995). Most African countries have now stopped, or are preparing to stop, vaccination and begin this process and the aim is to achieve global eradication by the year 2010 (Rweyemamu, 1999). If successful, rinderpest may thus be the first animal virus disease to achieve the status of global eradication.

Vaccination is also used to control distemper in domestic dogs, ferrets and farmed mink. However, the logistical problems posed by vaccination of wild animals are very great and it may be impossible to eradicate diseases caused by CDV, PDV and CeMV, particularly since there are so many different susceptible host species for these viruses. In addition, the vaccines developed for domestic species are not always attenuated for wildlife. In some susceptible species, such as foxes, black-footed ferrets and the lesser panda, the current CDV vaccines can cause a severe, life-threatening infection (Bush *et al.*, 1976; Carpenter *et al.*, 1976; Henke, 1997; Sutherland-Smith *et al.*, 1997). In the long term, the newly developed reverse genetics techniques for rescuing morbilliviruses (Conzelmann, 1998; Nagai, 1999) will allow a more rational approach to virus attenuation and vaccine design. It should be possible to identify the gene(s) which effects virulence in different species and alter it by direct genetic manipulation. In the short term, situations where vaccination against these diseases with less effective subunit vaccines might be feasible would be for captive and zoo animals and for sick marine mammals, such as seals, housed in rehabilitation centres before they are released back into the wild (Visser *et al.*, 1992). The ultimate aim would be to develop effective recombinant vaccines for morbilliviruses which could be delivered orally to wildlife species, as was successfully achieved for foxes and other wildlife species with the vaccinia/rabies recombinant vaccine (Pastoret & Brochier, 1996).

HIDDEN VIRUS

In 1994, an unexpected outbreak of rinderpest was diagnosed in wildlife, mainly buffalo (*Syncerus caffer*) and lesser kudu (*Tragelaphus imberbis*), in the Tsavo National Parks in the south of Kenya, away from the usual route of infection from Sudan in the north-west of the country. The virus spread in wildlife from November 1994 until May 1995, also reaching the Amboseli and Meru National Parks. The disease then appeared in Nairobi National Park in October 1996 in eland (*Taurotragus oryx*) and buffalo. In the preceding 10 years, the only known endemic source of rinderpest in Africa was in southern Sudan, from which point infection regularly entered adjacent countries (Uganda, Kenya and Ethiopia) by transhumance and with animals traded for slaughter.

The viruses isolated from cattle during the period 1984–1994 were all very closely related to each other and were of African lineage 1. A second virus type, African lineage 2, had been widespread in eastern Africa in the 1950s and 1960s but appeared to have died out after 1983 as no virus of this type was isolated in the period 1984–1994 (Wamwayi *et al.*, 1995; Forsyth & Barrett, 1995). However, molecular analysis showed that the Kenyan wildlife virus was different from the virus that had been overtly circulating in Africa and was clearly of lineage 2, thus indicating the presence of another, previously hidden, endemic focus in the region (Barrett *et al.*, 1998). The prototype African lineage 2 virus (RGK/1) was isolated in the same area of Kenya in 1962 from a giraffe suffering from rinderpest (Liess & Plowright, 1964). The 1994–1996 Kenyan wildlife virus isolates were most closely related to RGK/1 with less than 4 % nucleotide difference in the areas of the genome sequenced. The endemic focus of the African lineage 2 virus remains unknown, but surveillance suggests north-eastern Kenya or southern Somalia as the most likely source. This area is remote and sparsely populated and it is possible that the virus had persisted there unnoticed for many years; however, there was one report of virus in an eland in 1974 (Provost, 1980). The genetic stability of these viruses over a 30 year period may appear surprising, given the mutability of RNA virus genomes; however, RPV is not unique among RNA viruses. In the case of vesicular stomatitis virus, there is no evidence for a molecular clock but rather for a stepwise evolutionary pattern, reflecting geography rather than time. If the environmental conditions remain stable within an ecological niche then these viruses should remain stable at a peak of genetic fitness and, as a result, a relative genetic stasis would be observed despite the high RNA polymerase error rate (Gillespie, 1993; Nichol *et al.*, 1993). All Asian virus isolates examined to date fall into a single, but different, lineage, known as lineage 3 (see Fig. 2).

A factor which may be important in the maintenance of rinderpest in a hidden or silent form is the existence of viruses that cause only mild or subclinical infections in domestic ruminants in endemic areas. Virus isolated from wildlife species (kudu and eland) during the epidemic proved to be extremely mild in domestic cattle, in contrast to its severity in buffalo, kudu and eland. In fact, wildlife acted as sentinels to indicate the presence of disease; in areas where susceptible wildlife was absent the disease was either not apparent or confused with other virus infections of cattle such as bovine virus diarrhoea or malignant catharral fever (Kock *et al.*, 1999). Strains of the virus that cause extremely mild disease in cattle have been recognized in East Africa since the 1940s, or even earlier (Lowe *et al.*, 1947; Robson *et al.*, 1959; Plowright, 1963). They are often misdiagnosed and so can persist unnoticed for many years and, because the incubation period for these viruses can be up to 15 days, they can persist in smaller cattle populations (Wamwayi *et al.*, 1995; Tille *et al.*, 1991). Other outbreaks where hidden foci of infection were suspected occurred on the Russian/Mongolian and the Russian/Chinese borders in 1991 and 1998, respectively. Vaccination continues to be carried out on the Russian side because they suspect that virus remains in circulation in these areas. However, the viruses involved in both incidents were identical and very closely related to the vaccine strain used in Russia. Although not definitely established, the most likely explanation of these outbreaks was incomplete attenuation (Anonymous, 2000).

PPRV, the other ruminant morbillivirus, has become increasingly important economically as a consequence of the success of the rinderpest eradication campaign. Although PPRV was first discovered in West Africa, that may not be its original source, bearing in mind that measles and rinderpest are both of Asiatic origin. It is possible that this virus was present for a long time in Asia but was previously confused with rinderpest. It was only with the development of molecular probes that the two diseases could be easily distinguished (Diallo *et al.*, 1989; Shaila *et al.*, 1989). Previously it was necessary to isolate virus and carry out expensive and time-consuming differential neutralizing tests and animal inoculation experiments, procedures not always possible in endemic countries. Whatever its origin, PPR first appeared as a recognized disease in Asia in the early 1990s and it is now found all across the Middle East and the Indian subcontinent, reaching as far as Nepal and Bangladesh (Nanda *et al.*, 1996). Four distinct lineages of PPRV have been described (Shaila *et al.*, 1996) and a single lineage, distinct from the viruses found in Africa, has been responsible for most of these outbreaks (Fig. 3). The

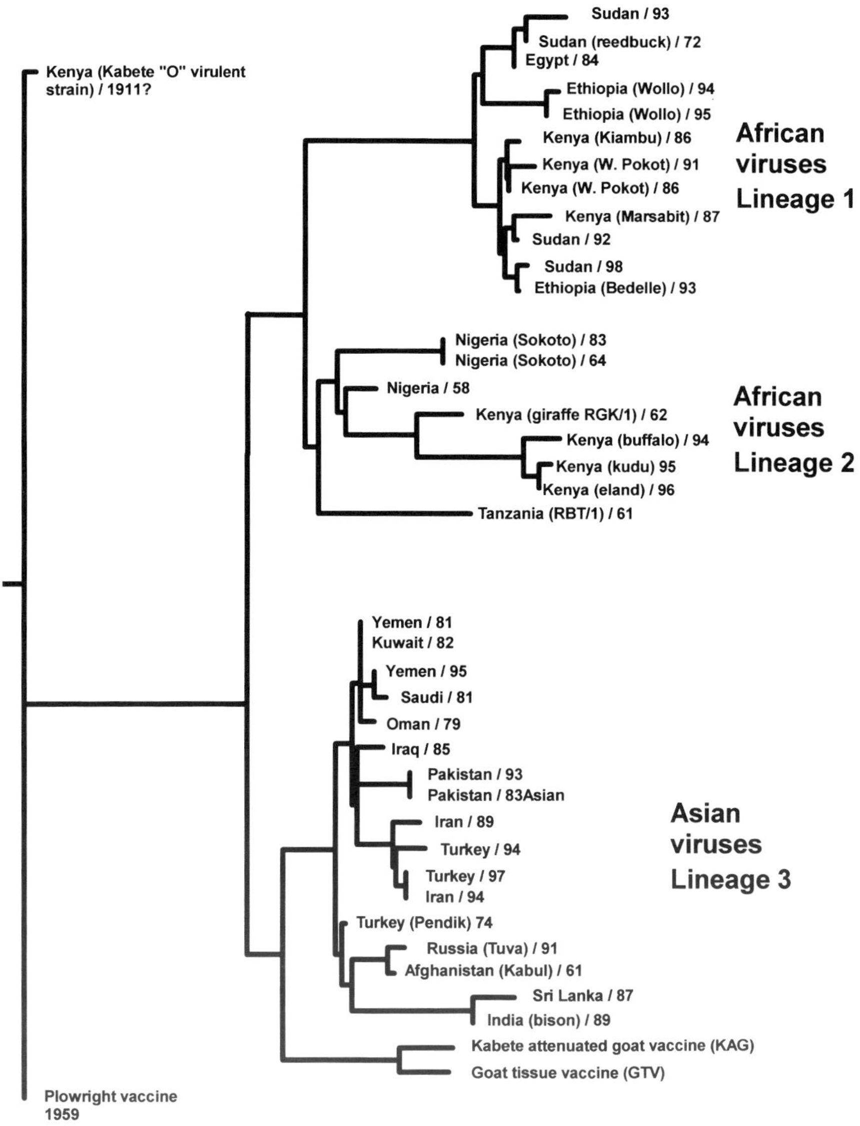

Fig. 2. Phylogenetic tree showing the relationships between the different isolates of RPV based on partial sequence data from the fusion (F) protein gene. A specific primer set was used to amplify a 372 bp DNA fragment as described by Forsyth & Barrett (1995). The tree was derived using the PHYLIP DNADIST and FITCH programs (Felsenstein, 1990). The branch lengths are proportional to the genetic distances between the viruses and the hypothetical common ancestor that existed at the nodes in the tree. The numbers on the figure represent the year of isolation.

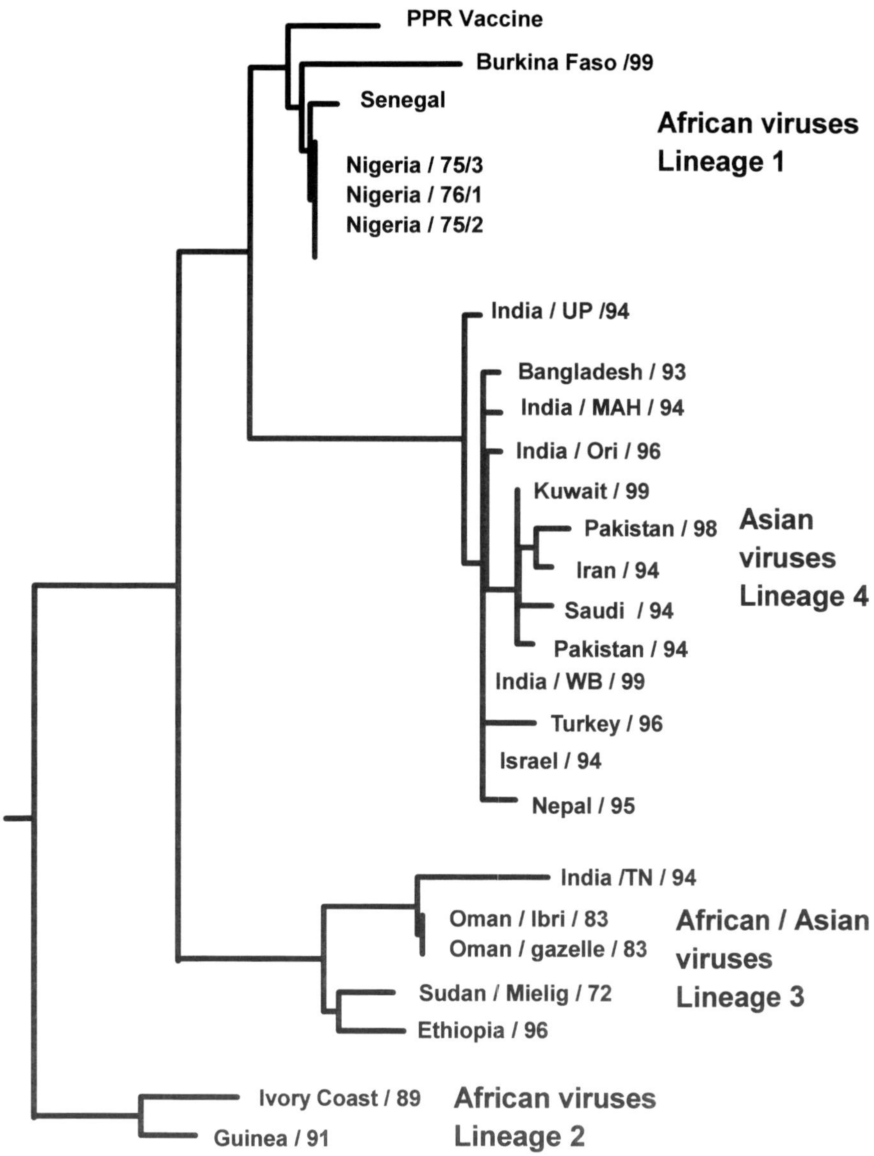

Fig. 3. Phylogenetic tree showing the relationships between the different isolates of PPRV based on partial sequence data from the fusion (F) protein gene. A specific primer set was used to amplify a 372 bp DNA fragment as described by Forsyth & Barrett (1995). The tree was derived as described in Fig. 2. The numbers on the figure represent the year of isolation.

full host range of this virus is unknown but several wild species of antelope died as a result of a zoo infection in the United Arab Emirates (Furley *et al.*, 1987). Cattle sero-positive for PPRV have been found in West Africa but there is no evidence that it can cause disease in cattle (Anderson & McKay, 1994).

HOST RANGE

While MV can only be maintained in human populations, many other primate species are susceptible and humans are a source of infection for endangered species such as the great apes. There is only one serotype of MV with no distinction between mild and pathogenic strains. Rinderpest probably can infect all artiodactyls but, depending on the species involved, different strains may vary in their pathogenicity and transmissibility. As mentioned above, the recent Kenyan lineage 2 virus is pathogenic in certain wildlife species but only causes mild disease in domestic cattle. Also warthog, which were considered to be highly susceptible to rinderpest infection (Plowright, 1982; Scott, 1970), were unaffected during the last Kenyan outbreak, even though they were foraging in the midst of sick and dying buffalo and eland. CDV causes a fatal disease in many species of carnivore, both wild and domesticated, and is globally distributed (Appel, 1987). In recent years, distemper has caused mortality in several species (lion, panther and hyaena) previously thought not to be susceptible. In 1992, the first fatal cases of CDV infection in big cats were observed in American zoo animals (Harder *et al.*, 1996). Shortly afterwards, CDV was responsible for the deaths of wild lions and hyaenas in the Serengeti Plain in Tanzania (Harder *et al.*, 1995; Haas *et al.*, 1996; Roelke-Parker *et al.*, 1996). It was thought, perhaps, that a new cat-adapted biotype of the virus had emerged or that co-infection with another virus, possibly feline immunodeficiency virus, which is prevalent in African big cats, might have exacerbated the disease in the lions. However, it is now clear that local strains of the virus were responsible for the deaths on the two continents and no correlation could be found between the presence of other viruses and mortality in lions (Roelke-Parker *et al.*, 1996; Haas *et al.*, 1996). A retrospective study on tissues from large cats which died in Swiss zoos over the period 1972–1992 found that 45 % had CDV-specific antigen in their tissues (Myers *et al.*, 1997), indicating that the disease has fatally infected large cats for much longer than was suspected. Failure to recognize the disease earlier in these species was probably the result of a lack of awareness of a possible viral aetiology and a lack of suitable diagnostic techniques in many laboratories. Sympatric domestic dogs were the most probable source of infection for the Serengeti lions and hyaenas as they were infected with a closely related virus (Roelke-Parker *et al.*, 1996). The dog population has risen alarmingly in that area in recent years and mostly they remain unvaccinated and a ready source of infection for wildlife. A similar threat to African wild hunting dogs and the Ethiopian wolf, two of the world's most endangered canids, is posed by contact with domestic dogs carrying distemper and rabies viruses (Gascoyne *et al.*, 1993; Laurenson *et al.*, 1998).

Morbilliviruses are not confined to hosts in the terrestrial environment and CDV can fatally infect seals (*Pinnipedia*) which form a suborder within the order *Carnivora*. During 1987/88, a great number of seals in Lake Baikal in Far Eastern Russia died as a result of a virus infection (Grachev *et al.*, 1989). Eventually the virus was shown to be CDV and was not related to the virus which caused the deaths of thousands of seals off the coasts of northern Europe during the summer and autumn of 1988 (see below). The most likely source of infection for the Lake Baikal seals was from land animals infected with the virus. Outbreaks of CDV are common in the large number of feral and domestic dogs around the lake, and annual hunting of the seals in the lake brings them into close contact with man and dogs. Wolves, ferrets, mink and bears could be another source of infection. A similar, but much more serious, CDV infection has recently occurred in seals in the Caspian Sea. The virus was first shown to be present in Caspian seals which suffered from an unusually high mortality in 1997 (Forsyth *et al.*, 1998), but in 2000 many more thousands of these animals died and CDV infection was again demonstrated (Kennedy *et al.*, 2000).

With hindsight, it is possible to ascribe the deaths of thousands of crabeater seals (*Lobodon carcinophagus*) which occurred in Antarctica in the 1950s to a CDV infection. At the time, the deaths were attributed to an acute viral infection which was not further characterized. The infection occurred near a scientific field station with unvaccinated sledge dogs. A recent serological survey of Antarctic seals showed a high prevalence of CDV-specific antibodies in the crabeater seals and so the virus appears to have become established in the population (Bengtson *et al.*, 1991). Since there are no native terrestrial carnivores in the Antarctic, it is likely that dogs were the original source of CDV on that continent and that they passed a CDV infection to the seals that resulted in their deaths. A ban is now in place on the importation of sledge dogs into Antarctica. Virus diseases carried by humans and domestic animals must therefore be considered as a threat to wildlife, particularly where endangered species are concerned, and control by vaccination or preventing contact is the most effective way of ensuring that wildlife will be safeguarded.

NEW MORBILLIVIRUSES

A morbillivirus with potentially severe ecological consequences for marine mammals was first identified in harbour and grey seals which died in large numbers off the coasts of northern Europe in 1988 (Osterhaus & Vedder, 1988). At first, based on antigenic similarities, the seal virus was thought to be CDV. The great interest generated by these new morbilliviruses meant that all available molecular resources were quickly employed to characterize the new viruses. Molecular probes (Mahy *et al.*, 1988), RT-PCR (Haas *et al.*, 1991) and monoclonal antibody (Cosby *et al.*, 1988) analyses showed that the seal virus was most closely related to, but different from, CDV, and it was then classified as a new member of the *Morbillivirus* genus. It was then named phocid distemper virus (PDV). The small size and scattered nature of European seal populations make it unlikely that they could maintain this virus in circulation and no disease with similar clinical signs has been seen at any of the seal rescue centres since 1989. The most likely source of infection for the European seals was contact with seals from the Arctic. This has been proposed on the grounds that morbillivirus, and more specifically PDV antibodies, were found in archival sera obtained from Arctic seals long before 1988 (Henderson *et al.*, 1992; Ross *et al.*, 1992). Arctic harp seals (*Phoca groenlandica*) migrated much further south to reach northern European waters in the year prior to the epizootic, probably as a result of climatic changes and overfishing around Greenland (Dietz *et al.*, 1989a, b). The harp seal population is extremely large, with an estimated four million individuals in Canadian waters alone, and is sufficient to maintain PDV in circulation. The full host range of PDV has not been fully determined but it can infect many species of seal (Duignan *et al.*, 1997). It is possible that future contact with Arctic seals could re-introduce the virus in epidemic form to European seals now that herd immunity has waned.

In 1988, morbilliviruses were also isolated from harbour porpoises (*Phocoena phocoena*) found off the British coast (Kennedy *et al.*, 1988; McCullough *et al.*, 1991). Shortly afterwards, in 1990, striped dolphins (*Stenella coeruleoalba*) in the Mediterranean Sea were also stricken by a morbillivirus infection (Domingo *et al.*, 1990). The porpoise and the dolphin morbilliviruses proved to be serologically closely related to each other but quite distinct from PDV; their nearest relative is PPRV (Haffar *et al.*, 1999; Visser *et al.*, 1993). Because the dolphin and porpoise viruses are as close genetically as different strains of RPV are to each other (Barrett *et al.*, 1993), they are considered to be strains of the same virus, CeMV.

CeMV appears to have a worldwide distribution. A retrospective serological study of cetacean species from along the North American coast revealed a high proportion positive for morbillivirus antibodies. The highest proportion of seropositive sera (92 %) was found in long-finned pilot whales (*Globicephala melas*) which were involved in mass strandings between 1982 and 1993 (Duignan *et al.*, 1995a, b). Morbillivirus-specific antibodies have been found in dolphins from regions as far apart as Canada, Texas and southern California (Duignan *et al.*, 1996; Reidarson *et al.*, 1998). The neutralizing titres of these sera were highest to dolphin and porpoise morbilliviruses. There is also molecular evidence (RT-PCR positive) for the presence of CeMV in tissues from sick dolphins from the west and east coasts of North America, the earliest positive sample dating from 1987 (Lipscomb *et al.*, 1994, 1996; Reidarson *et al.*, 1998; Schulman *et al.*, 1997), and from a pilot whale recently stranded in Delaware Bay in the USA. Sequence analysis of the latter virus grouped it with those obtained from porpoises and dolphins in Europe (Taubenberger *et al.*, 2000). Serological evidence has been obtained for morbillivirus infection in cetacean species in other oceans. Recent surveys of marine mammals found morbillivirus-specific antibodies in the south-west Atlantic, in the east-, west- and south-west Pacific and the Indian oceans (Van Bressem *et al.*, 1998, 2000). The high proportion of pilot whales showing evidence of infection has led to the suggestion that this species may act as the vector for the spread of CeMV (Duignan *et al.*, 1995b). Pilot whales are present in sufficient numbers, have a close social structure and a widespread pelagic distribution and are known to associate with many different cetacean species. Long-finned pilot whales are commonly seen in the western Mediterranean, where the 1990 dolphin epizootic was first observed. It is important to obtain virus isolates from regions other than the Atlantic Ocean to study the extent of geographic variation within this virus type.

EVOLUTION

Because of their very high level of antigenic relatedness and sequence similarity, the different morbilliviruses must all have evolved from one common ancestor. Rinderpest has been proposed as the 'archevirus' of the group since it reacts with a wide range of monoclonal antibodies produced against other morbilliviruses (Norrby *et al.*, 1985). Animal populations of the size needed to produce the constant supply of susceptible hosts required for a morbillivirus to survive would most likely have been large herds of ruminants roaming the plains of Asia, the historic source of rinderpest infection. Each known morbillivirus is only able to cause serious disease in one order of mammals; for example, MV is pathogenic for primates, RPV for artiodactyls and CDV for carnivores. However, MV is genetically closest to RPV and it is likely that, when man settled in sufficiently large communities to maintain a morbillivirus, close contact with infected cattle resulted in rinderpest, or a rinderpest-like infection, being passed to humans. Such a virus may then have evolved to acquire pathogenic characteristics for man and other primates to become MV.

Carnivores of various species preying on infected ruminants may similarly have become infected with the progenitor virus and subsequently this virus could have evolved to produce CDV. The two morbilliviruses of carnivores, CDV and PDV, are very closely related and it is probable that PDV evolved quite recently from CDV. Arctic seals may have been infected with CDV several hundreds (or several thousands) of years ago by contact with terrestrial carnivores which can carry the virus (wolves, foxes, dogs, polar bears). It is possible that in the future PDV could reverse direction and become a disease of terrestrial mammals; there has been one report that PDV caused a distemper-like infection in mink in Denmark at a farm in the immediate vicinity of diseased seals (Blixenkrone-Møller *et al.*, 1990).

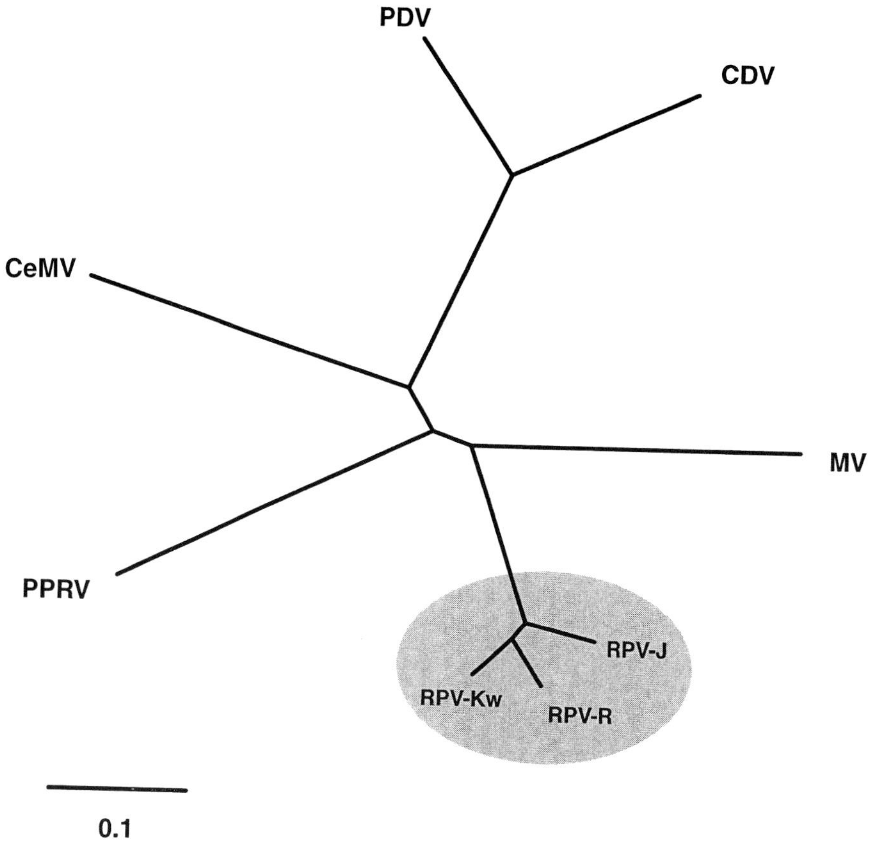

Fig. 4. Phylogenetic tree showing the relationships between the different morbilliviruses based on the coding sequence for the nucleocapsid (N) protein gene. Three strains of RPV were analysed: RPV-R, the RBOK vaccine; RPV-Kw, a field strain from Kuwait; and RPV-J, the lapinized vaccine strain from Japan. The tree was derived as described in Fig. 2. The bar represents nucleotide changes per position. Accession numbers: MV, Z66517; RPV-Kw, Z34262; RPV-J, E06018; RPV-R, X68311; CDV, AF014953; PPRV, X74443; PDV, X75717; CeMV, X75961.

The origin of the CeMV is more problematical and it is difficult to speculate on the origin of this virus since there is no natural contact between cetacean species and terrestrial species. The close serological relationship between CeMV and PPRV may just be due to chance and, given the great sequence diversity, it is unlikely to have evolved from either PDV or CDV. It is possible, with increasing awareness of the importance of virus diseases, that other morbilliviruses will be found in the future and techniques are now available which will allow their rapid identification and characterization. The phylogenetic relationships between the different morbilliviruses are shown in Fig. 4.

CONSEQUENCES

The losses due to rinderpest in domestic livestock can be devastating. Some 200 million cattle are estimated to have died in north-western Europe in the 18th century and as many as half a million cattle died during the last epidemic in Britain in the 1860s. The effect of the disease on the economies of western Europe was the main reason for establishing the first veterinary schools in the late 18th century (Wilkinson, 1984). The introduction of rinderpest into sub-Saharan Africa in the 1880s caused the so-called Great African Pandemic, which had a dramatic effect on native pastoralist people. In Ethiopia, it caused the great famine of 1888–1892 and in Kenya, people such as the Masai were so greatly affected that they never regained their dominance over other tribes. The subsequent famine, and outbreaks of diseases such as smallpox associated with population displacements, greatly reduced the human population and made it easier for the eventual colonization of the region (Barrett & Rossiter, 1999). Over 5 million cattle died over the 10 years it took for the virus to spread across the continent. During the second continental pandemic in the early 1980s, more than a million cattle may have died, despite the availability of an effective vaccine.

Direct losses which result from the high morbidity and mortality associated with infection are more than just the loss of meat and milk; in many parts of Africa and Asia cattle dung is the main source of fuel and fertilizer and oxen the main source of draught power. Indirect losses result from reduced international trade, decreased sales during quarantine restrictions and the high costs of control by vaccination in endemic and surrounding areas (Rossiter, 1994; Rossiter *et al.*, 1998). The effects in endemic areas are less severe than in virgin soil epidemics; nevertheless, there is continual mortality in young stock. In southern Sudan, one of the two remaining endemic foci in Africa, the virus affects nearly all unprotected young stock with an estimated case mortality of up to 40 %. For countries in the Horn of Africa, the export of livestock primarily to the Middle East is a major source of income and this is lost when rinderpest strikes (Rossiter, 1994). In West Africa, the presence of PPR is one of the major constraints limiting increased production of small ruminants. In recent years, this disease has spread across the Middle East and the Indian subcontinent and affected the livelihoods of many small farmers, particularly in India, Pakistan, Bangladesh and Nepal. Small ruminants form a major part of the rural economies of developing countries, and are particularly important for the less prosperous sections of farming communities.

In addition to the social and economic consequences associated with rinderpest infection, the introduction of virus can be catastrophic for wildlife. In Europe, the decline of the bison was attributed to rinderpest, and deer died in significant numbers during the last major outbreak in Britain in 1965. During the African pandemic of the 1890s, rinderpest killed more than 90 % of the domestic cattle and buffaloes and other highly susceptible species were also severely affected (Plowright, 1982; Scott, 1970). The loss of so many hosts may have led to the elimination of the tsetse fly from parts of southern Africa (Stevenson-Hamilton, 1957). It is also thought that the great pandemic and subsequent circulation of virus in equatorial Africa resulted in several anomalies in the distribution of certain antelope species; for example, in East Africa the greater kudu and bongo are restricted to small mountain top communities. An aerial census after the recent outbreaks in wildlife in the Tsavo National Parks in Kenya indicated that the disease was probably responsible for the loss of over 6000 of the estimated 11 000 buffaloes and perhaps 90 % of the lesser kudu (Kock *et al.*, 1999). The most dramatic demonstration of the impact that rinderpest can have on wildlife was the explosive increase in the population sizes of wildebeest and buffalo in the Serengeti Mara ecosystem after its eradication from cattle (Sinclair, 1973). This was also the most convincing proof that rinderpest is maintained by domestic cattle and not by wildlife, an essential precondition if a vaccination strategy for the eradication of rinderpest is to succeed. Were the virus to be maintained in wildlife it would be impossible, because of the numbers of animals involved and the lack of a suitable vaccine for oral delivery, to eradicate the disease by vaccination.

A similar situation exists with regard to distemper in wildlife; the source of CDV infection for the lions and hyaenas in the Serengeti was most probably the domestic dog. These have increased in numbers in recent years and are generally not vaccinated against CDV. The deaths of seals in Lake Baikal and in the Caspian Sea from CDV infection might have been prevented if the virus had been controlled in domestic dogs. Caspian Sea seals are an endangered species and are already severely threatened by hunting for meat and by a reduction in breeding efficiency caused by highly toxic pollutants in their environment (Allchin *et al.*, 1998). However, the most endangered seal species is the Mediterranean monk seal. Their numbers have declined alarmingly in the past 100 years, mainly as a result of hunting, and the CeMV has recently been isolated from dead monk seals in both the Mediterranean and the Atlantic (van de Bildt *et al.*, 2000). In one of the few remaining colonies of these seals located off the Mauritanian coast in 1997, an estimated 70 % of the population died within a month, a figure representing about one-third of the total world population, leaving only a few hundred remaining (Harwood, 1998). Fishermen in the area had reported sighting many dead dolphins before the seal deaths occurred, indicating the likely source of the virus. However, since the epidemiology did not fit a typical morbillivirus pattern, mortality being almost exclusively confined to sea-going adults, it is disputed whether or not this virus caused anything but an asymptomatic infection. Another argument against the CeMV as the main cause of death is that seals are classified in a different order (*Pinnipedia*) to whales and dolphins and, since infection with a morbillivirus generally causes disease in only one order of mammals, an alternative cause of the seal deaths has been suggested. A toxic algal bloom, which occurred at the same time, is considered by some the most likely cause of the high mortality among the adult seals in the Atlantic (Hernandez *et al.*, 1998). No mass deaths were reported in the Mediterranean monk seals when the virus was found in isolated dead animals. Virus infection without overt disease signs is possible; manatees (order *Sirenia*) sampled off the Florida coast were found to be serologically positive for CeMV but no virus-related deaths have been reported (Duignan *et al.*, 1995c). The monk seals would most probably be susceptible to PDV or CDV infections (Osterhaus *et al.*, 1992) and another mass mortality from whatever cause would certainly be enough to be wipe out the remaining population.

CONCLUSIONS

Throughout history, the social, economic and ecological consequences of morbillivirus infections have been severe and they continue to this day, despite the availability of modern vaccines and diagnostic tools. It will require a strong determination to continue the fight to eradicate measles and rinderpest and not to relax vigilance under economic or social pressures when it appears that these viruses no longer pose a threat or that the goal of eradication has been achieved. Already there are signs that measles is not considered a serious disease in parts of western Europe and vaccination levels are dropping in many countries. This is partly due to worries about vaccine safety, but complacency about the seriousness of the disease could mean that measles again becomes a significant cause of childhood mortality in developed countries. Similarly, the lack of significant mortality in domestic cattle with lineage 2 RPV in eastern Africa could lead to a situation where the virus is no longer considered a threat. This could have potentially disastrous consequences since the basis of the attenuation for cattle is not understood and, given the mutability of RNA viruses, the virus could revert to virulence for cattle. Not all lineage 2 viruses are attenuated and the giraffe isolate (RGK/1), which is closest in sequence to the recently isolated lineage 2 viruses from Kenya, is highly pathogenic in cattle (Liess & Plowright, 1964).

Research must continue to improve the safety and effectiveness of some morbillivirus vaccines. The current CDV vaccines are not suitable for use in wildlife species and so zoo animals cannot be effectively protected against distemper infection. The measles vaccine, unlike the rinderpest vaccine, has some immunosuppressive effects (Fireman *et al.*, 1969; Lund *et al.*, 2000). In addition, the live attenuated morbillivirus vaccines cannot be administered before about 9 months of age due to interference by maternal antibodies. New types of vaccine are required that can protect children and young animals from morbillivirus infections in the presence of passive maternal immunity. Another development which should be of great benefit to control and eradication programmes for RPV and PPRV is the availability of genetically marked vaccines. Their use would allow vaccinated animals to be distinguished from those that have recovered from natural infection and allow the detection of wild-type virus in the face of varying levels of herd vaccination (Walsh *et al.*, 2000). Vaccines which can be delivered in food or water would be ideal for controlling these diseases in wild species but little progress has been made in this area. In addition, improved diagnostic tests are required. More rapid and sensitive diagnostic aids designed for pen-side use would improve the ability of field veterinarians and stockmen to stop the spread of RPV and PPRV once an outbreak has occurred. Finally, more stringent measures must be taken to control morbillivirus infections in domestic species in order to lessen the risk of infection in wildlife.

REFERENCES

Allchin, C., Barrett, T., Duck, C., Eybatov, T., Forsyth, M., Kennedy, S. & Wilson, S. (1998). Surveys of Caspian seals in the Apsheron Peninsula region and residue and pathology analyses of dead seal tissues. In *Studies on the Present Status of Marine Biological Resources of the Caspian Sea*, pp. 101–118. Bio-resources Network Meeting, Bordeaux, November 1997. Edited by H. Dumont, S. Wilson & B. Wazniewicz. Washington: World Bank.

Anderson, E. C., Hassan, E., Barrett, T. & Anderson, J. (1990). Observations on the pathogenicity and transmissibility for sheep and goats of the strain of virus isolated during the rinderpest outbreak in Sri Lanka in 1987. *Vet Microbiol* **21**, 309–318.

Anderson, J. & McKay, J. A. (1994). The detection of antibodies against peste des petits ruminants virus in cattle, sheep and goats and the possible implications to rinderpest control programmes. *Epidemiol Infect* **112**, 225–231.

Anderson, R. M. & May, R. M. (1991). *Infectious Diseases of Humans: Dynamics and Control*. Oxford: Oxford University Press.

Anonymous (2000). Rinderpest. *EMPRES: Transboundary Anim Dis Bull* **13**, 2–9 (www.fao.org/empres).

Appel, M. J. G. (1987). Canine distemper virus. In *Virus Infections of Vertebrates*, vol. 1, *Virus Infections of Carnivores*, pp. 133–159. Amsterdam: Elsevier.

Appel, M. J. G., Shek, W. R., Shesberadaran, H. & Norrby, E. (1984). Measles virus and inactivated canine distemper virus induce incomplete immunity to canine distemper. *Arch Virol* **82**, 73–82.

Barrett, T. & Rossiter, P. B. (1999). Rinderpest: the disease and its impact on humans and animals. *Adv Virus Res* **53**, 89–110.

Barrett, T., Visser, I. K. G., Mamaev, L., Van Bressem, M.-F. & Osterhaus, A. D. M. E. (1993). Dolphin and porpoise morbilliviruses are genetically distinct from phocine distemper virus. *Virology* **193**, 1010–1012.

Barrett, T., Forsyth, M., Inui, K., Wamwayi, H., Kock, R., Wambula, J., Mwanzia, J. & Rossiter, P. (1998). Rediscovery of the second African lineage of rinderpest virus: its epidemiological significance. *Vet Rec* **142**, 669–671.

Barton, A. (1956). Plaques and contagions in antiquity. *J Am Vet Assoc* **129**, 503–505.

Bengtson, J. L., Boveng, P., Franzèn, U., Have, P., Heide-Jørgensen, M.-P. & Härkönen, T. J. (1991). Antibodies to canine distemper virus in Antarctic seals. *Mar Mamm Sci* **71**, 85–87.

Black, F. L. (1991). Epidemiology of paramyxoviridae. In *The Paramyxoviruses*, pp. 509–536. Edited by D. Kingsbury. New York: Plenum.

Blixenkrone-Møller, M., Svansson, V., Örvell, C. & Have, P. (1990). Phocid distemper virus – a threat to terrestrial mammals? *Vet Rec* **127**, 263–264.

Bush, M., Montpelier, J., Brownstein, D., James, A. E. & Appel, M. J. G. (1976). Vaccine induced canine distemper in a lesser panda. *J Am Vet Med Assoc* **169**, 959–960.

Carmichael, A. G. (1997). Measles: the red menace. In *Plague, Pox and Pestilence*, pp. 80–85. Edited by K. F. Kipple. London: Weidenfeld and Nicholson.

Carpenter, J. W., Appel, M. J., Erickson, R. C. & Novilla, N. (1976). Fatal vaccine-induced canine distemper virus infection in black-footed ferrets. *J Am Vet Med Assoc* **169**, 961–964.

Chant, K., Chan, R., Smith, M., Dwyert, D. E., Kirkland, P. & the NSW Expert Group (1998). Probable human infection with a newly described virus in the family *Paramyxoviridae*. *Emerg Infect Dis* **4**, 273–275.

Chua, K. B., Bellini, W. J., Rota, P. A. & 19 other authors (2000). Nipah virus: a recently emergent deadly paramyxovirus. *Science* **288**, 1432–1435.

Conzelmann, K.-K. (1998). Nonsegmented negative-strand RNA viruses: genetics and manipulation of viral genomes. *Annu Rev Genet* **32**, 123–162.

Cosby, S. L., McQuaid, S., Duffy, N. & 9 other authors (1988). Characterization of a seal morbillivirus. *Nature* **336**, 115–116.

Diallo, A., Barrett, T., Barbron, M., Shaila, M. S. & Taylor, W. P. (1989). Differentiation of rinderpest and peste des petits ruminants viruses using specific cDNA clones. *J Virol Methods* **23**, 127–136.

Dietz, R., Ansen, C. T., Have, P. & Heide-Jirgensen, M.-P. (1989a). Clue to seal epizootic? *Nature* **338**, 627.

Dietz, R., Heide-Jorgensen, M.-P. & Harkonen, T. (1989b). Mass deaths of harbor seals (*Phoca vitulina*) in Europe. *Ambio* **18**, 258–264.

Domingo, M., Ferrer, L., Pumarola, M., Marco, A., Plana, J., Kennedy, S., McAliskey, M. & Rima, B. K. (1990). Morbillivirus in dolphins. *Nature* **348**, 21.

Duignan, P. J., House, C., Geraci, J. R., Early, G., Walsh, M. T., St Aubin, D. J., Koopman, H. & Rhinelart, H. (1995a). Morbillivirus infection in cetaceans of the Western Atlantic. *Vet Microbiol* **44**, 241–249.

Duignan, P. J., House, C., Geraci, J. R. & 8 other authors (1995b). Morbillivirus infection in two species of pilot whales (*Globicephala sp.)* from the western Atlantic. *Mar Mamm Sci* **11**, 150–162.

Duignan, P. J., House, C., Walsh, M. T., Campbell, T., Bossart, G. D., Duffy, N., Fernandes, P. J., Rima, B. K., Wright, S. & Gerachi, J. R. (1995c). Morbillivirus infections in manatees. *Mar Mamm Sci* **11**, 441–451.

Duignan, P. J., House, C., Odell, D. K., Wells, R. S., Hansen, L. J., Walsh, M. T., St Aubin, D. J., Rima, B. K. & Geraci, J. R. (1996). Morbillivirus infection in bottlenose dolphins: evidence for recurrent epizootics in the Western Atlantic and Gulf of Mexico. *Mar Mamm Sci* **12**, 499–515.

Duignan, P. J., Nielsen, O., House, C., Kovacs, K. M., Duffy, N., Early, G., Sadove, S., St Aubin, D. J., Rima, B. K. & Geraci, J. R. (1997). Epizootology of morbillivirus infection in harp, hooded and ringed seals from the Canadian Arctic and Western Atlantic. *J Wildl Dis* **33**, 719.

Felsenstein, J. (1990). PHYLIP manual version 3·3. University Herbarium, University of California, Berkeley, California.

Fireman, P., Friday, G. & Kumate, J. (1969). Effect of measles vaccine on immunologic responsiveness. *Pediatrics* **43**, 264–272.

Forsyth, M. & Barrett, T. (1995). Evaluation of polymerase chain reaction for the detection of rinderpest and peste des petits ruminants viruses for epidemiological studies. *Virus Res* **39**, 151–163.
Forsyth, M., Kennedy, S., Wilson, S., Eybatov, T. & Barrett, T. (1998). Canine distemper in a Caspian seal. *Vet Rec* **143**, 662–664.
Furley, C. W., Taylor, W. P. & Obi, T. U. (1987). An outbreak of peste des petits ruminants in a zoological collection. *Vet Rec* **121**, 443–447.
Gargadennec, L. & Lalanne, A. (1942). PPR. *Bull Serv Zootech Epizoot Afr Occident Fr* **5**, 16–21.
Gascoyne, S. C., King, A. A., Laurenson, M. K., Borner, M., Schildger, B. & Barrat, J. (1993). Aspects of rabies infection and control in the conservation of the African wild dog (*Lycaon pictus*) in the Serengeti Region, Tanzania. *Onderstepoort J Vet Res* **60**, 415–420.
Gibbs, E. P., Taylor, W. P., Lawman, M. J. P. & Bryant, J. (1979). Classification of peste des petits ruminants virus as the fourth member of the genus morbillivirus. *Intervirology* **11**, 268–274.
Gillespie, J. H. (1993). Episodic evolution of RNA viruses. *Proc Natl Acad Sci U S A* **90**, 10411–10412.
Grachev, M. A., Kumarev, V. P., Mamaev, L. V. & 13 other authors (1989). Distemper virus in Baikal seals. *Nature* **338**, 209.
Haas, L., Subbarao, S. M., Harder, T., Liess, B. & Barrett, T. (1991). Detection of phocid distemper virus RNA in seal tissues using slot hybridisation and the polymerase chain reaction amplifying assay: genetic evidence that the virus is distinct from canine distemper virus. *J Gen Virol* **72**, 825–832.
Haas, L., Hofer, H., East, M., Wohlsein, P., Liess, B. & Barrett, T. (1996). Epizootic of canine distemper virus infection in Serengetti spotted hyaenas (*Crocuta crocuta*). *Vet Microbiol* **49**, 147–152.
Haffar, A., Libeau, G., Moussa, A., Minet, C. & Diallo, A. (1999). The matrix protein gene sequence analysis reveals close relationship between peste des petits ruminants (PPRV) and dolphin morbillivirus. *Virus Res* **64**, 69–75.
Harder, T. C., Kenter, M., Appel, M. J. G., Roelke-Parker, M. E., Barrett, T. & Osterhaus, A. D. M. E. (1995). Phylogenetic evidence of canine distemper virus in Serengetti's lions. *Vaccine* **13**, 521–523.
Harder, T. C., Kenter, M., Vos, H., Siebelink, K., Huisman, W., van Amerongen, G., Orvell, C., Barrett, T., Appel, M. J. G. & Osterhaus, A. D. M. E. (1996). Morbilliviruses isolated from diseased captive large felids: pathogenicity for domestic cats and comparative molecular analysis. *J Gen Virol* **77**, 397–405.
Harwood, J. (1998). What killed the monk seals? *Nature* **393**, 17–18.
Henderson, G., Trudgett, A., Lyons, C. & Ronald, K. (1992). Demonstration of antibodies in archival sera from Canadian seals reactive with a European isolate of phocine distemper virus. *Sci Total Environ* **115**, 93–98.
Henke, S. E. (1997). Effects of modified live-virus canine distemper vaccines in grey foxes. *J Wildl Rehabil* **20**, 3–7.
Hernandez, M., Robinson, I., Aguilar, A., Gonzalez, L. M., Lopez-Jurado, L. F., Reyero, M. I., Cacho, E., Franco, J., Lopez-Rodas, V. & Costas, E. (1998). Did algal toxins cause monk seal mortality? *Nature* **393**, 28–29.
Kennedy, S., Smyth, J. A., Cush, P. F., McCullough, S. J., Allan, G. M. & McQuaid, S. (1988). Viral distemper found in porpoises. *Nature* 336, **21.**

Kennedy, S., Kuiken, T., Jepson, P. D. & 10 other authors (2000). Canine distemper virus identified as cause of recent mass mortality in Caspian seals (*Phoca caspica*). *Emerg Infect Dis* **6**, (in press).

Kock, R. A., Wambua, J., Mwanzia, J., Rossiter, P., Wamwayi, H., Ndungu, E. K., Barrett, T., Kock, N. D. & Rossiter, P. B. (1999). Rinderpest epidemic in wild ruminants in Kenya 1993–1997. *Vet Rec* **145**, 275–283.

Laurenson, K., Sillero-Zubiri, C., Thompson, H., Shiferaw, F., Thirgood, S. & Malcolm, J. (1998). Disease as a threat to endangered species: Ethiopian wolves, domestic dogs and canine pathogens. *Anim Conserv* **4**, 273–280.

Liess, B. & Plowright, W. (1964). Studies on the pathogenesis of rinderpest in experimental cattle. 1. Correlation of clinical signs, viraemia and virus excretion by various routes. *J Hyg* **62**, 81–100.

Lipscomb, T. P., Schulman, F. Y., Moffett, D. & Kennedy, S. (1994). Morbilliviral disease in Atlantic bottlenose dolphins (*Tursiops truncatus*) from the 1987–1988 epizootic. *J Wildl Dis* **30**, 567–571.

Lipscomb, T. P., Kennedy, S., Moffett, D., Krafft, A., Klaunberg, B. A., Lichy, J. H., Regan, G. T., Worthy, G. A. & Taubenberger, J. K. (1996). Morbilliviral epizootic in bottlenose dolphins of the Gulf of Mexico. *J Vet Diagn Investig* **8**, 283–290.

Lowe, H. J., Wilde, J. K. H., Lee, R. P. & Stuchbery, H. M. (1947). An outbreak of an aberrant type of rinderpest in Tanganyika Territory. *J Comp Pathol* **57**, 175–183.

Lund, B. T., Tiwari, A., Galbraith, S., Baron, M. D., Morrison, W. I. & Barrett, T. (2000). Vaccination of cattle with attenuated rinderpest virus stimulates $CD4^+$ T cell responses with broad viral antigen specificity. *J Gen Virol* **81**, 2137–2146.

McCullough, S. J., McNeilly, F., Allan, G. M., Kennedy, S., Smyth, J. A., Cosby, S. L., McQuaid, S. & Rima, B. K. (1991). Isolation and characterisation of a porpoise morbillivirus. *Arch Virol* **118**, 247–252.

Mack, R. (1970). The great African cattle plague epidemic of the 1890s. *Trop Anim Health Prod* **2**, 210–219.

Mahy, B. W. J., Barrett, T., Evans, S., Anderson, E. C. & Bostock, C. J. (1988). Characterisation of a seal morbillivirus. *Nature* **336**, 115.

Murray, K., Selleck, P., Hooper, P. & 8 other authors (1995). A morbillivirus that caused fatal disease in horses and humans. *Science* **268**, 94–97.

Myers, D. L., Zurbriggen, A., Lutz, H. & Pospishil, A. (1997). Distemper: not a new disease in lions and tigers. *Clin Diagn Lab Immunol* **4**, 180–184.

Nagai, Y. (1999). Paramyxovirus replication and pathogenesis: reverse genetics transforms understanding. *Rev Med Virol* **9**, 83–99.

Nanda, Y. P., Chatterjee, A., Purohit, A. K. & 10 other authors (1996). The isolation of peste des petits ruminants virus from Northern India. *Vet Microbiol* **51**, 207–216.

Nichol, S. T., Rowe, J. E. & Fitch, W. M. (1993). Punctuated equilibrium and positive Darwinian evolution in vesicular stomatitis virus. *Proc Natl Acad Sci U S A* **90**, 10424–10428.

Norrby, E., Sheshberadaran, H., McCullough, K. C., Carpenter, W. C. & Orvell, C. (1985). Is rinderpest virus the archevirus of the *Morbillivirus* genus? *Intervirology* **23**, 228–232.

Osterhaus, A. D. M. E. & Vedder, E. J. (1988). Identification of virus causing recent seal deaths. *Nature* **335**, 20.

Osterhaus, A. D. M. E., Visser, I. K. G., Deswart, R. L., Van Bressem, M.-F., Van De Bildt, M. W. G., Orvell, C., Barrett, T. & Raga, J. A. (1992). Morbillivirus threat to Mediterranean monk seals? *Vet Rec* **130**, 141–142.

Pastoret, P.-P. & Brochier, B. (1996). The development and use of a vaccinia-rabies recombinant oral vaccine for the control of wildlife rabies; a link between Jenner and Pasteur. *Epidemiol Infect* **116**, 235–240.

Plowright, W. (1963). Some properties of strains of rinderpest virus recently isolated in East Africa. *Res Vet Sci* **4**, 96–108.

Plowright, W. (1982). The effects of rinderpest and rinderpest control on wildlife in Africa. *Symp Zool Soc Lond* **50**, 128.

Provost, A. (1980). Queries about rinderpest in African wild animals. In *Wildlife Disease Research and Economic Development*, pp. 19–20. Edited by L. Karstad, B. Nestel & M. Graham. Proceedings of a Workshop held in Kabete, Kenya. Ottawa: International Development Research Centre.

Reidarson, T. H., McBain, J., House, C., King, D. P., Stott, J. L., Krafft, A., Taubenberger, J. K., Heyning, J. & Lipscomb, T. P. (1998). Morbillivirus infection in stranded common dolphins from the Pacific Ocean. *J Wildl Dis* **34**, 771–776.

Renshaw, R. W., Glaser, A. L., Campen, H., Van Weiland, F. & Dubovi, E. J. (2000). Identification and phylogenetic comparison of Salem virus, a novel paramyxovirus of horses. *Virology* **270**, 417–429.

Robson, J., Arnold, R. M., Plowright, W. & Scott, G. R. (1959). The isolation from an eland of a strain of rinderpest virus attenuated for cattle. *Bull Epizoot Dis Afr* **7**, 97–102.

Roelke-Parker, M. E., Munson, L., Packer, C. & 12 other authors (1996). A canine distemper virus epidemic in Serengeti lions (*Panthera leo*). *Nature* **379**, 441–445.

Ross, P. S., Visser, I. K. G., Broeders, H. W. J., Bildt, M. W. G., van de Bowen, W. D. & Osterhaus, A. D. M. E. (1992). Antibodies to phocine distemper virus in Canadian seals. *Vet Rec* **130**, 514–516.

Rossiter, P. B. (1994). Rinderpest. In *Infectious Diseases of Livestock with Special Reference to Southern Africa*, vol. 2, pp. 735–757. Edited by J. A. W. Coetzer, G. R. Thompson, R. C. Tustin & N. P. Kriek. Capetown: Oxford University Press.

Rossiter, P. B., Hussain, M., Raja, R. H., Moghul, W., Khan, Z. & Broadbent, D. W. (1998). Cattle plague in Shangri-La: observations on a severe outbreak of rinderpest in northern Pakistan 1994–1995. *Vet Rec* **143**, 39–42.

Rweyemamu, M. M. (1999). Rinderpest. *EMPRES: Transboundary Anim Dis Bull* **12**, 2–5 (www.fao.org/empres).

Rweyemamu, M. M. & Cheneau, Y. (1995). Strategy for the global rinderpest eradication programme. *Vet Microbiol* **44**, 369–376.

Sabin, A. B. (1992). My last will and testament on rapid elimination and ultimate global eradication of poliomyelitis and measles. *Pediatrics* **90**, 162–169.

Schneider-Schaulies, J., Niewiesk, S., Schneider-Schaulies, S. & ter Meulen, V. (1999). Measles virus in the CNS: the role of viral and host factors for the establishment and maintenance of a persistent infection. *J Neurovirol* **5**, 613–622.

Schulman, F. Y., Lipscomb, T. P., Moffett, D., Krafft, A. E., Lichy, J. H., Tsai, M. M., Taubenberger, J. K. & Kennedy, S. (1997). Histologic, immunohistochemical and polymerase chain reaction studies of bottlenose dolphins from the 1987–1988 United States Atlantic coast epizootic. *Vet Pathol* **34**, 288–295.

Scott, G. R. (1970). Rinderpest. In *Infectious Diseases of Wild Mammals*, pp. 20–35. Edited by J. W. Davis, L. H. Karstad & D. O. Trainer. Ames, IA: Iowa State University.

Scott, G. R. (1981). Rinderpest and peste des petits ruminants. In *Virus Diseases of Food Animals*, vol. 2, pp. 401–432. Edited by E. P. J. Gibbs. London: Academic Press.

Scott, G. R. (1996). The history of rinderpest in Britain: part one: 809–1799. *State Vet J* **6**, 8–10.
Scott, G. R. (1997). The history of rinderpest in Britain: part two: 1800 to date. *State Vet J* **7**, 9–13.
Shaila, M. S., Purushothaman, V., Bhavasar, D., Venugopal, K. & Venkatesan, R. A. (1989). PPR of sheep in India. *Vet Rec* **125**, 602.
Shaila, M. S., Shamaki, D., Forsyth, M., Diallo, A., Goatley, L., Kitching, P. & Barrett, T. (1996). Geographic distribution and epidemiology of peste des petits ruminants viruses. *Virus Res* **43**, 149–153.
Sinclair, A. R. E. (1973). Population increases of buffalo and wildebeest in the Serengeti. *East Afr Wildl J* **11**, 93–107.
Stevenson-Hamilton, J. (1957). Tsetse fly and the rinderpest epidemic of 1896. *S Afr J Sci* **58**, 216–218.
Sutherland-Smith, M. R., Rideout, B. A., Mikolon, A. B., Appel, M. J. G., Morris, P. J., Shima, A. & Janssen, D. J. (1997). Vaccine-induced canine distemper in European mink, *Mustela lutreola*. *J Zoo Wildl Med* **28**, 312–318.
Taubenberger, J. K., Tsai, M., Atkin, T. J., Fanning, T. G., Krafft, A. E., Moeller, R. B., Kodsi, S. E., Mense, M. G. & Lipscomb, T. P. (2000). Molecular genetic evidence of a novel morbillivirus in a long-finned pilot whale (*Globicephalus melas*). *Emerg Infect Dis* **6**, 42–45.
Taylor, W. P., Bhat, P. N. & Nanda, Y. P. (1995). The principles and practice of rinderpest eradication. *Vet Microbiol* **44**, 359–367.
Tille, A., Lefevre, P.-C., Pastoret, P.-P. & Thiry, E. (1991). A mathematical model of rinderpest infection in cattle populations. *Epidemiol Infect* **107**, 441–452.
Van Bressem, M.-F., Van Waerebeek, K., Fleming, M. & Barrett, T. (1998). Serological evidence of morbillivirus infection in small cetaceans from the Southeast Pacific. *Vet Microbiol* **59**, 89–98.
Van Bressem, M.-F., Van Waerebeek, K., Jepson, P. & 14 other authors (2000). An insight into the epidemiology of dolphin morbillivirus worldwide. *Vet Microbiol* (in press).
Van de Bildt, M. W. G., Martina, B. E. E., Vedder, E. J. & 8 other authors (2000). Identification of morbilliviruses of probable cetacean origin in carcasses of Mediterranean monk seals (*Monachus monachus*). *Vet Rec* **146**, 691–694.
Visser, I. K. G., Vedder, E. J., Van de Bildt, M. W. G., Örvell, C., Barrett, T. & Osterhaus, A. D. M. E. (1992). Canine distemper virus ISCOMs induce protection in harbour seals against phocid distemper but still allow subsequent infection with phocid distemper virus-1. *Vaccine* **10**, 435–438.
Visser, I. K. G., Van Bressem, M.-F., De Swart, R. L. & 9 other authors (1993). Characterisation of morbillivirus isolated from dolphins and porpoises in Europe. *J Gen Virol* **74**, 631–641.
Walsh, E. P., Baron, M. D., Rennie, L. F., Monaghan, P., Anderson, J. & Barrett, T. (2000). Recombinant rinderpest vaccines expressing membrane anchored proteins as genetic markers: evidence for exclusion of marker protein from the virus envelope. *J Virol* **74**, 10165–10175.
Wamwayi, H. M., Fleming, M. & Barrett, T. (1995). Characterisation of African isolates of rinderpest virus. *Vet Microbiol* **44**, 151–163.
Wild, F. T., Vidalain, P. O., Servet-Delprat, C. & Rabourdin-Combe, C. (2000). Vers l'éradication de la rougeole? *Med/Sci* **16**, 87–93.
Wilkinson, L. (1984). Rinderpest and mainstream infectious disease concepts in the eighteenth century. *Med Hist* **28**, 129–150.

Structure–function analysis of prion protein

Charles Weissmann,[1] Doron Shmerling,[1†] Daniela Rossi,[1] Antonio Cozzio,[1‡] Ivan Hegyi,[2] Marek Fischer,[1§] Rainer Leimeroth[1||] and Eckhard Flechsig[1]

[1]MRC Prion Unit/Neurogenetics, Imperial College School of Medicine at St Mary's, London W2 1PG, UK

[2]Institut für Neuropathologie, Universitätsspital Zürich, 8091 Zürich, Switzerland

[†]Current address: Core Technologies Dept, Novartis Pharma AG, 4002 Basel, Switzerland.

[‡]Current address: Dept of Pathology, Stanford University School of Medicine, Stanford, CA 94304, USA.

[§]Current address: Dept für Innere Medizin, Abt. für Infektionskrankheiten, Universitätsspital Zürich, Zürich 8091, Switzerland.

[||]Current address: Dept of Cell Biology, ETH, Hönggerberg, 8093 Zürich, Switzerland.

INTRODUCTION

PrP, the prion protein, plays a central role in the pathogenesis of transmissible spongiform encephalopathies such as scrapie or bovine spongiform encephalopathy (BSE) (Prusiner, 1996, 1998; Weissmann, 1999; Weissmann *et al.*, 1996). The normal form of PrP, designated PrP^C, is encoded by a single-copy gene (Basler *et al.*, 1986) and is expressed in the brain of healthy and prion-infected organisms to about the same extent (Chesebro *et al.*, 1985; Oesch *et al.*, 1985). There is overwhelming evidence that a modified form of PrP^C, which we designate PrP* (Weissmann, 1991), is the principal if not the only component of the infectious agent, or prion, and that it is devoid of nucleic acid. The 'protein-only' hypothesis (Griffith, 1967) states that the abnormal form of PrP propagates by interacting with PrP^C and converting it into a likeness of itself (Prusiner, 1989, 1996; Weissmann *et al.*, 1996). It has been proposed that a partially protease-resistant, aggregated form of PrP, named PrP^{Sc} or PrP-res, is congruent with PrP* (Prusiner, 1989).

Mice devoid of PrP develop and behave normally (Büeler *et al.*, 1992) but are resistant to prion disease (Büeler *et al.*, 1993; Manson *et al.*, 1994; Sailer *et al.*, 1994; Sakaguchi *et al.*, 1995). Moreover, introduction of PrP transgenes into such *Prnp*$^{0/0}$ mice restores susceptibility to scrapie (Fischer *et al.*, 1996), thus paving the way for structure–function analysis of PrP.

In this chapter, we review our deletion analysis of PrP with regard to its ability to mediate scrapie pathogenesis and prion replication as well as some aspects of its conjectured natural function. PrP with deletions up to residue 93, but not to residue 106, still

SGM symposium 60: New challenges to health: the threat of virus infection.
Editors G. L. Smith, W. L. Irving, J. W. McCauley, D. J. Rowlands. Cambridge University Press. ISBN 0 521 80614 3 ©SGM 2001.

sustained prion replication and development of scrapie, showing that the octarepeats are not required for these processes. Ablation of PrP in itself does not lead to disease and thus casts no light on its natural function; however, expression in PrP knockout mice of PrP truncated to residue 121 or 134 leads to ataxia and ablation of the cerebellar granule or Purkinje cell layer, depending on the promoter used for expression. This deleterious effect is abrogated by simultaneous expression of full-length PrP, suggesting that a pathway in which PrP is normally involved is being disrupted by its truncated counterpart. Interestingly, PrP knockout mice with deletions in the PrP gene which extend beyond the upstream boundary of the third exon also suffer from a cerebellar syndrome, due to overexpression in the brain of doppel (German for 'double'; Dpl), a protein similar to the pathogenic, truncated PrP.

VECTORS FOR THE EXPRESSION OF PrP IN THE CNS

We examined several PrP-encoding constructs for their efficacy in expressing PrP in the CNS and restoring susceptibility of PrP knockout mice to scrapie. It had previously been shown that wild-type mice harbouring multiple copies of a 40 kb *Prnp*-containing cosmid overexpressed PrP (Hsiao *et al.*, 1990; Scott *et al.*, 1989). Because such large constructs are difficult to modify and handle, we developed a shorter vector for the expression of PrP in the CNS.

The mouse PrP gene contains an upstream intron of 2 kb and a downstream intron of 6–12 kb (Westaway *et al.*, 1994a). We generated a PrP-encoding construct from which the large intron ('half-genomic construct' or phgPrP) or both introns ('cDNA construct' or pPrPcDNA) were deleted and which contained 6 kb of 5′ and 2.2 kb of 3′ flanking sequence (Fig. 1a). These constructs, as well as *Prnp*-containing cosmid DNA (cosmid cos6.I/LnJ-4; Westaway *et al.*, 1991), were introduced into $Prnp^{0/0}$ mice by nuclear injection. All but one of the six lines carrying half-genomic *Prnp* transgenes tested expressed PrP. In contrast, the cDNA construct, in all eight lines tested, even in those with very high transgene copy numbers (>150 copies), did not give rise to detectable levels of PrP protein in the brain. The one line of cosmid transgenic mice examined showed efficient PrP expression, as expected from previous work (Westaway *et al.*, 1991). These results showed that the presence of at least one intron is essential for the efficient expression of a PrP-encoding transgene under the control of its own promoter, as is the case for at least some other transgenes (Brinster *et al.*, 1988).

Two 'half-genomic' (*tga19*/+; *tga20*/+) mouse lines and one cosmid mouse line (*tgc35*/+), all with similar gene copy numbers of about 30–40, overexpressed PrP in the brain at levels ranging from three- to fivefold (*tga19*/+ and *tgc35*/+, respectively) to six- to sevenfold (*tga20*/+) relative to wild-type $Prnp^{+/+}$ mice. As shown by *in situ* hybridization, expression patterns were similar except that PrP RNA was not detected in the Purkinje cells of animals carrying half-genomic transgenes (Fig. 2), while it was abundant in Purkinje cells of wild-type and cosmid transgenic mice (Fischer *et al.*, 1996). This suggests that one or more control elements essential for the expression of PrP in Purkinje cells are absent from the half-genomic construct.

PrP knockout mice containing the cosmid transgene, but not those with the half-genomic construct, gradually developed ataxia as early as 4–6 months after birth, as reported previously (Westaway *et al.*, 1994b). Zürich I mice bearing this cosmid express in their brains mRNAs derived from the recently discovered *Prnd* gene (Moore *et al.*, 1999). The cognate protein Dpl is believed to be responsible for the ataxic syndrome (Moore *et al.*, 1999; Silverman *et al.*, 2000), as described below.

Inoculation with scrapie prions of mice expressing the half-genomic construct (*tga20/+*) led to typical scrapie symptoms after 68 days, as compared to 166 days for wild-type mice. Mice transgenic for the cosmid (*tgc35/+*) had an incubation time of 130 days. Incubation times were inversely correlated with expression levels of PrP^{C} (Büeler *et al.*, 1994; Fischer *et al.*, 1996; Scott *et al.*, 1989).

SUSCEPTIBILITY TO SCRAPIE OF PrP KNOCKOUT MICE EXPRESSING AMINO-PROXIMALLY TRUNCATED PrP

Transgenes encoding a set of PrP molecules with progressively longer amino-proximal deletions were introduced into PrP knockout mice and these were challenged with mouse-adapted scrapie prions. In all cases, the expression level of truncated PrP was twice or more that of full-length PrP. PrP with deletions up to residue 93 still sustained prion replication and development of scrapie, albeit with incubation times that became increasingly longer with more extensive deletions (Fig. 1b). PrP truncated to position 93, which lacked the five copper-binding octarepeats, resulted in prion titres and protease-resistant PrP levels about 30-fold lower than in wild-type mice. Brains of terminally ill animals showed no histopathology typical of scrapie; however, in the spinal cord, infectivity, gliosis and motor neurone loss were as in scrapie-infected wild-type controls. PrP truncated to position 106 was no longer able to sustain scrapie infection. Amplification of the number of octarepeats in PrP beyond the usual five is associated with human familial prion diseases (Collinge, 1997) and the number of repeats appears to determine the type of cerebellar amyloid deposits (Vital *et al.*, 1998). Nonetheless, PrP completely devoid of the repeats can still sustain prion propagation and scrapie disease, albeit with modified pathology and clinical presentation (Flechsig *et al.*, 2000). Because PrP which, in addition to an amino-terminal deletion to position 88, lacked the region 141–176, corresponding to the first α-helix, can also sustain prion replication and pathogenesis (Supattapone *et al.*, 1999), part or all of the region between 93 and 141 must be essential for these functions.

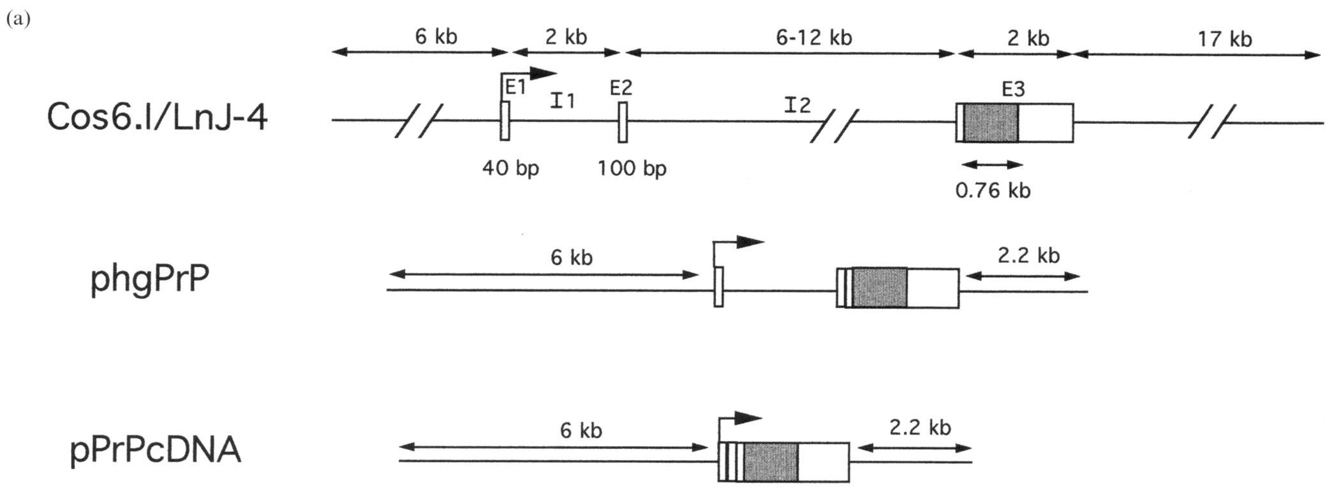

Fig. 1. Expression of mutant and wild-type PrP. (a) Maps of constructs encoding wild-type PrP. cos6.I/LnJ-4, a cosmid-derived clone comprising three exons (E1, E2, E3), two introns (I1, I2), and 6 kb of 5′ and 18 kb of 3′ flanking regions; phgPrP, as before, but lacking intron 2 and with only 2.2 kb of 3′ flanking region; pPrPcDNA, as phgPrP, but lacking both introns.

(b)

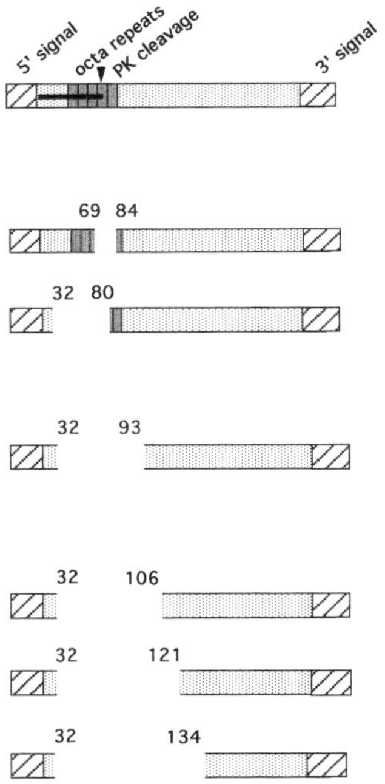

PrP type	Population	PrP levels*	Scrapie susceptibility *(incubation time)*	Cerebellar syndrome onset (weeks)	terminal (months)
Wild		1	yes *(171 ± 11 d)*	no	
"	tga20/tga20	10	yes *(62 ± 4 d)*	no	
Δ69-84	e19h/+	3-4	yes *(72 ± 16 d)*	no	
Δ32-80	d11/+	6	yes *(103 ± 11 d)*	no	
"	d12/+	10-12	yes *(89 ± 12 d)*	no[&]	
Δ32-93	C4/C4	4	yes *(232 ± 21 d)*	no	
"	C15/C15	3.8	yes *(239 ± 45 d)*	no	
Δ32-106	D32	3	no	no	
Δ32-121	E11	6	no	6-8	4-9
Δ32-134	F11	2	no	5	3-6
"	F35	2	no	3-4	2-3

* Relative to wild-type
& Mice homozygous for the transgene also showed no abnormalities

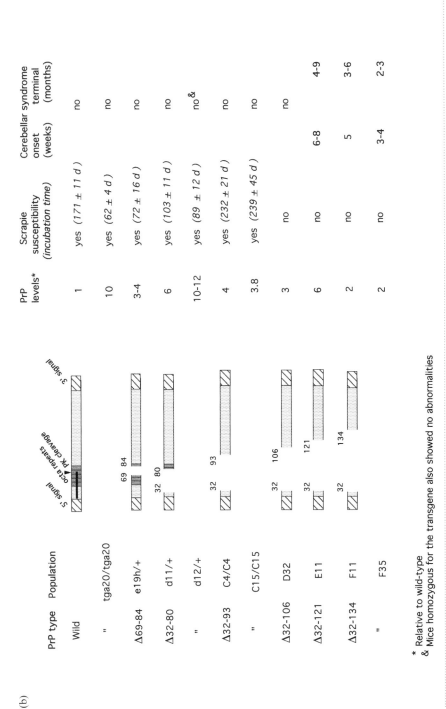

Fig. 1. *(cont.)*
(b) Wild-type PrP and PrP with amino-proximal deletions were expressed in Zürich I PrP knockout mice.

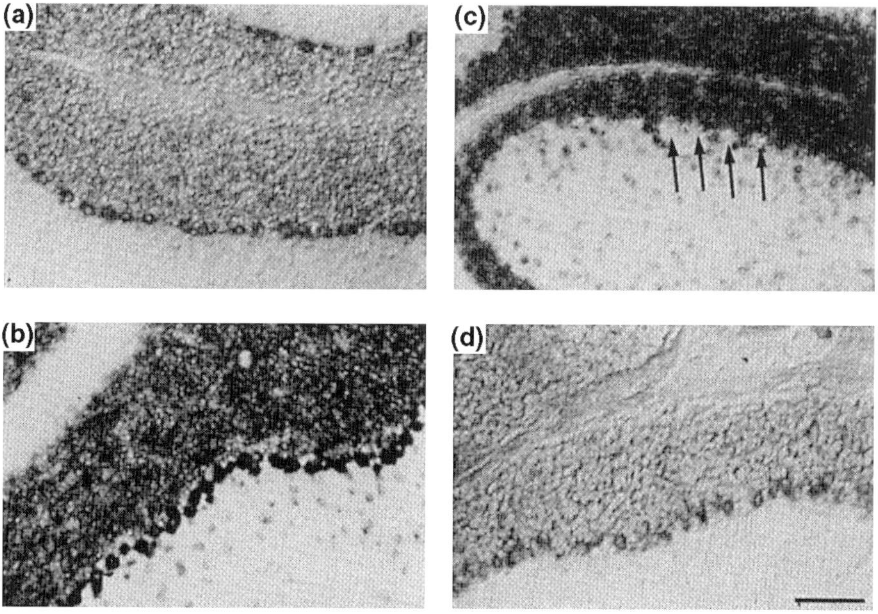

Fig. 2. PrP knockout mice transgenic for the 'half-genomic' PrP construct phgPrP express PrP mRNA in granule cells but not in Purkinje cells. Sagittal sections through the cerebellum of wild-type mice (a), and PrP Zürich I knockout mice expressing the *Prnp* cosmid cos6.I/LnJ-4 (*tgc35*/+) (b) or the half-genomic PrP construct *hgPrnp* (*tga19*/+) (c, d) were hybridized *in situ* with digoxigenin-labelled riboprobes detecting PrP (a–c) or neuron-specific enolase (NSE) RNA (d). Note the absence of a PrP RNA-specific signal in Purkinje cells of *tga19*/+ mice (arrows) (c) even though NSE mRNA can be detected in them in a parallel section (d). Bar, 310 μm. Modified from Fischer *et al.* (1996).

ATAXIC SYNDROME IN PrP KNOCKOUT MICE EXPRESSING PrP TRUNCATED TO POSITIONS 121 OR 134

Uninfected PrP knockout mice expressing PrP truncated to positions 121 or 134, but not PrP with shorter deletions, exhibited severe ataxia and depletion of the granule cell layer of the cerebellum as early as 1–3 months after birth (Fig. 3). The defect was completely abolished by concomitant expression of a single wild-type PrP gene, although the level of the truncated PrP remained unchanged (Shmerling *et al.*, 1998). Co-expression of PrP truncated to position 93 also abolished the syndrome, while co-expression of PrP truncated to position 106 retarded but did not prevent onset of the ataxic syndrome (E. Flechsig & C. Weissmann, unpublished results).

Expression or overexpression of a mutated PrP is often pathogenic, due to aggregate formation and/or abnormal intracellular localization (Chiesa *et al.*, 1998; Hegde *et al.*, 1999; Muramoto *et al.*, 1997), but usually co-expression of the normal protein does not abrogate the deleterious phenotype (Chiesa *et al.*, 2000). Why does absence of PrP

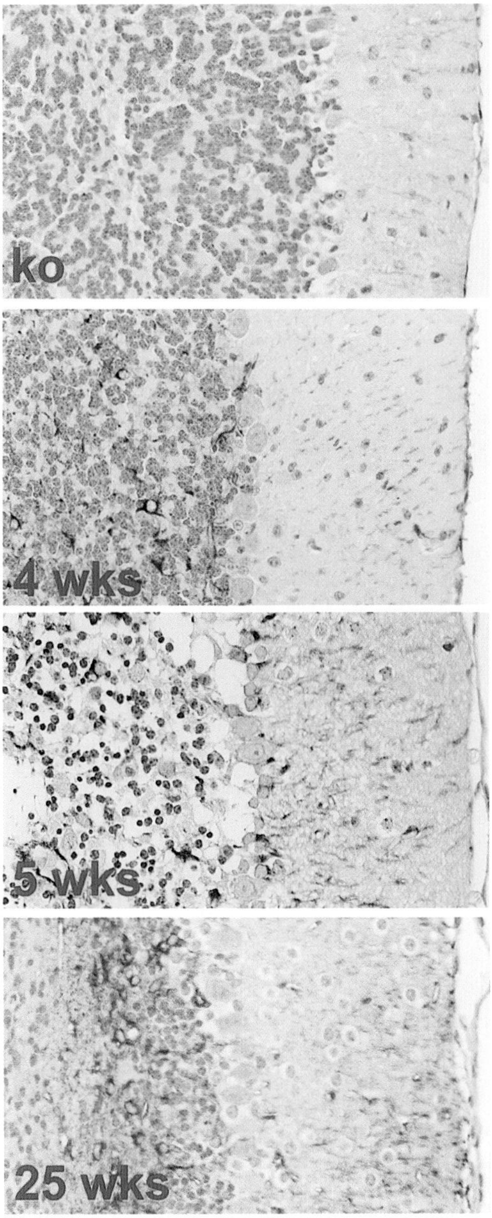

Fig. 3. Cerebella of Zürich I PrP knockout mice expressing PrPΔ32–134, at various times after birth. Sections were immunostained for glial fibrillary acidic protein (GFAP). At 4 weeks of age (4 wks), the internal granular layer is still largely intact and similar to the non-transgenic *Prnp*$^{0/0}$ control (ko), but incipient astrogliosis is observed. In 5-week-old mice (5 wks), there is prominent diffuse astrogliosis of the whole cerebellar cortex and coarse vacuolation of the granular cell layer. Several apoptotic figures are visible in the granular cell layer. At 25 weeks (25 wks), the granular layer has collapsed and shows extreme cell loss as well as gliosis. The Purkinje cells appear intact throughout this process.

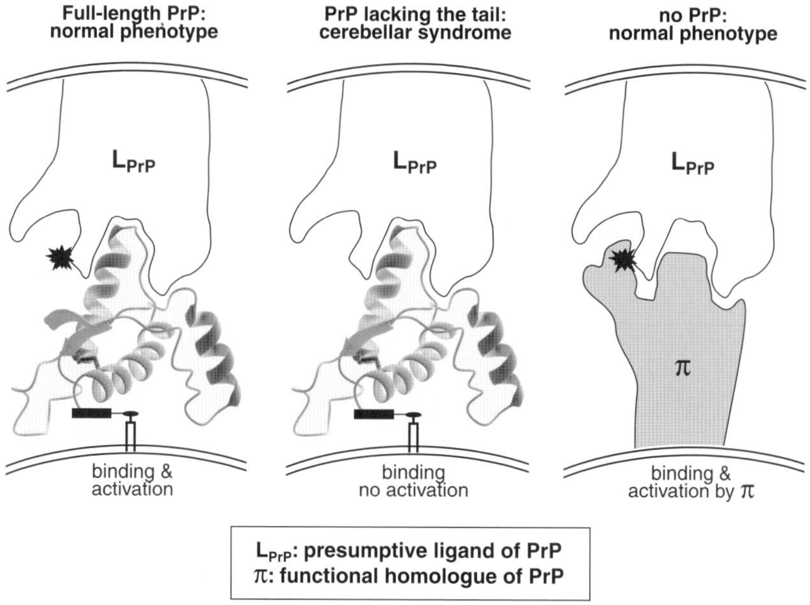

Fig. 4. Model to explain the pathogenic effect of truncated PrP. PrP, consisting of a globular part comprising three alpha helices and a flexible tail (Riek *et al.*, 1997), interacts with a presumed ligand, 'L_{PrP}', via these two domains, thereby eliciting a signal (black star burst). The same signal is elicited by the interaction of L_{PrP} with π, a conjectural protein which has the functional properties of PrP, explaining why PrP knockout mice show no phenotype. PrP lacking the flexible tail can interact with L_{PrP} without eliciting a signal and competes efficiently with π, thus acting as a dominant inhibitor of the latter. The postulated signal could act through PrP if the deleterious effect is cell autonomous or through L_{PrP} if it is not. If PrP is co-expressed with truncated PrP, it displaces the latter and restores the signal. Modified from Shmerling *et al.* (1998).

fail to cause ataxia while PrP truncated to position 134 does, and why does co-expression of full-length PrP restore the normal phenotype? We propose (Fig. 4) that there is a ligand for PrP, designated L_{PrP}, whose interaction with PrP elicits a signal required for maintenance of certain cell types, such as granule cells. This signal could be mediated through either PrP or L_{PrP}. Furthermore, we postulate that a functional homologue of PrP, which we designate π, can interact with L_{PrP} and elicit the maintenance signal in PrP knockout mice. We assume that PrP truncated to position 134 can bind to L_{PrP} more tightly than π but is unable to elicit the maintenance signal, thus causing cell death. Co-expression of PrP abrogates the effect of the truncated PrP because it binds more tightly to L_{PrP}.

Interestingly, when expressed under the direction of the Purkinje-cell-specific L7 promoter (E. Flechsig, R. Leimeroth & I. Hegyi, unpublished results), PrP truncated to position 134 causes Purkinje cell degeneration and ataxia. Here too, co-expression of a single wild-type PrP gene abrogates the pathogenic phenotype. Clearly, Purkinje cells are susceptible to the same pathogenic effect of truncated PrP as granule cells, but they are spared when the half-genomic vector is used because, as stated above, this vector does not direct expression in Purkinje cells (Fig. 2).

DELETIONS IN THE *Prnp* LOCUS LEADING TO ATAXIA

The essential role of PrP in prion disease was first revealed by showing that PrP knockout mice (Büeler *et al.*, 1992) were resistant to disease and failed to propagate mouse scrapie prions (Büeler *et al.*, 1993). These so-called Zürich I *Prnp*$^{0/0}$ mice, as well as mice from a later PrP knockout line designated Edinburgh *Prnp*$^{-/-}$ (Manson *et al.*, 1994), were clinically healthy, although they showed discrete neurophysiological changes and demyelination of peripheral nerves on ageing. However, mice of a third PrP knockout line, Nagasaki *Prnp*$^{-/-}$ (Sakaguchi *et al.*, 1996), came down with ataxia and loss of cerebellar Purkinje cells at 6–12 months of age; because the disease was prevented by introduction of a PrP-encoding cosmid, it was attributed to PrP ablation (Nishida *et al.*, 1999). A comparison of the knockout strategies (Fig. 5) showed that in the Zürich I mice two-thirds of the open reading frame (ORF) were replaced by a neomycin-resistance cassette, in the Edinburgh mice a similar cassette was inserted into the ORF, while in the Nagasaki mice not only the ORF, but also 0.9 kb of intron 2, the 5′ non-coding region and 0.45 kb of the 3′ non-coding region were deleted. In none of these lines could PrP or a fragment thereof be detected. It therefore seemed likely that the removal of the flanking sequences and not the ablation of PrP was responsible for the phenotype (Weissmann, 1996). Two additional PrP knockout lines, Zürich II (Rossi *et al.*, 2001) and Rcm0 (Moore *et al.*, 1999; Silverman *et al.*, 2000), in which the PrP ORF and its flanking regions were replaced by a *loxP* sequence and an HPRT (hypoxanthine phosphoribosyltransferase) cassette, respectively, confirmed the observations on the Nagasaki line.

Moore *et al.* (1999) provided surprising insight into the origin of the phenotype. Sequencing a *Prnp*-containing murine cosmid revealed an ORF encoding a 179-residue protein, christened Dpl, 16 kb downstream of the PrP gene (Fig. 6a). The predicted protein showed about 25 % identity with the carboxy-proximal two-thirds of PrP. Fig. 6(b) indicates that Dpl may contain three α-helices and an S–S linkage between the 2nd and 3rd helix, as does PrP (Riek *et al.*, 1997). Strikingly, however, Dpl lacks sequences homologous to the amino-proximal copper-binding octarepeat region and to the highly conserved region 112–126 of PrP, which is part of the region essential for its capacity to sustain prion replication, as described above (Schatzl *et al.*, 1995). Dpl mRNA is expressed at high levels in testis, less in other peripheral organs, but at very low levels in the brain of wild-type mice. However, there is a relatively high expression level of Dpl-specific RNA in the brain of Nagasaki, Rcm0 (Moore *et al.*, 1999) and

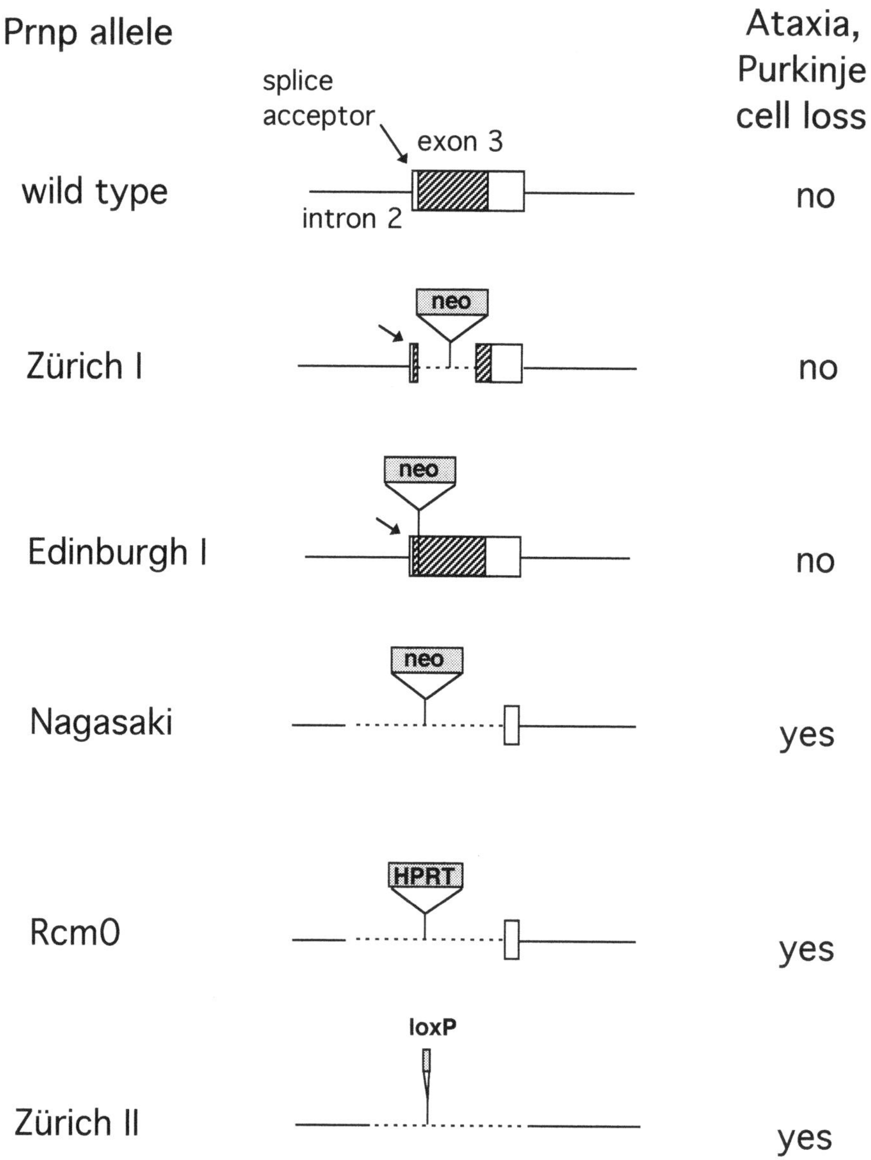

Fig. 5. Various strategies employed to disrupt the *Prnp* locus. The dotted line indicates the deleted DNA segment and the dotted box an inserted sequence. Hatched boxes, open reading frame; white boxes, non-coding regions; neo, neomycin phosphotransferase; HPRT, hypoxanthine phosphoribosyltransferase; *loxP*, a 34 bp recombination site from phage P1.

Zürich II mice (Rossi *et al.*, 2001), all of which present with ataxia and Purkinje cell death, but not in that of Zürich I or Edinburgh mice. Analysis of brain-derived cDNAs indicates that in wild-type mice, Dpl mRNA is very weakly expressed, mainly from a promoter upstream of exon 1 of *Prnd* (the gene encoding Dpl), whilst the strong expression in Nagasaki and Rcm0 mice is due to a chimeric RNA which originates at the *Prnp* promoter, runs all the way past the Dpl ORF and is processed by one or more splicing events that link the 3′ end of the second PrP exon directly or indirectly to the Dpl-encoding exon (Fig. 6a) (Moore *et al.*, 1999). This intergenic splicing, which is also detected at very low levels in wild-type mice, is greatly enhanced in the ataxic mice because the splice acceptor site upstream of the PrP-encoding exon is deleted, thus diverting the splice to a downstream acceptor site. The chimeric mRNA is expressed undiminished in Nagasaki mice 'cured' by the introduction of a PrP-expressing cosmid.

Why should overexpression of Dpl cause ataxia and concurrent overexpression of PrP restore normal function? As described above, expression of PrP truncated to position 134 in Zürich I $Prnp^{0/0}$ mice leads to ataxia and degeneration of the cerebellar granule cell layer within weeks of birth. Moreover, introduction of a single intact PrP allele prevented the disease. Because Dpl resembles the truncated PrP, it might cause disease by the same mechanism as proposed for the latter (Fig. 4). Furthermore, because expression of Dpl in the brain is directed by the PrP promoter, which is very active in Purkinje cells (Fig. 2a, b) (Fischer *et al.*, 1996), the resulting phenotype resembles that found with truncated PrP expressed under the direction of the Purkinje-cell-specific L7 promoter (see above) rather than the one resulting from its expression from the half-genomic vector.

CONCLUDING REMARKS

The normal function of PrP has not yet been elucidated. Because the octarepeats of PrP bind Cu^{2+} (Hornshaw *et al.*, 1995) and the Cu^{2+} complex of PrP shows superoxide dismutase activity (Brown & Besinger, 1998), roles in copper ion transport (Brown *et al.*, 1997) and resistance to oxygen stress have been proposed (Brown *et al.*, 1999). On the other hand, Cu^{2+} may serve mainly to impart a defined tertiary structure to PrP (Stockel *et al.*, 1998) and thereby support some other function, for example signalling. A number of possible ligands of PrP have been proposed (Edenhofer *et al.*, 1996; Graner *et al.*, 2000; Martins *et al.*, 1997; Rieger *et al.*, 1997; Yehiely *et al.*, 1997); however, none of them have yet been shown to be functionally significant. We believe that PrP truncated to position 121 or 134 interferes with the putative signalling function of PrP; complementation with PrP carrying other deletions or mutations allows us to map the regions critical for PrP function without precise knowledge of this function.

Finally, it is worth re-emphasizing the pitfalls that may beset the interpretation of knockout experiments. In the instance described, deletion of the PrP-encoding exon

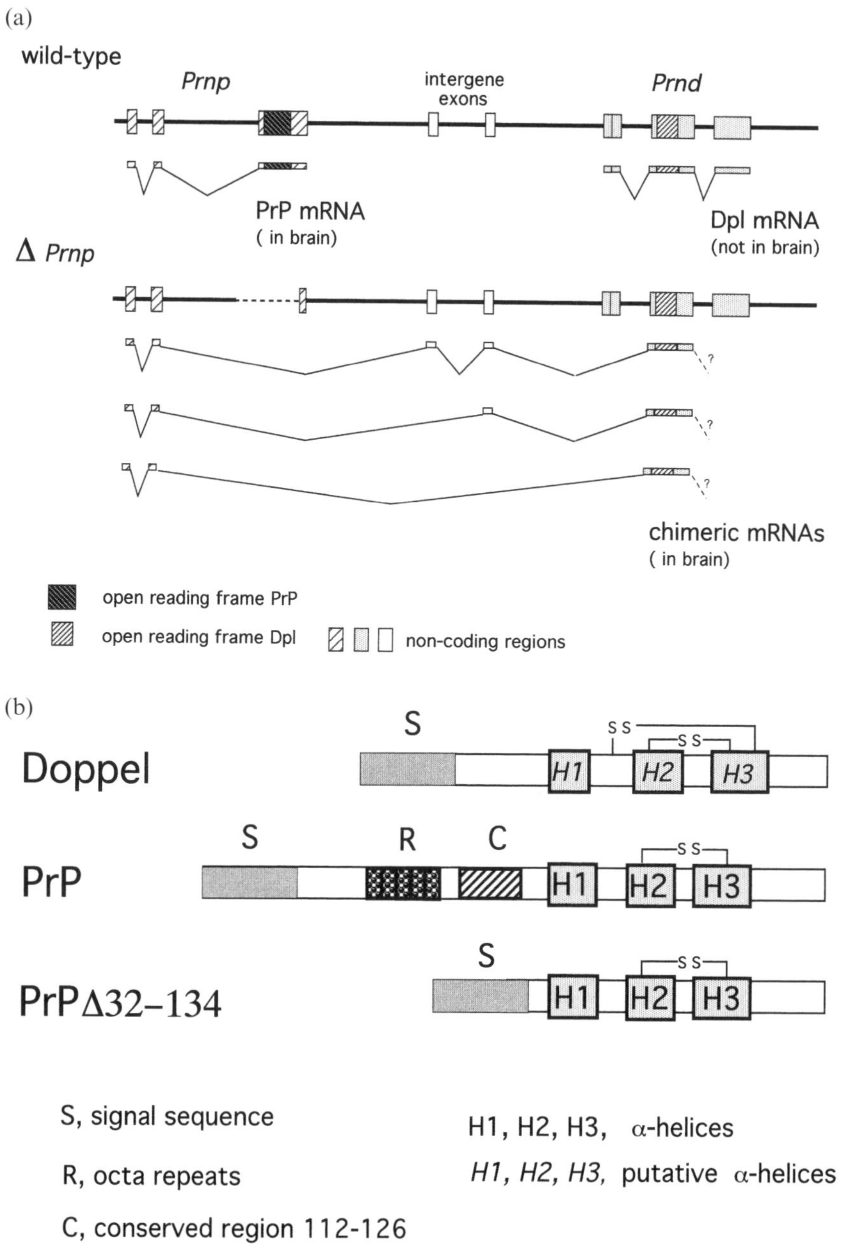

Fig. 6. *Prnp* and *Prnd* loci and their expression. (a) Top: Coding and non-coding exons of *Prnp* and *Prnd* and intergenic exons of unknown function in the wild-type allele. Bottom: Deletion of the ORF-containing exon of *Prnp* and its flanking regions entails the formation of several chimeric mRNAs in which the first two exons of *Prnp* are spliced directly or indirectly to the doppel (Dpl)-encoding exon of *Prnd*. (b) Comparison of predicted domains of Dpl with full-length PrP and PrP with residues 32–134 deleted. Modified from Moore *et al.* (1999).

and its flanking regions gave rise to a severe phenotype which could be reversed by introduction of an intact PrP-expressing transgene, the classical experiment correlating a phenotype with ablation of a gene. Nonetheless, in this case the conclusion was misleading, because the ataxic phenotype did not result from the deletion of the PrP ORF but from the incidental up-regulation of a deleterious gene product whose pathogenicity could be offset by wild-type PrP.

REFERENCES

Basler, K., Oesch, B., Scott, M., Westaway, D., Wälchli, M., Groth, D. F., McKinley, M. P., Prusiner, S. B. & Weissmann, C. (1986). Scrapie and cellular PrP isoforms are encoded by the same chromosomal gene. *Cell* **46**, 417–428.

Brinster, A. L., Allen, J. M., Behringer, R. R., Gelinas, R. E. & Palmiter, R. D. (1988). Introns increase transcriptional efficiency in transgenic mice. *Proc Natl Acad Sci U S A* **85**, 836–840.

Brown, D. R. & Besinger, A. (1998). Prion protein expression and superoxide dismutase activity. *Biochem J* **334**, 423–429.

Brown, D. R., Qin, K., Herms, J. W. & 10 other authors (1997). The cellular prion protein binds copper in vivo. *Nature* **390**, 684–687.

Brown, D. R., Wong, B. S., Hafiz, F., Clive, C., Haswell, S. J. & Jones, I. M. (1999). Normal prion protein has an activity like that of superoxide dismutase. *Biochem J* **344**, 1–5.

Büeler, H., Fischer, M., Lang, Y., Bluethmann, H., Lipp, H.-P., DeArmond, S. J., Prusiner, S. B., Aguet, M. & Weissmann, C. (1992). Normal development and behaviour of mice lacking the neuronal cell-surface PrP protein. *Nature* **356**, 577–582.

Büeler, H., Aguzzi, A., Sailer, A., Greiner, R. A., Autenried, P., Aguet, M. & Weissmann, C. (1993). Mice devoid of PrP are resistant to scrapie. *Cell* **73**, 1339–1347.

Büeler, H., Raeber, A., Sailer, A., Fischer, M., Aguzzi, A. & Weissmann, C. (1994). High prion and PrPSc levels but delayed onset of disease in scrapie-inoculated mice heterozygous for a disrupted PrP gene. *Mol Med* **1**, 19–30.

Chesebro, B., Race, R., Wehrly, K. & 9 other authors (1985). Identification of scrapie prion protein-specific messenger RNA in scrapie-infected and uninfected brain. *Nature* **315**, 331–333.

Chiesa, R., Piccardo, P., Ghetti, B. & Harris, D. A. (1998). Neurological illness in transgenic mice expressing a prion protein with an insertional mutation. *Neuron* **21**, 1339–1351.

Chiesa, R., Drisaldi, B., Quaglio, E., Migheli, A., Piccardo, P., Ghetti, B. & Harris, D. A. (2000). Accumulation of protease-resistant prion protein (PrP) and apoptosis of cerebellar granule cells in transgenic mice expressing a PrP insertional mutation. *Proc Natl Acad Sci U S A* **97**, 5574–5579.

Collinge, J. (1997). Human prion diseases and bovine spongiform encephalopathy (BSE). *Hum Mol Genet* **6**, 1699–1705.

Edenhofer, F., Rieger, R., Famulok, M., Wendler, W., Weiss, S. & Winnacker, E. L. (1996). Prion protein PrPc interacts with molecular chaperones of the Hsp60 family. *J Virol* **70**, 4724–4728.

Fischer, M., Rülicke, T., Raeber, A., Sailer, A., Moser, M., Oesch, B., Brandner, S., Aguzzi, A. & Weissmann, C. (1996). Prion protein (PrP) with amino-proximal deletions restoring susceptibility of PrP knockout mice to scrapie. *EMBO J* **15**, 1255–1264.

Flechsig, E., Shmerling, D., Hegyi, I., Raeber, A. J., Fischer, M., Cozzio, A., von Mering, C., Aguzzi, A. & Weissmann, C. (2000). Prion protein (PrP) devoid of the octapeptide repeat region restores susceptibility to scrapie in PrP knockout mice. *Neuron* **27**, 399–408.

Graner, E., Mercadante, A. F., Zanata, S. M. & 10 other authors (2000). Cellular prion protein binds laminin and mediates neuritogenesis. *Brain Res Mol Brain Res* **76**, 85–92.

Griffith, J. S. (1967). Self-replication and scrapie. *Nature* **215**, 1043–1044.

Hegde, R. S., Tremblay, P., Groth, D., DeArmond, S. J., Prusiner, S. B. & Lingappa, V. R. (1999). Transmissible and genetic prion diseases share a common pathway of neurodegeneration. *Nature* **402**, 822–826.

Hornshaw, M. P., McDermott, J. R. & Candy, J. M. (1995). Copper binding to the N-terminal tandem repeat regions of mammalian and avian prion protein. *Biochem Biophys Res Commun* **207**, 621–629.

Hsiao, K. K., Scott, M., Foster, D., Groth, D. F., DeArmond, S. J. & Prusiner, S. B. (1990). Spontaneous neurodegeneration in transgenic mice with mutant prion protein. *Science* **250**, 1587–1590.

Manson, J. C., Clarke, A. R., McBride, P. A., McConnell, I. & Hope, J. (1994). PrP gene dosage determines the timing but not the final intensity or distribution of lesions in scrapie pathology. *Neurodegeneration* **3**, 331–340.

Martins, V. R., Graner, E., Garcia-Abreu, J., de Souza, S. J., Mercadante, A. F., Veiga, S. S., Zanata, S. M., Moura Neto, V. & Brentani, R. R. (1997). Complementary hydropathy identifies a cellular prion protein receptor. *Nat Med* **3**, 1376–1382.

Moore, R. C., Lee, I. Y., Silverman, G. L. & 18 other authors (1999). Ataxia in prion protein (PrP)-deficient mice is associated with upregulation of the novel PrP-like protein doppel. *J Mol Biol* **292**, 797–817.

Muramoto, T., DeArmond, S. J., Scott, M., Telling, G. C., Cohen, F. E. & Prusiner, S. B. (1997). Heritable disorder resembling neuronal storage disease in mice expressing prion protein with deletion of an alpha-helix. *Nat Med* **3**, 750–755.

Nishida, N., Tremblay, P., Sugimoto, T. & 12 other authors (1999). A mouse prion protein transgene rescues mice deficient for the prion protein gene from purkinje cell degeneration and demyelination. *Lab Investig* **79**, 689–697.

Oesch, B., Westaway, D., Wälchli, M. & 9 other authors (1985). A cellular gene encodes scrapie PrP 27-30 protein. *Cell* **40**, 735–746.

Prusiner, S. B. (1989). Scrapie prions. *Annu Rev Microbiol* **43**, 345–374.

Prusiner, S. B. (1996). Molecular biology and genetics of prion diseases. *Cold Spring Harbor Symp Quant Biol* **61**, 473–493.

Prusiner, S. B. (1998). Prions. *Proc Natl Acad Sci U S A* **95**, 13363–13383.

Rieger, R., Edenhofer, F., Lasmézas, C. I. & Weiss, S. (1997). The human 37-kDa laminin receptor precursor interacts with the prion protein in eukaryotic cells. *Nat Med* **3**, 1383–1388.

Riek, R., Hornemann, S., Wider, G., Glockshuber, R. & Wüthrich, K. (1997). NMR characterization of the full-length recombinant murine prion protein, *m*PrP(23-231). *FEBS Lett* **413**, 282–288.

Rossi, D., Cozzio, A., Flechsig, E., Klein, M. A., Rülicke, T., Aguzzi, A. & Weissmann, C. (2001). Onset of ataxia and Purkinje cell loss in PrP knockout mice is inversely correlated with expression level of doppel mRNA in brain. *EMBO J* (in press).

Sailer, A., Büeler, H., Fischer, M., Aguzzi, A. & Weissmann, C. (1994). No propagation of prions in mice devoid of PrP. *Cell* **77**, 967–968.

Sakaguchi, S., Katamine, S., Shigematsu, K. & 10 other authors (1995). Accumulation of proteinase K-resistant prion protein (PrP) is restricted by the expression level of normal PrP in mice inoculated with a mouse-adapted strain of the Creutzfeldt-Jakob disease agent. *J Virol* **69**, 7586–7592.

Sakaguchi, S., Katamine, S., Nishida, N. & 11 other authors (1996). Loss of cerebellar Purkinje cells in aged mice homozygous for a disrupted PrP gene. *Nature* **380**, 528–531.

Schatzl, H. M., Da, C. M., Taylor, L., Cohen, F. E. & Prusiner, S. B. (1995). Prion protein gene variation among primates. *J Mol Biol* **245**, 362–374.

Scott, M., Foster, D., Mirenda, C. & 9 other authors (1989). Transgenic mice expressing hamster prion protein produce species-specific scrapie infectivity and amyloid plaques. *Cell* **59**, 847–857.

Shmerling, D., Hegyi, I., Fischer, M. & 10 other authors (1998). Expression of amino-terminally truncated PrP in the mouse leading to ataxia and specific cerebellar lesions. *Cell* **93**, 203–214.

Silverman, L. G., Qin, K., Moore, R. C., Yang, Y., Mastrangelo, P., Tremblay, P., Prusiner, S. B., Cohen, F. E. & Westaway, D. (2000). Doppel is an N-glycosylated GPI-anchored protein: expression in testis and ectopic production in the brains of *Prnp*$^{0/0}$ mice predisposed to Purkinje cell loss. *J Biol Chem* **275**, 26834–26841.

Stockel, J., Safar, J., Wallace, A. C., Cohen, F. E. & Prusiner, S. B. (1998). Prion protein selectively binds copper(II) ions. *Biochemistry* **37**, 7185–7193.

Supattapone, S., Bosque, P., Muramoto, T. & 11 other authors (1999). Prion protein of 106 residues creates an artificial transmission barrier for prion replication in transgenic mice. *Cell* **96**, 869–878.

Vital, C., Gray, F., Vital, A., Parchi, P., Capellari, S., Petersen, R. B., Ferrer, X., Jarnier, D., Julien, J. & Gambetti, P. (1998). Prion encephalopathy with insertion of octapeptide repeats: the number of repeats determines the type of cerebellar deposits. *Neuropathol Appl Neurobiol* **24**, 125–130.

Weissmann, C. (1991). Spongiform encephalopathies. The prion's progress. *Nature* **349**, 569–571.

Weissmann, C. (1996). PrP effects clarified. *Curr Biol* **6**, 1359.

Weissmann, C. (1999). Molecular genetics of transmissible spongiform encephalopathies. *J Biol Chem* **274**, 3–6.

Weissmann, C., Fischer, M., Raeber, A., Büeler, H., Sailer, A., Shmerling, D., Rülicke, T., Brandner, S. & Aguzzi, A. (1996). The role of PrP in pathogenesis of experimental scrapie. *Cold Spring Harbor Symp Quant Biol* **61**, 511–522.

Westaway, D., Mirenda, C. A., Foster, D. & 9 other authors (1991). Paradoxical shortening of scrapie incubation times by expression of prion protein transgenes derived from long incubation period mice. *Neuron* **7**, 59–68.

Westaway, D., Cooper, C., Turner, S., Da, C. M., Carlson, G. A. & Prusiner, S. B. (1994a). Structure and polymorphism of the mouse prion protein gene. *Proc Natl Acad Sci U S A* **91**, 6418–6422.

Westaway, D., DeArmond, S. J., Cayetano, C. J., Groth, D., Foster, D., Yang, S. L., Torchia, M., Carlson, G. A. & Prusiner, S. B. (1994b). Degeneration of skeletal muscle, peripheral nerves, and the central nervous system in transgenic mice overexpressing wild-type prion proteins. *Cell* **76**, 117–129.

Yehiely, F., Bamborough, P., Da Costa, M., Perry, B. J., Thinakaran, G., Cohen, F. E., Carlson, G. A. & Prusiner, S. B. (1997). Identification of candidate proteins binding to prion protein. *Neurobiol Dis* **3**, 339–355.

Endogenous retroviruses and xenotransplantation

Jonathan P. Stoye

Division of Virology, National Institute for Medical Research, The Ridgeway, Mill Hill, London NW7 1AA, UK

THE PROMISE OF XENOTRANSPLANTATION

Every year, thousands of patients with chronic endstage organ failure will die while on the waiting lists for potentially life-saving transplant surgery. Many others will not even make the waiting lists, as they are judged less likely to benefit from scarce human organs. The transplantation of foetal neural cells or foetal islets shows great potential for the treatment of individuals with Parkinson's disease or Type I diabetes, but again the number of potential beneficiaries from such therapies is limited by the availability of donor tissue.

Over the years, such shortages have prompted a number of investigators to explore the possibility of using non-human sources of organs and cells for transplantation into humans. Until now, such efforts have met with little success, but a number of recent developments in biomedical science have appeared to confirm the potential of such approaches (Cooper & Lanza, 2000). Over the coming years, such efforts are likely to become much more frequent as increasing numbers of physicians and surgeons seek to tap into the potential of techniques involving the transplantation of cells, tissues or whole organs from other species into humans, procedures collectively known as xenotransplantation.

SOURCE ANIMALS FOR XENOTRANSPLANTATION

To date, most animal to human xenotransplantation protocols have featured the use of non-human primates as source animals because the close evolutionary relationship to humans might be expected to minimize problems of rejection and to maximize the

SGM symposium 60: New challenges to health: the threat of virus infection.
Editors G. L. Smith, W. L. Irving, J. W. McCauley, D. J. Rowlands. Cambridge University Press. ISBN 0 521 80614 3

probability of physiological compatibility between graft and recipient. However, opinion has now turned firmly against the use of non-human primates, partly on ethical and partly on practical and safety grounds (see US Food and Drug Administration Guidance Notice at http://www.fda.gov/cber/gdlns/xenoprim.txt). Attention within the xenotransplantation community now focuses on the use of the pig. Consequently, this will be the only source animal considered in this chapter.

CHALLENGES FOR XENOTRANSPLANTATION

For xenotransplantation to find widespread use, three scientific/medical challenges must be overcome. First, the problem of xenograft rejection must be solved. Second, the transplant must function properly in its new setting. Third, introduction of novel infectious micro-organisms into the human population via the transplant must be prevented. This chapter will focus on the potential infectious disease threat posed by one group of infectious agents, the porcine endogenous retroviruses (PERVs), but will start with a brief summary of the current status in all three areas.

It is thought that there are four barriers to the xenotransplantation of organs: hyperacute rejection, acute vascular rejection (also known as delayed xenograft rejection), acute cellular rejection and chronic rejection (Bach, 1998; Platt & Lin, 1998). For many years, hyperacute rejection appeared an insurmountable barrier. Within minutes of transplantation, the new organ would turn black and would be destroyed very rapidly. Research showed that this process was mediated by the binding of naturally occurring human antibodies reactive with α-galactosyl residues on the engrafted tissue followed by complement activation (Lin *et al.*, 1997). Hyperacute rejection can now be prevented, at least in pig to monkey models, by preventing antibody binding or complement activation. The favoured solution appears to be the use of transgenic pigs carrying regulators of human complement, for example decay-accelerating factor, on the surfaces of their cells (Cozzi *et al.*, 1997). Acute vascular rejection can also be overcome using the same pigs plus very high doses of drugs suppressing antibody synthesis or splenectomy (Bhatti *et al.*, 1999; Schmoeckel *et al.*, 1999). Although these protocols are probably too severe to contemplate for human trials, such results are encouraging in suggesting that acute vascular rejection can be overcome. Acute cellular rejection and chronic rejection are the rejection mechanisms currently encountered in allotransplantation; though we do not know whether the analogous processes in xenotransplantation will be more or less severe, there is optimism that they can be managed by immunosuppressive drugs and retransplantation, respectively.

Even assuming that these immunological barriers can be overcome, it remains an open question when xenografts will fulfil their intended functions in the new host. There are many levels, both anatomical and physiological in nature, at which incompatibility

Table 1. Examples of infectious agents transmitted in allotransplants

See Michaels & Simmons (1994).

HIV-1/HIV-2
Hepatitis B virus
Hepatitis C virus
Epstein–Barr virus
Human herpesvirus 6
Cytomegalovirus
Rabies virus
Toxoplasma gondii
Prions

might theoretically be manifested (Hammer, 1998). Single gene problems could undoubtedly be overcome with further transgenic manipulations, but this approach would probably not be feasible with multiple incompatibilities. In general, it seems most likely that transplants involving organs or cells serving only a single mechanical or physiological function, for example heart or islets, will work better than those with multiple biochemical roles, such as liver or kidney. Unfortunately, there are no satisfactory model systems to resolve these issues; it is only in long-term clinical trials that such questions will be answered in a definitive fashion.

The third challenge concerns the risk of infectious disease (Fishman, 1994; Michaels & Simmons, 1994). There are two kinds of hazards to be considered. The first is transmission of known pathogens to the transplant recipient. The second and much more worrying concern is that xenotransplantation would allow the introduction of previously unknown or undetected pathogens into the general population. Organ transplantation represents a perfect means for transmission of infectious agents. Not only are the primary barriers (skin, intestine) to infection bypassed, but the recipient is also heavily immunosuppressed and therefore unable to respond to a wide variety of opportunistic infections by agents carried by the transplanted organ (Table 1) or acquired from environmental sources. Infection is therefore a major complication of allotransplantation.

It is hoped that xenotransplantation might help reduce the risk associated with known micro-organisms. Allotransplants are usually unscheduled; when an organ becomes available there is relatively little time to assess the infectious disease status of the organ before its physical condition begins to deteriorate. By contrast, xenotransplants will be carefully planned with source animals bred according to defined protocols designed to produce specific-pathogen-free animals (Fishman, 1998). At the same time, such protocols should also eliminate most, if not all, unknown infectious agents from the source herds. However, the techniques used to exclude pathogens are not perfect; agents which are capable of infecting foetuses congenitally, for example herpes viruses, or which are passed in the germ line, for example endogenous retroviruses, will not be eliminated. The existence and properties of such agents has therefore become one of the most controversial areas of xenotransplantation research.

INTRODUCTION TO ENDOGENOUS RETROVIRUSES

The replication cycle of retroviruses involves an obligate step of viral DNA integration into the host-cell genome to form the provirus (Coffin, 1996). Infection of germ cells would therefore be expected to provide an ideal means for germ-line colonization. Examination of normal genomic DNA reveals the presence of multiple copies of integrated proviruses which are inherited in Mendelian fashion. These are called endogenous retroviruses. Because of their intimate relationship with their host, such elements have been studied intensively over the past 30 years. In any given species, there are likely to be upwards of 10 different groups of endogenous retroviral element present in copy numbers ranging from one or a few to a thousand or more (Boeke & Stoye, 1997).

Endogenous retroviruses were initially studied because of their close structural similarity to cancer-causing exogenous retroviruses. Although there are a few examples of 'spontaneous' tumours, for example murine thymomas and mammary carcinomas (Rosenberg & Jolicoeur, 1997), that result from the activation of endogenous proviruses, most endogenous retroviruses appear benign, at least in the species in which they have made their home. Most have been present in the germ line for hundreds of thousands or millions of years (Goodchild *et al.*, 1993); during this time they have accumulated sufficient mutations to render them incapable of replication. Others have entered the germ line more recently and retain replication potential (Kozak & Silver, 1985), but are restrained by host adaptations. Good examples of the latter phenomenon are mutations in the cellular receptors for virus that can block reinfection (Marin *et al.*, 1999). Thus the mouse genome contains multiple endogenous proviruses encoding xenotropic viruses which grow well on non-murine cells but are incapable of infecting murine cells. A number of investigators have unwittingly rediscovered this phenomenon by isolating murine viruses from human tumour cells passaged through mice (Tralka *et al.*, 1983). It would be of considerable interest to examine the pathogenic effect of murine xenotropic viruses if functional receptors were to be restored to mice by transgenesis.

Little is known about the pathogenic properties of endogenous retroviruses following cross-species transmission to new hosts. However, based on experiences with a variety of micro-organisms, it would not be surprising if replication to high titre in a naïve host led to disease. The case of gibbon ape leukaemia virus (GALV) provides some support for this idea. GALV causes haematopoietic tumours in gibbons and can be transmitted both horizontally and vertically in gibbons (Teich *et al.*, 1982). The origin of GALV is unknown. It is clearly not endogenous to gibbons, but bears some similarity to endogenous retroviruses isolated from mice derived from SE Asia (Lieber *et al.*, 1975b). It is therefore tempting to conclude that GALV is derived from such a harmless resident of the mouse germ line and that cross-species transmission was associated with a gain in pathogenicity. However, until the presumptive source virus is identified and characterized, this conclusion is somewhat tentative. Very recently, an endogenous virus closely related to GALV has been found in koala bears, where it may play a role in immunological malignancies (Hanger *et al.*, 2000). The presence of such a virus in koalas, but not other marsupials, might also be explained by a relatively recent cross-species transmission associated with an apparent gain in pathogenicity.

ENDOGENOUS RETROVIRUSES AS A THREAT IN XENOTRANSPLANTATION

Consideration of the basic properties of endogenous retroviruses and their exogenous cousins suggests that they might represent hazards for xenotransplantation (Smith, 1993; Stoye & Coffin, 1995; Stoye *et al.*, 1998). They have little or no effect on their hosts and therefore their presence might remain undetected; they are inherited in the germ line and are therefore impossible to remove by most techniques; if they cause disease it is likely to follow a long asymptomatic latent period during which time spread from transplant recipients into the general population could occur.

For PERVs to pose a serious threat to the health of the general public, the following chain of events must occur (Stoye, 1998). First, endogenous retroviruses capable of replicating in human cells must be encoded by proviruses present in the germ line of pigs. Second, such proviral loci must be present in the herds of pigs taken as source animals for xenotransplantation. Third, these proviruses must be expressed in the cells, tissues or organs taken for transplantation. Fourth, these viruses must be capable of infecting the transplant recipient. Fifth, virus replication and growth to high titre in the recipient must take place. Sixth, this process must lead to disease. Seventh, spread to others must occur. The rest of this chapter will be structured to consider the evidence for or against each link in this chain.

PROPERTIES OF PERVs

Pig cell lines were first tested for retrovirus production in the early 1970s. A number of established lines were shown to produce reverse-transcriptase-positive, C-type particles (Armstrong *et al.*, 1971; Strandström *et al.*, 1974; Todaro *et al.*, 1974). Labelled probes prepared with viral RNA hybridized with normal pig liver DNA indicated that these viruses were endogenous to the pig (Todaro *et al.*, 1974). Host range studies showed that at least two of the viral isolates were capable of replication on pig cells (Lieber *et al.*, 1975a). Productive infection of cells from other species was not observed; however, unsurprisingly in light of our current knowledge, limited rescue of murine sarcoma virus from a variety of transformed, non-producer cells was reported (Lieber *et al.*, 1975a). Over the next 20 years, relatively little attention was devoted to pig retroviruses. Two reports documented virus production associated with malignant transformation of lymphocytes (Frazier, 1985; Suzuka *et al.*, 1985). In one case, the isolated virus, Tsukuba-1, was cloned (Suzuka *et al.*, 1986), but little further characterization was performed. The relationship, if any, between these viruses and malignancy has not been determined.

Concerns about the potential activation of endogenous retroviruses in xenotransplantation led to a re-examination of the host range of viruses produced by two of the previously studied pig kidney lines, PK-15 and MPK (Patience *et al.*, 1997). Confirming previous reports, the MPK virus is ecotropic. It grew only on cells of the species from which it was isolated (pig) but not on non-porcine cells like mink or human. By contrast, PK-15 cells, which had previously been thought to release non-infectious virus, were found to produce virus that could infect human 293 cells as well as mink and pig cells. Two further properties of PK-15 virus immediately became obvious. First, though it can be serially passaged on 293 cells, titres remain low (<500 infectious units ml^{-1}). Indeed, infection of certain cell lines can only be accomplished by co-cultivation. Second, PERV infection of certain human cell lines as well as human peripheral blood mononuclear cells tends to be non-productive – integrated virus DNA can be found in infected cells but serial passage is not possible. These properties, which have been confirmed by subsequent investigations (Wilson *et al.*, 2000), presumably explain the 'non-infectivity' of PK-15 virus in the earlier studies.

The host range of the virus produced by PK-15 cells suggested that either it was a mixture of two viruses (ecotropic and xenotropic) or it represented a polytropic virus. To examine this question, the host-range-determining envelope gene(s) of virus from PK-15-infected 293 cells were cloned and sequenced. Two related but clearly distinguishable groups of clones were obtained, with the greatest diversity occurring within the regions predicted to specify receptor binding, implying that they interact with different receptors (Le Tissier *et al.*, 1997). Subsequent studies showed that they correspond to the envelope genes of two separate classes, PERV-A and PERV-B, of polytropic virus, both capable of infecting a variety of cell types (Takeuchi *et al.*, 1998). A third class of PERV, PERV-C, was identified based on the sequences of a full-length viral cDNA clone isolated from miniswine lymphocytes and the Tsukuba-1 virus (Akiyoshi *et al.*, 1998).

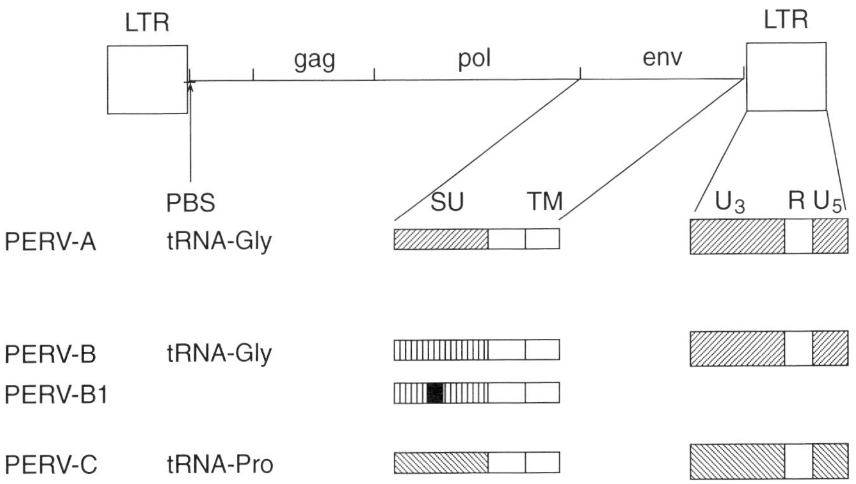

Fig. 1. Structures of the different classes of PERV. PERV-A, -B, -C show significant differences in the tRNA primer binding site for reverse transcriptase, in the *env* gene and within the U3 and U5 regions of the LTR. PERV-B1 (Galbraith *et al.*, 1997) contains two frameshift mutations in *env*, preserving an open reading frame but altering 22 codons; it is not known whether the resulting Env protein is functional. SU, surface protein; TM, transmembrane protein.

The ecotropic virus identified in MPK cells appears to belong to the PERV-C class (Takeuchi *et al.*, 1998). PCR (Martin *et al.*, 1998a), Southern hybridization analysis (Akiyoshi *et al.*, 1998; Le Tissier *et al.*, 1997; Patience *et al.*, 1997) and isolation of genomic PERV clones (Rogel-Gaillard *et al.*, 1999) have provided ample evidence that all three classes of PERV are present in normal pig DNA, directly demonstrating the endogenous nature of this class of retrovirus.

A number of molecular clones of intact PERV elements have now been sequenced (Akiyoshi *et al.*, 1998; Czauderna *et al.*, 2000; Galbraith *et al.*, 1997). These permit a reasonably detailed comparison between the different classes of virus (Fig. 1). They appear very similar in the viral *gag* and *pol* genes and the 3′ half of *env*, but there are extensive differences within the 5′ half of *env*, the region encoding the host range determinants on the viral surface protein (Battini *et al.*, 1995). Within the LTRs (long terminal repeats), the R region is similar between the three classes, but the U3 and U5 regions of PERV-C differ significantly from those of PERV-A and -B. Since the U3 region of a retrovirus typically includes its promoter/enhancer elements (Rabson & Graves, 1997), this might suggest that PERV-C proviruses will show different expression patterns compared to PERV-A and -B. Unusually, PERV-A and PERV-B would be

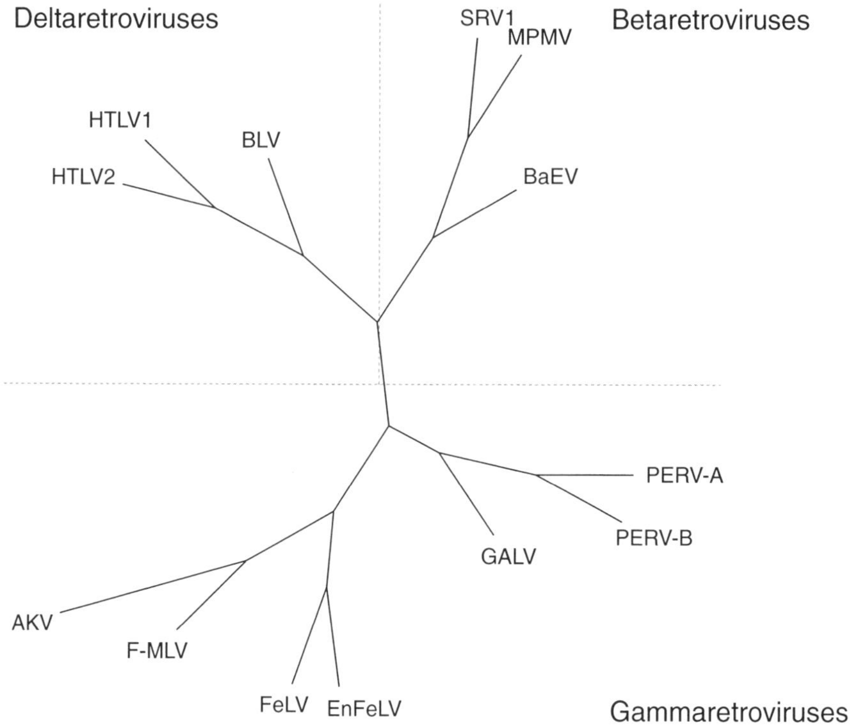

Fig. 2. Phylogenetic relationship of PERV-A and PERV-B. An unrooted tree illustrating the relationship of PERV-A and PERV-B to other retroviruses assessed using the sequence of the transmembrane protein. AKV, AKR endogenous ecotropic virus; F-MLV, Friend murine leukaemia virus; FeLV, exogenous feline leukaemia virus; enFeLV, endogenous feline leukaemia virus; BaEV, baboon endogenous virus; SRV1, simian retrovirus 1; MPMV, Mason–Pfizer monkey virus; BLV, bovine leukosis virus; HTLV1, human T-cell leukaemia virus type I; HTLV2, human T-cell leukaemia virus type II. BaEV is a recombinant beta/gammaretrovirus (van der Kuyl *et al.*, 1996).

predicted to utilize a tRNA-Gly for priming reverse transcriptase, whereas PERV-C utilizes a more conventional tRNA-Pro. It is not clear how much variation will be seen between different members of these families – a detailed analysis of multiple pig genomic clones will be required to address this issue – but initial comparisons suggest that there may be more variation than that seen in a similar study of endogenous murine leukaemia viruses (Stoye & Coffin, 1987). Phylogenetic analysis places the PERVs within the gammaretrovirus family, the closest relative being GALV (Fig. 2; Martin *et al.*, 1999).

A variety of strategies have been employed to look for endogenous retroviruses (Boeke & Stoye, 1997). The identification of PERV-A, -B and -C exemplifies the simplest approach, namely looking for virus production. Among alternative approaches, perhaps the easiest is to use PCR to amplify conserved regions of the viral polymerase gene (Shih *et al.*, 1989; Tristem *et al.*, 1996). In this way, at least eight more families of retroviral proviruses have been identified in pigs and related species (Patience *et al.*, 2001). It seems that most, if not all, of these novel elements are more ancient than the PERVs and have therefore accumulated inactivating mutations. Nevertheless, a more detailed analysis of these retroviral families would appear warranted.

GENETIC DISTRIBUTION OF PERVs

Endogenous retroviruses are inherited through the germ line. They cannot therefore be eliminated by techniques designed to prevent mother to infant transmission via infection that have been used to generate specific-pathogen-free pigs. Rather, it would be necessary to identify, or create, virus-free stocks and then expand them by selective breeding, an approach typified by the breeding of chickens free of endogenous avian leukosis virus (Astrin *et al.*, 1979).

The simplest scenario would arise if some varieties of pigs lacked specific classes of PERVs. Using class-specific PCR primers, a number of different kinds of pig were tested for various PERVs. Unfortunately, all tested breeds of pigs (Martin *et al.*, 1998a; Patience *et al.*, 2001) as well as their nearest relatives contain both PERV-A and PERV-B, the two classes of human tropic PERV. Ironically, pigs lacking PERV-C have been identified, but since this virus appears to be ecotropic, this observation appears of little importance with respect to xenotransplantation.

A more sophisticated approach involves the identification and analysis of specific proviral loci. Members of a given retroviral family are highly related to one another and it is therefore relatively difficult to generate locus-specific probes directed against viral sequences. However, because each provirus will have different host sequences flanking its integration site, it is possible to use these sequences to define different loci. DNA digested with an enzyme that cuts once at a conserved site within the provirus will generate a number of bands on a Southern blot, analysed with a viral probe, with discrete sizes depending on the position of the nearest restriction site in the flanking DNA. This strategy has been used to define and map over 100 non-ecotropic endogenous murine leukaemia viruses (Frankel *et al.*, 1990). Blots illustrating this approach using PERV-A- and PERV-B-specific probes (Le Tissier *et al.*, 1997) are shown in Fig. 3. Counting the bands gives an estimate of the number of proviruses of each class present in each breed tested. Bands with identical mobilities, indicated with arrows in Fig. 3, suggest the presence of identical proviruses in the different breeds. From this analysis, we can see that it would be very difficult, if not impossible, to breed pigs lacking all PERV-A and PERV-B

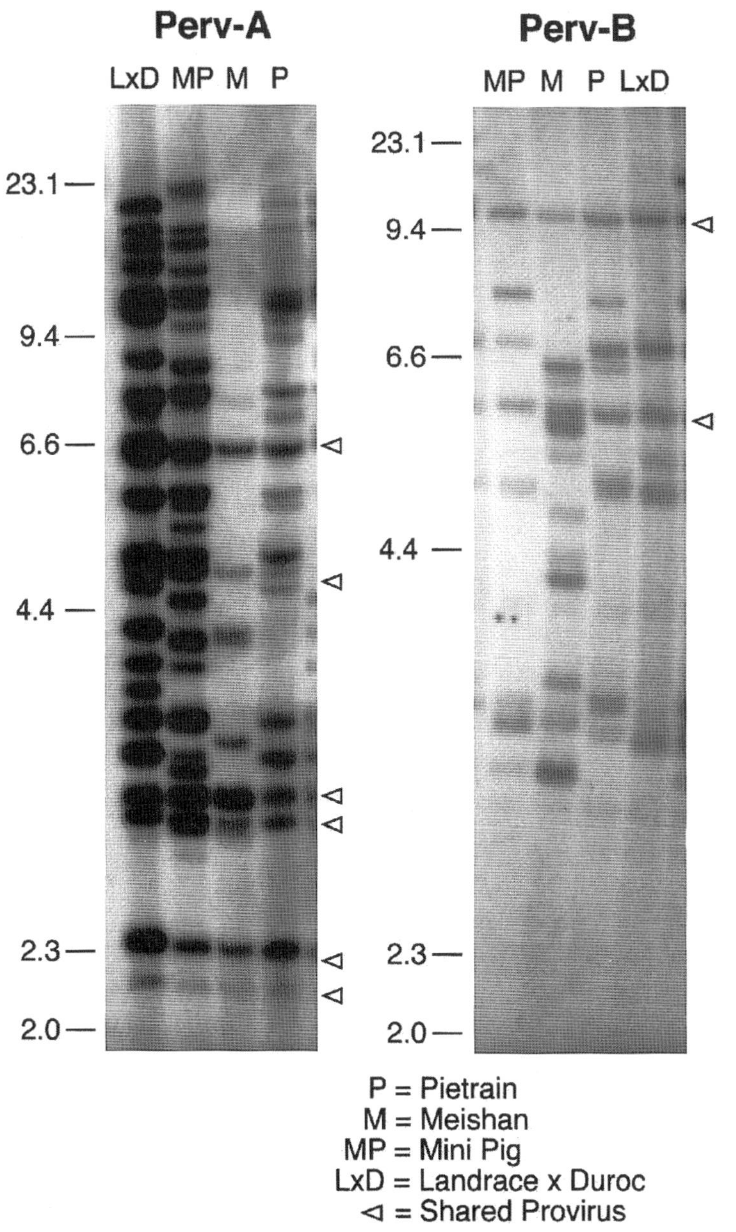

Fig. 3. Southern analysis of the number and genetic distribution of PERV viruses. PERV-A, *Nsi*I-digested genomic DNA probed with a PERV-A-specific probe; PERV-B, *Bam*HI-digested genomic DNA probed with a PERV-B-specific probe. DNA samples: LxD, Landrace X Duroc F1; M, Meishan; P, Pietrain; MP, miniswine. Arrowheads indicate apparently equivalent proviruses. Reprinted by permission from *Nature* (Le Tissier *et al.*, 1997; **389**, 681–682) copyright 1997 Macmillan Magazines Ltd. See original paper for descriptions of the probes.

proviruses. However, this is not necessarily the end of the story. By analogy with murine and avian systems, it is highly likely that the majority of proviruses detected do not encode infectious proviruses. Although not a trivial undertaking, it should be possible to identify infectious proviruses (by molecular cloning followed by transfection assays) and consider breeding them out. Even if these proviruses are not polymorphic, specific gene knock-outs in cell culture followed by pig cloning using nuclear transfer (Onishi *et al.*, 2000; Polejaeva *et al.*, 2000) would appear feasible and could be attempted if the need were considered great enough.

EXPRESSION OF PERVs

To examine the possibility of xenotransplantation-associated infection, a number of investigators have looked for PERV RNA expression in fresh porcine tissues by RT-PCR and Northern hybridization analysis. A variety of tissues, including several that are candidates for transplantation, apparently express PERV-A and PERV-B mRNA (Takeuchi *et al.*, 1998). Infectious virus production from stimulated porcine lymphocytes has also been detected (Wilson *et al.*, 1998). Interestingly, the recovered virus appeared to be a PERV-A/PERV-C recombinant (Wilson *et al.*, 2000). Primary aortal endothelial cells produce both PERV-A and PERV-C (Martin *et al.*, 1998a). By contrast, foetal neural cells, which have already been used in clinical trials for the treatment of Parkinson's disease, are negative (Schumacher *et al.*, 2000).

One concern for the safety of xenotransplantation is whether these patterns of expression will remain stable following transplantation. Although the study of endogenous proviral expression has attracted considerable attention over the past 30 years (Boeke & Stoye, 1997), the identification of factors controlling the expression of specific proviruses remains problematic. The sequence of the viral promoter is clearly important, as are *cis*-acting flanking sequences, but global host controls such as methylation or differentiation specific *trans*-acting factors also appear to play a role. Thus in the transplant setting, where porcine cells may be subject to a range of novel stimuli, the expression patterns of specific proviruses may change. The extent of this potential problem remains to be evaluated.

IN VIVO REPLICATION OF PERVs

Taken together, the data described in the preceding three sections make it highly likely that most, if not all, porcine xenotransplants will show the spontaneous expression of one or more human tropic retroviruses. But will they give rise to infection *in vivo*? To try to answer this question, a number of groups have looked for signs of PERV infection by very sensitive PCR and serological techniques in individuals exposed to living porcine tissue. To date, no evidence demonstrating virus transmission has been reported in over 180 patients (Table 2).

Table 2. Patients exposed to living pig tissues

Number	Procedure	Reference
100	Extracorporeal splenic perfusion through pig spleen	Paradis *et al.* (1999)
28	Extracorporeal perfusion through bioartificial liver	Paradis *et al.* (1999); Pitkin & Mullon (1999)
15	Pig skin grafts	Paradis *et al.* (1999)
14	Pancreatic islet cell transplantation	Paradis *et al.* (1999); Heneine *et al.* (1998)
11	Pig foetal mesencephalic tissue transplant	Schumacher *et al.* (2000)
2	Extracorporeal kidney perfusion	Paradis *et al.* (1999); Patience *et al.* (1998)
1	Extracorporeal liver perfusion	Paradis *et al.* (1999)

One major practical complication of such studies is to distinguish the true products of new viral infections from the endogenous proviruses present in any surviving pig cells. Survival of low numbers of transplanted donor cells in the recipient is a phenomenon known as microchimerism. Very low level positive signals for PERV were observed in a significant fraction of the patients whose blood was perfused through pig spleens (Paradis *et al.*, 1999). In all cases where there was sufficient DNA for testing, porcine centromeric or mitochondrial DNA could be detected, implying the presence of microchimerism. Formally, very low levels of non-productive infection accompanied by microchimerism cannot be ruled out; however, negative results in follow-up samples from these patients argue against a spreading infection. As an alternative approach, serology was used to try to detect immune responses to PERVs. Four out of 160 patients gave positive results on Western blots against Gag but not Env antigens, a figure which is similar to that seen with control populations. Two of the four were positive in pretreatment samples; such samples were not available from the other two (K. Paradis. Presentation to the Xenotransplantation Advisory Sub-Committee of the Food and Drug Administration, Washington, January 2000). Taken together, these data argue against viral infections in these patients.

These results appear reassuring, but should not be regarded as an immediate green light for all xenotransplantation procedures. The risk of virus transmission is likely to depend on the degree of immunosuppression in recipients as well as the nature of the transplant and extent of exposure. These factors will vary from procedure to procedure and were poorly quantified in the majority of these retrospective studies. It would certainly seem unwise to base risk/benefit analyses for the most risky procedures on some of the cases described above. None of these cases involved the use of transgenic pigs carrying regulators of human complement; since a mechanism akin to hyperacute rejection efficiently lyses viruses carrying the α-Gal epitope (Takeuchi *et al.*, 1996), the transmission of PERVs to humans may be significantly enhanced using transgenic pigs. It should also be remembered that the AIDS epidemic may well have its origin in a tiny number of cross-species transmissions (Gao *et al.*, 1999); thus even a PERV transmission rate of one in a thousand or lower may be of concern if spread from the initial recipient is reasonably efficient.

PATHOGENICITY OF ENDOGENOUS RETROVIRUSES

A number of exogenous gammaretroviruses are pathogenic; examples include viruses capable of causing malignancies, immunodeficiencies and neurological diseases (Rosenberg & Jolicoeur, 1997). Typically, these diseases involve extensive virus replication and many are characterized by a long latent period. If virus replication is blocked by effective immune responses, disease is usually prevented. Thus many viruses will only induce disease when animals are exposed as neonates, but not when infected in later life. Altering the immune status of adult animals may also influence the result of infection (Donahue *et al.*, 1992).

Relatively few studies to examine the pathogenesis of endogenous viruses in other species have been performed. Such studies have yielded mainly negative results (Levy, 1978), but there is now considerable interest in asking whether the PERVs have pathogenic potential. Unfortunately from the point of view of such studies, non-human primates do not appear susceptible to productive infection (Martin *et al.*, 1998b), perhaps because they lack the appropriate receptor proteins (Takeuchi *et al.*, 1998). Meaningful safety trials cannot therefore be combined with efficacy testing. Other systems remain to be explored. There are preliminary observations that PERV-B replication will occur in adult guinea pigs (D. Onions. Presentation to the Xenotransplantation Advisory Sub-Committee of the Food and Drug Administration, Washington, January 2000) and it will be extremely interesting to examine the sites of viral replication as well as effects of immunosuppression in these animals. Other investigators have initiated a variety of small animal studies, but it is still too early to draw any conclusions from these experiments.

RISK ASSESSMENT

Given the presence of inherited retroviruses in source tissues, xenotransplantation will always carry with it some risk of cross-species transmission. Though the chance of this hazard being expressed may be vanishingly small, it will never be zero. If political imperatives demand that procedures be risk-free, xenotransplantation can have little future. If, however, we are willing to accept a limited degree of risk, as we must be to introduce any new medical procedure, then major advances in the treatment of a number of conditions appear possible. Provided we know the potential enemy, and steps can be taken to manage the threat, then it does not seem unreasonable to proceed, but with due caution. Protocol-by-protocol risk/benefit analyses are necessary and great care must be taken not to extrapolate from one procedure to another.

Certain cell therapies where either virus production does not take place, or infection can be prevented effectively using semi-permeable membranes (Nyberg *et al.*, 2000), appear to offer little or no significant threat. Clinical trials are on-going and evidence for or against clinical efficacy will hopefully become available in the not too distant future. Other procedures, including organ transplantation, still appear more risky and evidence for or against the possible transmission of PERVs is still lacking. It is to be hoped that during the time required for developing an acceptable solution to the problem of acute vascular rejection, more can be learned about the likelihood and consequences of PERV transmission.

In most cases (but with a few notable exceptions such as islet treatment of diabetics), a low chance of retrovirally acquired disease several years after transplantation would not contribute significantly to the risk/benefit analysis for the individual patient. The great concern in all cases is the possibility of onward transmission into the general human population. The limited growth potential of the current form of PERVs is reassuring in this regard. The great imponderable is whether some change in the virus could occur, by either mutation or recombination with another endogenous or exogenous retrovirus, to give rise to a virus with enhanced replicative potential. Since retrovirus genetic change requires active virus replication, such events appear unlikely but can never be formally ruled out. Further investigations in this area would appear to be indicated.

REFERENCES

Akiyoshi, D. E., Denaro, M., Zhu, H., Greenstein, J. L., Banerjee, P. & Fishman, J. A. (1998). Identification of a full-length cDNA for an endogenous retrovirus of miniature swine. *J Virol* **72**, 4503–4507.

Armstrong, J. A., Porterfield, J. S. & de Madrid, A. T. (1971). C-type virus particles in pig kidney cell lines. *J Gen Virol* **10**, 195–198.

Astrin, S. M., Buss, E. G. & Hayward, W. S. (1979). Endogenous viral genes are non-essential in the chicken. *Nature* **282**, 339–341.

Bach, F. H. (1998). Xenotransplantation: problems and prospects. *Annu Rev Med* **49**, 301–310.

Battini, J. L., Danos, O. & Heard, J. M. (1995). Receptor-binding domain of murine leukemia virus envelope glycoproteins. *J Virol* **69**, 713–719.

Bhatti, F. N., Schmoeckel, M., Zaidi, A., Cozzi, E., Chavez, G., Goddard, M., Dunning, J. J., Wallwork, J. & White, D. J. (1999). Three-month survival of HDAF transgenic pig hearts transplanted into primates. *Transplant Proc* **31**, 958.

Boeke, J. D. & Stoye, J. P. (1997). Retrotransposons, endogenous retroviruses, and the evolution of retroelements. In *Retroviruses*, pp. 343–435. Edited by J. M. Coffin, S. H. Hughes & H. E. Varmus. Cold Spring Harbor, NY: Cold Spring Harbor Laboratory.

Coffin, J. M. (1996). Retroviridae: the viruses and their replication. In *Fields Virology*, 3rd edn, pp. 1767–1847. Edited by B. N. Fields, D. M. Knipe, P. M. Howley, R. M. Chanock, J. L. Melnick, T. P. Monath, B. Roizman & S. E. Strauss. Philadelphia: Lippincott–Raven.

Cooper, D. K. C. & Lanza, R. P. (2000). *Xeno: The Promise of Transplanting Animal Organs into Humans*. Oxford & New York: Oxford University Press.

Cozzi, E., Tucker, A. W., Langford, G. A. & 9 other authors (1997). Characterization of pigs transgenic for human decay-accelerating factor. *Transplantation* **64**, 1383–1392.

Czauderna, F., Fischer, N., Boller, K., Kurth, R. & Tönjes, R. R. (2000). Establishment and characterization of molecular clones of porcine endogenous retroviruses replicating on human cells. *J Virol* **74**, 4028–4038.

Donahue, R. E., Kessler, S. W., Bodine, D. & 10 other authors (1992). Helper virus induced T cell lymphoma in nonhuman primates after retroviral mediated gene transfer. *J Exp Med* **176**, 1125–1135.

Fishman, J. A. (1994). Miniature swine as organ donors for man: strategies for prevention of xenotransplant-associated infections. *Xenotransplantation* **1**, 47–57.

Fishman, J. A. (1998). Infection and xenotransplantation: developing strategies to minimize risk. *Ann N Y Acad Sci* **862**, 52–66.

Frankel, W. N., Stoye, J. P., Taylor, B. A. & Coffin, J. M. (1990). A genetic linkage map of endogenous murine leukemia viruses. *Genetics* **124**, 221–236.

Frazier, M. E. (1985). Evidence for retrovirus in minature swine with radiation-induced leukemia or metaplasia. *Arch Virol* **83**, 83–97.

Galbraith, D. N., Haworth, C., Lees, G. M. & Smith, K. T. (1997). Q-One Biotech Ltd, Immuntran Ltd. *International Patent Application. WO 97/40167*.

Gao, F., Bailes, E., Robertson, D. L. & 9 other authors (1999). Origin of HIV-1 in the chimpanzee *Pan troglodytes troglodytes*. *Nature* **397**, 436–441.

Goodchild, N. L., Wilkinson, D. A. & Mager, D. L. (1993). Recent evolutionary expansion of a subfamily of RTVL-H human endogenous retrovirus-like elements. *Virology* **196**, 778–788.

Hammer, C. (1998). Physiological obstacles after xenotransplantation. *Ann N Y Acad Sci* **862**, 19–27.

Hanger, J. J., Bromham, L. D., McKee, J. J., O'Brein, T. M. & Robinson, W. F. (2000). The nucleotide sequence of koala (*Phascolarctos cinereus*) retrovirus: a novel type C endogenous virus related to Gibbon Ape Leukemia Virus. *J Virol* **74**, 4264–4272.

Heneine, W., Tibell, A., Switzer, W. M., Sandstrom, P., Vazquez Rosales, G., Mathews, A., Korsgren, O., Chapman, L. E., Folks, T. M. & Groth, C. G. (1998). No evidence of infection with porcine endogenous retrovirus in recipients of porcine islet-cell xenografts. *Lancet* **352**, 695–699.

Kozak, C. & Silver, J. (1985). The transmission and activation of endogenous retroviral genomes. *Trends Genet* **1**, 331–334.

van der Kuyl, A. C., Dekker, J. T. & Gouldsmit, J. (1996). Baboon endogenous virus evolution and ecology. *Trends Microbiol* **4**, 455–459.

Le Tissier, P., Stoye, J. P., Takeuchi, Y., Patience, C. & Weiss, R. A. (1997). Two sets of human-tropic pig retrovirus. *Nature* **389**, 681–682.

Levy, J. A. (1978). Xenotropic type C retroviruses. *Curr Top Microbiol Immunol* **79**, 111–213.

Lieber, M. M., Sherr, C. J., Benveniste, R. E. & Todaro, G. J. (1975a). Biologic and immunologic properties of porcine type C viruses. *Virology* **66**, 616–619.

Lieber, M. M., Sherr, C. J., Todaro, G. J., Benveniste, R. E., Callahan, R. & Coon, H. G. (1975b). Isolation from the asian mouse *Mus caroli* of an endogenous type C virus related to infectious primate type C viruses. *Proc Natl Acad Sci U S A* **72**, 2315–2319.

Lin, S. S., Kooyman, D. L., Daniels, L. J. & 11 other authors (1997). The role of natural anti-Galα1-3Gal antibodies in hyperacute rejection of pig-to-baboon cardiac xenotransplants. *Transpl Immunol* **5**, 212–218.

Marin, M., Tailor, C. S., Nouri, A., Kozak, S. L. & Kabat, D. (1999). Polymorphisms of the cell surface receptor control mouse susceptibilities to xenotropic and polytropic leukemia viruses. *J Virol* **73**, 9362–9368.

Martin, J., Herniou, E., Cook, J. & Tristem, M. (1999). Interclass transmission and phyletic host range tracking in murine leukemia virus-related retroviruses. *J Virol* **73**, 2442–2449.

Martin, U., Kiessig, V., Blusch, J. H., Haverich, A., von der Helm, K., Herden, T. & Steinhoff, G. (1998a). Expression of pig endogenous retrovirus by primary porcine endothelial cells and infection of human cells. *Lancet* **352**, 692–694.

Martin, U., Steinhoff, G., Kiessig, V., Chikobava, M., Anssar, M., Morschheuser, T., Lapin, B. & Haverich, A. (1998b). Porcine endogenous retrovirus (PERV) was not transmitted from transplanted porcine endothelial cells to baboons in vivo. *Transpl Int* **11**, 247–251.

Michaels, M. G. & Simmons, R. L. (1994). Xenotransplant-associated zoonoses. *Transplantation* **57**, 1–7.

Nyberg, S. L., Hibbs, J. R., Hardin, J. A., Germer, J. J., Platt, J. L., Paya, C. V. & Weisner, R. H. (2000). Influence of human fulminant hepatic failure sera on endogenous retroviral expression in pig hepatocytes. *Liver Transplant* **6**, 76–84.

Onishi, A., Iwamoto, M., Akita, T., Mikawa, S., Takeda, K., Awata, T., Hanada, H. & Perry, A. C. (2000). Pig cloning by microinjection of fetal fibroblast nuclei. *Science* **289**, 1188–1190.

Paradis, K., Langford, G., Long, Z., Heneine, W., Sandstrom, P., Switzer, W. M., Chapman, L. E., Lockey, C., Onions, D. & Otto, E. (1999). Search for cross-species transmission of porcine endogenous retrovirus in patients treated with living pig tissue. *Science* **285**, 1236–1241.

Patience, C., Takeuchi, Y. & Weiss, R. A. (1997). Infection of human cells by an endogenous retrovirus of pigs. *Nat Med* **3**, 282–286.

Patience, C., Patton, G. S., Takeuchi, Y., Weiss, R. A., McClure, M. O., Rydberg, L. & Breimer, M. E. (1998). No evidence of pig DNA or retroviral infection in patients with short-term extracorporeal connection to pig kidneys. *Lancet* **352**, 699–701.

Patience, C., Switzer, W. M., Takeuchi, Y., Griffiths, D. J., Goward, M. E., Heneine, W., Stoye, J. P. & Weiss, R. A. (2001). Multiple groups of retroviral genomes in pigs and related species. *J Virol* (in press).

Pitkin, Z. & Mullon, C. (1999). Evidence of absence of porcine endogenous retrovirus (PERV) infection in patients treated with a bioartificial liver support system. *Artif Organs* **23**, 829–833.

Platt, J. L. & Lin, S. S. (1998). The future promises of xenotransplantation. *Ann N Y Acad Sci* **862**, 5–18.

Polejaeva, I. A., Chen, S.-H., Vaught, T. D. & 9 other authors (2000). Cloned pigs produced by nuclear transfer from adult somatic cells. *Nature* **407**, 405–409.

Rabson, A. B. & Graves, B. J. (1997). Synthesis and processing of viral RNA. In *Retroviruses*, pp. 205–262. Edited by J. M. Coffin, S. H. Hughes & H. E. Varmus. Cold Spring Harbor, NY: Cold Spring Harbor Laboratory.

Rogel-Gaillard, C., Bourgeaux, N., Billault, A., Vaiman, M. & Chardon, P. (1999). Construction of a swine BAC library: application to the characterization and mapping of porcine type C endoviral elements. *Cytogenet Cell Genet* **85**, 205–211.

Rosenberg, N. & Jolicoeur, P. (1997). Retroviral pathogenesis. In *Retroviruses*, pp. 475–585. Edited by J. M. Coffin, S. H. Hughes & H. E. Varmus. Cold Spring Harbor, NY: Cold Spring Harbor Laboratory.

Schmoeckel, M., Bhatti, F. N. K., Zaidi, A., Cozzi, E., Chavez, G., Wallwork, J., White, D. J. G. & Friend, P. J. (1999). Splenectomy improves survival of HDAF transgenic pig kidneys. *Transplant Proc* **31**, 961.

Schumacher, J. M., Ellias, S. A., Palmer, E. P. & 12 other authors (2000). Transplantation of embryonic porcine mesencephalic tissue in patients with PD. *Neurology* **54**, 1042–1050.

Shih, A., Misra, R. & Rush, M. G. (1989). Detection of multiple, novel reverse transcriptase coding sequences in human nucleic acids: relation to primate retroviruses. *J Virol* **63**, 64–75.

Smith, D. M. (1993). Endogenous retroviruses in xenografts. *N Engl J Med* **328**, 142–143.

Stoye, J. P. (1998). No clear answers on safety of pigs as tissue donor source. *Lancet* **352**, 666–667.

Stoye, J. P. & Coffin, J. M. (1987). The four classes of endogenous murine leukemia viruses: structural relationships and potential for recombination. *J Virol* **61**, 2659–2669.

Stoye, J. P. & Coffin, J. M. (1995). The dangers of xenotransplantation. *Nat Med* **1**, 1100.

Stoye, J. P., Le Tissier, P., Takeuchi, Y., Patience, C. & Weiss, R. A. (1998). Endogenous retroviruses: a potential problem for xenotransplantation? *Ann N Y Acad Sci* **862**, 67–74.

Strandström, H., Veijalainen, P., Moennig, V., Hunsmann, G., Schwartz, H. & Schäfer, W. (1974). C-type particles produced by a permanent cell line from a leukemic pig. I. Origin and properties of the host cells and some evidence for the occurrence of C-type-like particles. *Virology* **57**, 175–178.

Suzuka, I., Sekiguchi, K. & Kodama, M. (1985). Some characteristics of a porcine retrovirus from a cell line derived from swine malignant lymphomas. *FEBS Lett* **183**, 124–128.

Suzuka, I., Shimizu, N., Sekiguchi, K., Hoshino, H., Kodama, M. & Shimotohno, K. (1986). Molecular cloning of unintegrated closed circular DNA of porcine retrovirus. *FEBS Lett* **198**, 339–343.

Takeuchi, Y., Porter, C. D., Strahan, K. M., Preece, A. F., Gustafsson, K., Cosset, F.-L., Weiss, R. A. & Collins, M. K. L. (1996). Sensitization of cells and retroviruses to human serum by (α1-3) galactosyltransferase. *Nature* **379**, 85–88.

Takeuchi, Y., Patience, C., Magre, S., Weiss, R. A., Banerjee, P. J., Le Tissier, P. & Stoye, J. P. (1998). Host-range and interference studies on three classes of pig endogenous retrovirus. *J Virol* **72**, 9986–9991.

Teich, N. M., Wyke, J., Mak, T., Bernstein, A. & Hardy, W. (1982). Pathogenesis of retrovirus-induced disease. In *Molecular Biology of Tumour Viruses*, 2nd edn, *RNA Tumour Viruses*, pp. 785–998. Edited by R. Weiss, N. Teich, H. Varmus & J. M. Coffin. Cold Spring Harbor, NY: Cold Spring Harbor Laboratory.

Todaro, G. J., Benveniste, R. E., Lieber, M. M. & Sherr, C. J. (1974). Characterization of a type C virus released from the porcine line PK(15). *Virology* **58**, 65–74.

Tralka, T. S., Yee, C. L., Rabson, A. B., Wivel, N. A., Stromberg, K. J., Rabson, A. S. & Costa, J. C. (1983). Murine type C retroviruses and intracisternal A-particles in human tumours serially passaged in nude mice. *J Natl Cancer Inst* **71**, 591–599.

Tristem, M., Kabat, P., Lieberman, L., Linde, S., Karpas, A. & Hill, F. (1996). Characterization of a novel murine leukemia virus-related subgroup within mammals. *J Virol* **70**, 8241–8246.

Wilson, C. A., Wong, S., Muller, J., Davidson, C. E., Rose, T. M. & Burd, P. (1998). Type C retrovirus released from porcine primary peripheral blood mononuclear cells infects human cells. *J Virol* **72**, 3082–3087.

Wilson, C. A., Wong, S., VanBrocklin, M. & Federspiel, M. J. (2000). Extended analysis of the in vitro tropism of porcine endogenous retrovirus. *J Virol* **74**, 49–56.

Gammaherpesviral infections and neoplasia in immunocompromised populations

Chris Boshoff

Departments of Oncology and Molecular Pathology, The CRC Viral Oncology Group, Wolfson Institute for Biomedical Research, Cruciform Building, Gower Street, University College London, London WC1E 6BT, UK

INTRODUCTION

A substantial burden of human cancer worldwide is attributable to infection. Viral infections account for approximately 15 % of all human cancers. Cervical cancer, hepatocellular carcinoma and Kaposi's sarcoma, all caused by viruses, are some of the most important tumours in sub-Saharan Africa.

The interface between infection and malignancy is highlighted by the cancers prevalent in immunocompromised patients (Table 1). Human immunodeficiency virus (HIV)-infected individuals are specifically prone to cancer caused by viruses, e.g. Epstein–Barr virus (lymphomas), papillomaviruses (squamous carcinomas of skin and ano-genital carcinoma) and Kaposi's sarcoma-associated herpesvirus (lymphomas, multicentric Castleman's disease and Kaposi's sarcoma). The tumours increased in HIV-infected individuals where a virus has not yet been described (Table 1) may nevertheless have a viral aetiology. Carcinogenesis is a multifactorial process and not all persons infected with oncogenic viruses will develop a cancer: in fact, only a fraction of infected individuals will develop a tumour, particularly in the absence of immunosuppression. Certain tumours with a known viral aetiology (e.g. nasopharyngeal and hepatocellular carcinoma) are not increased in AIDS, indicating that viral infection and immunosuppression, without co-factors, are not enough to precipitate those specific cancers. Efficacious immunization against the primary infection would virtually eliminate the occurrence of the tumour with which the virus is associated. A successful vaccine against hepatitis B is available, whereas vaccines against Epstein–Barr virus and human papillomaviruses are currently in clinical studies.

SGM symposium 60: New challenges to health: the threat of virus infection.
Editors G. L. Smith, W. L. Irving, J. W. McCauley, D. J. Rowlands. Cambridge University Press. ISBN 0 521 80614 3 ©SGM 2001.

Table 1. Tumours increased in patients with AIDS

KSHV, Kaposi's sarcoma-associated herpesvirus; EBV, Epstein–Barr virus; HPV, human papillomavirus.

Cancer type	Relative risk*	Viral link
Kaposi's sarcoma	310.0	KSHV
Non-Hodgkin's lymphoma	113.0	EBV
Angiosarcoma	36.7	?
Anal cancer†	31.7	HPV
Leukaemias other than lymphoid and myeloid	11.0	?
Hodgkin's disease	7.6	?EBV
Leiomyosarcoma and other soft tissue sarcomas‡	7.2	?EBV
Multiple myeloma	4.5	?
Primary brain cancer§	3.5	Polyomaviruses
Testicular seminoma or malignant germinoma	2.9	?

*Adapted from Goedert *et al.* (1998).

†Anal carcinoma is increased in homosexual men, even prior to HIV.

‡EBV sequences described in smooth-muscle tumours from children with AIDS, including monoclonal EBV episomes in some cases.

§Predominantly malignant glioma or astrocytoma. The detection of papovavirus sequences in brain tumours remains unconfirmed.

This review will focus on the neoplasia associated with the gammaherpesviruses Epstein–Barr virus (EBV) and Kaposi's sarcoma-associated herpesvirus (KSHV).

GAMMAHERPESVIRUSES

'herpein' Greek, to creep or crawl ... refers to the characteristic skin lesions caused by herpes simplex and herpes zoster infections.

O'er ladies lips, who straight on kisses dream,
Which oft the angry Mab with blisters plagues,
Because their breaths with sweetmeats tainted are.
Shakespeare
Romeo and Juliet c. 1595

Nearly 100 herpesviruses have been identified and almost all mammalian species have been shown to be infected by at least one member of the family. The known herpesviruses share a common virion architecture and four critical biological properties. (1) All herpesviruses encode a large variety of enzymes involved in nucleic acid metabolism,

DNA synthesis and protein processing. (2) The synthesis of viral DNA and assembly of the capsid occur mainly in the nucleus of infected cells. (3) Production of infectious virus progeny is generally accompanied by destruction of the infected cell (lytic infection). (4) Herpesviruses have the ability to become latent and persist for life in their hosts. Latent infection occurs in specific cell types which vary between the different viruses. The latent viral genomes take the form of circular episomes with only a fraction of viral genes expressed.

The subfamily *Gammaherpesvirinae* includes the genera *Lymphocryptovirus* and *Rhadinovirus*. Viruses of this subfamily are large double-stranded DNA viruses which infect and persist in lymphocytes. They are also characterized by their capacity to induce cell proliferation *in vivo*, resulting in transient or chronic lymphoproliferative disorders. Gammaherpesviruses are widely disseminated in nature, causing infection and disease in many species (Fig. 1).

EPSTEIN–BARR VIRUS

Epstein–Barr virus (EBV) is the prototype of gammaherpesviruses. *In vitro*, EBV infection of primary B lymphocytes efficiently induces continuous proliferation and transformation into permanent cell lines (called lymphoblastoid cell lines or LCLs). These LCLs normally contain multiple episomal copies of the EBV genome (Adams & Lindahl, 1975). Of the approximately 100 viral genes, only 13 genes are expressed in LCLs, including those encoding the six nuclear proteins (EBNAs 1–6), three membrane proteins (LMPs 1–3) and two non-translated RNAs (EBER 1 and 2) (reviewed by Rickinson & Kieff, 1996). These transformation-associated viral proteins regulate the maintenance of episomal viral DNA and viral gene expression, drive cellular proliferation directly and via the transactivation of cellular oncogenes, and block apoptosis.

EBV-infected B lymphocytes are highly immunogenic and elicit powerful cytotoxic T-cell (CTL) responses (Rickinson & Moss, 1997). EBV-specific CTLs are targeted against human leukocyte antigen (HLA) class 1 associated peptides derived from the EBNAs and the latent membrane proteins (LMPs), with the notable exception of EBNA-1. The choice of the viral target depends on the HLA phenotype of the responder (de Campos-Lima *et al.*, 1993). A sufficient variety of immunogenic peptides can be presented by the HLA spectrum to provide immunosurveillance against uncontrolled growth of virally transformed immunoblasts.

Although EBV is a highly transforming virus, only a fraction of EBV-infected individuals will actually develop EBV-associated tumours, and despite the interactions between autologous EBV-infected B cells and $CD8^+$ T lymphocytes, EBV persists in

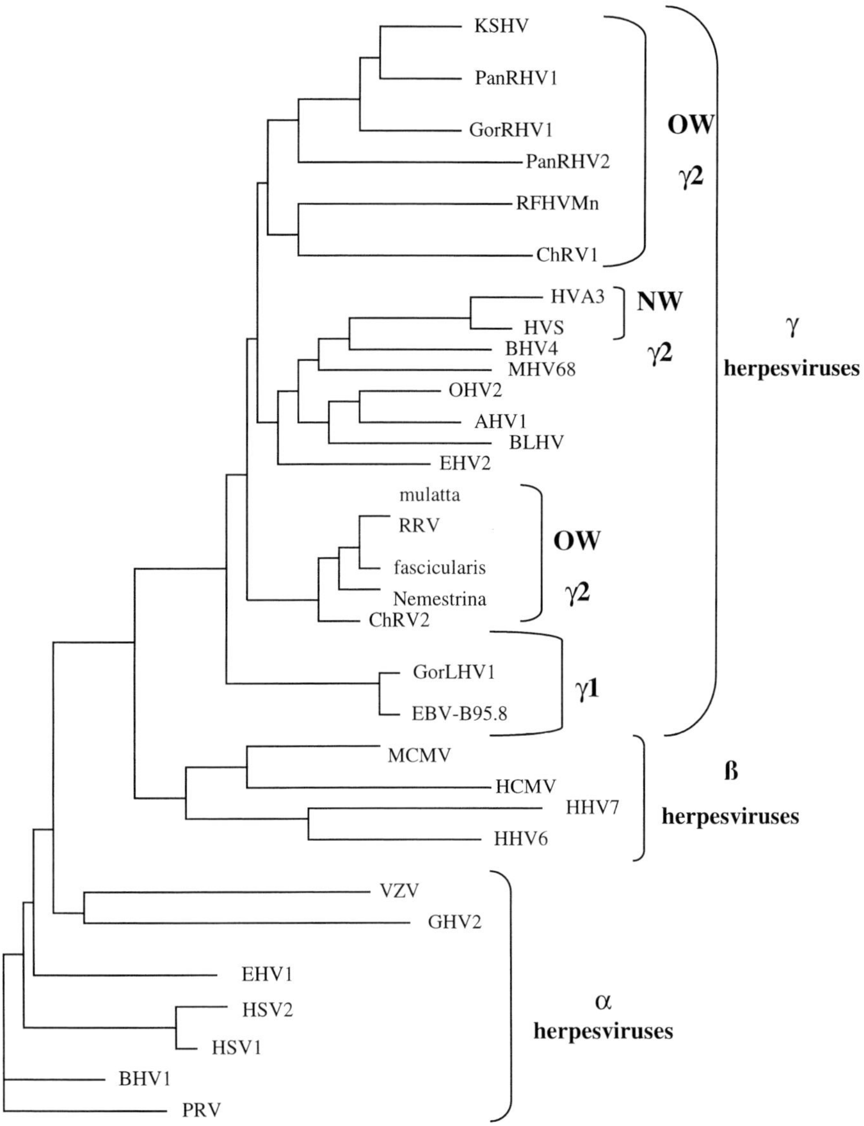

Fig. 1. Expanding family of herpesviruses (adapted from Lacoste *et al.*, 2000). Phylogenetic tree of related herpesviruses: HHV, human herpesvirus; HVS, herpesvirus saimiri; RRV, rhadino rhesus herpesvirus; RFHV, retroperitoneal fibrosis herpesvirus (macaques); BHV, bovine herpesvirus; AHV, alcelaphine herpesvirus; GorRHV, gorilla rhadinoherpesvirus; ChRV, African green monkey rhadinovirus; PanRHV, pan rhadinoherpesvirus (chimpanzees).

B lymphocytes. These two apparent paradoxes are solved by the down-regulation of all growth-transforming-associated viral proteins, which include those known to elicit CTL responses, in persistently infected B lymphocytes (Klein, 1994). Only EBNA-1 (EBV nuclear antigen 1) is expressed in these cells. EBNA-1 is essential to maintain the stability and proliferation of the viral episome (unintegrated viral DNA) (Yates *et al.*, 1985) and does not evoke an immune response.

The other EBNAs and the LMPs are required for the transcriptional regulation, growth transformation and immortalization of infected B cells (Rickinson & Kieff, 1996). EBNA-1 is required for the controlled replication of the EBV genome and also *trans*-activates other viral latency-specific EBV promoters. EBNA-1 is expressed in all known latently infected cells, irrespective of the cellular phenotype. EBNA-2 transcriptionally *trans*-activates the viral LMP-1 and 2A as well as cellular genes, such as the protooncogene c-*fgr* and the gene encoding the B-cell activation antigen CD23, that are believed to play a role in EBV-induced B-cell growth transformation. EBNAs 3, 4 and 6 (also called EBNAs 3A, 3B and 3C) are encoded by three genes that are tandemly placed in the EBV genome (Rickinson & Kieff, 1996). Because of their sequence similarity these proteins are likely to have similar functions in latent EBV infection and transformation. LMP-1 activates human B lymphocytes, can protect LCLs from apoptosis through the induction of bcl-2 and is defined as a viral oncogene, because it can transform cultured rodent fibroblasts and induce tumours in SCID mice.

Diseases in the immunocompetent host

Burkitt's lymphoma (BL). The recognition of BL and of its associations with malaria and EBV infection is one of the great achievements of 20th century medicine. Post World War II, the Irish surgeon Denis Burkitt returned to Africa and studied a peculiar lymphoma common in African children (Burkitt, 1958). He noticed the specific geographical distribution. His lecture tours on this topic brought him into contact with Anthony Epstein (at the Middlesex Hospital), who suspected an infective agent was the culprit. Epstein volunteered to look for viruses from BL cell lines by electron microscopy and discovered the new herpesvirus (Epstein *et al.*, 1964, 1965).

Only EBNA-1 is expressed in BL cells (Klein, 1994). These cells thus represent the resting persistently infected B cells in the circulation. However, all BL cells are in cell cycle and rapidly proliferating due to the universally present translocation of the c-*myc* gene (Bernheim *et al.*, 1981; Manolov & Manolova, 1972). This translocation brings c-*myc* under the control of an immunoglobulin gene locus, and it is no longer down-regulated when proliferating cells enter a resting phase. This translocation 'accident' may therefore switch-on resting EBNA-1-expressing B cells to start proliferating.

The distribution of BL in Africa coincides with the distribution of hyperendemic malaria. The association is further supported by the fact that the incidence of both diseases decreases with chloroquine prophylaxis. The immune dysregulation caused by malaria, including chronic Th2 stimulation of B cells and chronic antigen exposure, contributes to B-cell hyperplasia. Chronic B lymphocyte hyperplasia increases the target cell population for aberrant recombinations. Malaria-triggered B-cell hyperplasia with subsequent c-*myc* immunoglobulin juxtaposition in conjunction with EBV-driven cell proliferation may be enough to precipitate endemic BL in Africa. The EBNA-1 latency programme and down-regulation of MHC class I contribute to immune evasion (Rickinson & Moss, 1997).

Nasopharyngeal carcinoma (NPC). All NPCs express EBNA-1, but not the immunogenic EBNA-2 to -6 proteins. Most cells also express LMP-1 and LMP-2. LMP-1 is not always expressed, and this may reflect differences in virus expression dependent on the state of cellular differentiation. It seems that the virus enters epithelial cells from permissively infected lymphocytes trafficking through lymphoid-rich epithelium. Evidence of latent infection in normal epithelium has not been detected. The establishment of predominantly non-permissive (latent) infection in epithelial cells could be the major event leading to NPC development. The expression of latent viral genes could then promote cellular proliferation and the progression from dysplasia to invasive carcinoma. Like BL, co-factors are probably required to precipitate NPC. Animal experimentation and epidemiological data indicate that consumption of large quantities of salted fish (e.g. in South East China) is one such important co-factor (Yu & Henderson, 1987; Yu *et al.*, 1989). Epidemiological data also support other factors such as the consumption of Chinese herbal teas, exposure to domestic woodfire and possibly genetic predisposition (Yu, 1991; Zheng *et al.*, 1994).

Hodgkin's disease (HD). The finding of clonal EBV genomes in Reed–Sternberg cells and the restricted pattern of latent viral gene expression in nearly 30–50 % of all HD cases throughout the world suggest that EBV is not simply a passenger in this disease (Armstrong *et al.*, 1992). The presence of EBV in Reed–Sternberg cells correlates with an increased expression of lymphocyte activation antigens, a decreased expression of the CD20 B-cell antigen and with expression of cytokines such as interleukin 10 (IL-10) and IL-6. Other evidence for an aetiological role of EBV in HD is provided from both cohort and case-control studies in which a positive association between history of infectious mononucleosis and subsequent HD is consistently reported. The exact role of EBV and other co-factors still needs to be elucidated. There is an increased incidence of HD in patients who have previously received curative cytotoxic therapy for various cancers. These Hodgkin's lymphomas are often associated with EBV infection and

Table 2. Neoplasms associated with EBV infection and the EBV latent proteins expressed in these tumours

Tumour	EBV gene expression
Burkitt's lymphoma	EBNA-1
Nasopharyngeal carcinoma (NPC)	EBNA-1
	LMP-1
	LMP-2A*
Hodgkin's disease	EBNA-1
	LMP-1, 2A and 2B
Post-transplant lymphoproliferative disease†	EBNA-1 to -6
	LMP-1, 2A and 2B

*Not always expressed in NPC.

†EBV gene expression pattern is similar to that seen in *in vitro* EBV-transformed B lymphoblastoid cell lines.

immunological factors. Genetic damage done by cytotoxic therapies probably plays a role in the pathogenesis of these cases.

Diseases in the immunosuppressed host

Post-transplant lymphoproliferative disorder (PTLD). PTLDs represent one of the most common complications of immunosuppression following organ transplantation and occur in around 1–10 % of cases. Predisposing factors include high cumulative doses of immunosuppressive drugs and primary EBV infection at, or post-dating, transplantation. In heart and lung transplant recipients, the risk of PTLD is ~10-fold higher in those who are EBV-seronegative at the time of transplant compared to those that are seropositive. PTLDs in EBV-seronegative patients often present with a polyclonal lymphadenopathy (infectious mononucleosis-like primary infection), whereas EBV-seropositive patients present with monoclonal lymphomas in extranodal sites like gut, brain and in the grafted organ. Similar lymphoproliferative disorders also occur in some congenital immunodeficiencies, e.g. X-linked lymphoproliferative syndrome. Phenotypically, PTLDs resemble *in vitro* transformed B lymphocytes, i.e. LCLs, where the tumour cells express all the latent viral genes (Table 2). These cells are able to proliferate because of a lack of CTL responses in the immunodeficient host and these tumours may regress when immunosuppression is discontinued (Rickinson & Moss, 1997). Antiviral agents (e.g. ganciclovir and acyclovir) may be added to the treatment of PTLD; however, their role is unclear and the latent viral proteins expressed in these tumour cells are not inhibited by these agents. The treatment of PTLD remains unsatisfactory and current studies are evaluating adoptive immunotherapies. One approach is to infuse EBV-specific cytotoxic T lymphocytes (Heslop & Rooney, 1997; Khanna *et al.*, 1999). In bone marrow transplant recipients, the healthy donor can act as a source of CTL. However, in solid organ transplant recipients, either autologous CTL must be grown from a stored pretransplant blood sample or allogeneic HLA matched CTL must be used.

CD20 is frequently expressed on the tumour cells and immunotherapy against this antigen, using rituximab (anti-CD20 antibody), is an effective and well-tolerated treatment option (Kuehnle *et al.*, 2000; Zompi *et al.*, 1999).

AIDS-associated lymphomas. Both post-transplant proliferative disorder-like lymphoma and Burkitt's-type lymphoma (BL) develop in HIV-infected patients. The overall incidence is about 3 %. AIDS-BL occurs relatively early in HIV infection and shows a peak at age 10–19 years. BL accounts for ~20 % of all HIV-related lymphomas. However, only some of these contain EBV DNA. Non-BL AIDS lymphoma is referred to as large cell lymphoma and >50 % of such tumours contain EBV DNA. Nearly 100 % of AIDS central nervous system (CNS) lymphomas contain EBV, and PCR detection of EBV in the cerebrospinal fluid can help in the differential diagnosis of CNS lesions in HIV-infected patients (Antinori *et al.*, 1999; Cinque *et al.*, 1993).

KAPOSI'S SARCOMA-ASSOCIATED HERPESVIRUS

For over 100 years, Kaposi's sarcoma (KS) remained a rare curiosity to clinicians and cancer researchers, until it shot to prominence as the sentinel of what we now call AIDS. In 1872, the Hungarian dermatologist Moriz Kaposi published the case histories of five middle-aged and elderly male patients in Vienna with *idiopathic multiple pigmented sarcomas* of the skin (Kaposi, 1872).

Classic KS occurs predominantly in elderly male patients of southern European ancestry (Franceschi & Geddes, 1995). A high frequency is also seen in Israel and other Middle Eastern countries. This form of the disease is generally not as aggressive as the form originally described by Kaposi, for unknown, possibly immunological, reasons.

In some equatorial countries of Africa, KS has existed for many decades, long preceding HIV (known as *endemic KS*) (Oettle, 1962). This form is found in younger patients as well as the elderly; the male : female ratio is >3 : 1. It is generally a more aggressive disease than classic KS, though less so than African AIDS-associated KS (Bayley, 1984). In particular, endemic KS in African children is often associated with lymph node involvement only and no skin lesions.

KS is also known to develop after organ transplantation (*post-transplant or iatrogenic KS*) (Penn, 1983). Patients of Mediterranean, Jewish or Arabian ancestry are also clearly over-represented among immunosuppressed patients who develop KS after a transplant (Franceschi & Geddes, 1995).

In 1981, the US Centers for Disease Control and Prevention (CDC) became aware of an increased occurrence of two rare diseases in young homosexual men from New York City (NY, USA) and California (Service, 1981): KS and *Pneumocystis carinii* pneumonia (PCP). This was the beginning of the AIDS epidemic and *AIDS-KS* is today the most common form of KS.

Studies of AIDS case surveillance support the pre-AIDS data on the existence of a sexually transmissible KS co-factor: KS occurs predominantly in homosexual and bisexual men with AIDS, less commonly in those acquiring HIV through heterosexual contact, and rarely in AIDS patients with haemophilia or in intravenous drug users (Beral, 1991; Beral *et al.*, 1990).

Chang *et al.* (1994) employed representational difference analysis (RDA) to identify sequences of a new herpesvirus [Kaposi's sarcoma-associated herpesvirus (KSHV) or human herpesvirus-8] in AIDS-KS biopsies. RDA relies on cycles of subtractive hybridization and PCR amplification to enrich and isolate rare DNA fragments that are present in only one of two otherwise identical populations of DNA. This is a powerful technique which can detect small differences between complex genomes, and has been used to identify DNA amplifications in tumour tissues and the lack of tumour suppressor genes in cancers (Lisitsyn *et al.*, 1993, 1995).

The KSHV genome consists of a ~140 kb long unique coding region (LUR) flanked by ~800 bp non-coding tandemly repeated units. Over 85 open reading frames (ORFs) have thus far been identified, including nearly 70 with sequence similarity to those of related gammaherpesviruses (Fig. 2). Novel ORFs not present in other herpesviruses were designated K1–K15, although many of these now appear to be present in related viruses.

KSHV encodes a number of cellular homologues, including a viral cyclin, bcl-2, IL-6, interferon regulatory factor (IRFs), FLICE inhibitory protein (FLIP) and chemokine homologues (vMIP I, II and III).

ORF-K1 is used to subtype KSHV (McGeoch & Davidson, 1999; Nicholas *et al.*, 1998): subtypes A, B, C and D have been identified which display between 15 and 30 % amino acid differences between their ORF-K1 coding regions. Within these four

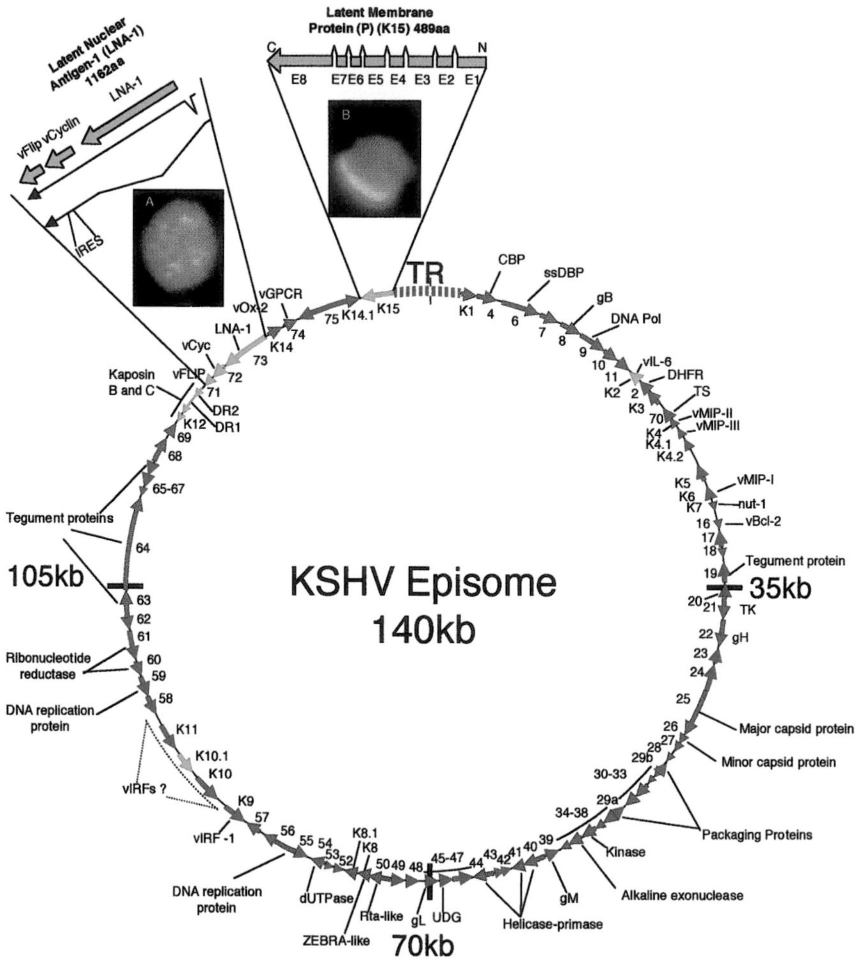

Fig. 2. Putative structure of the KSHV episome (from Sharp & Boshoff, 2000, and reproduced with permission). Red and blue arrows indicate directions of lytic transcripts; green arrows indicate latent transcripts.

subtypes, over 13 clades have now been described (Zong *et al.*, 1999). These subtypes have close associations with the geographic and ethnic background of individuals. Subtype B is found almost exclusively in patients from Africa, subtype C in individuals from the Middle East and Mediterranean Europe, subtype A in western Europe and North America and subtype D has only so far been described in individuals from the Pacific Islands (Hayward, 1999; Zong *et al.*, 1999). So far, no subtype appears to correlate with a specific disease entity or with a more aggressive course for KS. The unusually high genetic divergence identified in ORF-K1 reflects some unknown powerful biological selection process acting specifically on this immunoglobulin-receptor-like signal transducing protein (Hayward, 1999; McGeoch & Davidson, 1999). This could be related to evolving mechanisms of virus evasion from the immune system among different populations.

KSHV latent proteins

In EBV, the latent nuclear proteins (EBNAs 1–6) and latent membrane proteins (LMPs 1–3) are essential for persistence of the episomal genome, maintenance of latency, initiating lytic viral infection, evasion of, or eliciting antiviral immune responses, and driving cellular proliferation (and therefore tumorigenesis). These EBV proteins have been shown to interact with or up-regulate cellular proteins involved in transformation (including p53, pRb, cyclin D, histone deacetylase and TRAF). A number of KSHV ORFs are also transcribed during latency (Dittmer *et al.*, 1998; Sarid *et al.*, 1998).

Latent nuclear antigen (ORF 73/LNA). ORF 73 of KSHV encodes a latent immunogenic nuclear protein (LNA) detected as nuclear speckling by immunofluorescence using anti-KSHV-positive sera on primary effusion lymphoma (PEL) cells (Kedes *et al.*, 1997; Kellam *et al.*, 1997; Rainbow *et al.*, 1997). ORF 73 is transcribed with the viral cyclin (see later) and FLIP homologues (Kellam *et al.*, 1997; Sarid *et al.*, 1999; Talbot *et al.*, 1999). LNA is expressed in all KS spindle cells latently infected with KSHV, in all the plasmablasts in KSHV-associated Castleman's disease and in all cells of PEL (Dupin *et al.*, 1999). Like EBNA-1, LNA is essential to maintain the KSHV episome (extra-chromosomal persistence) (Ballestas *et al.*, 1999). Furthermore, LNA tethers KSHV DNA to chromosomes during mitosis to allow the segregation of viral episomes to all progeny cells (Ballestas *et al.*, 1999). LNA therefore maintains a stable episome during mitosis. LNA has also been shown to interact with p53 and with the retinoblastoma protein (Radkov *et al.*, 2000).

v-cyclin. Cellular cyclins are critical components of the cell cycle: cyclins are regulatory subunits of a specific class of cellular kinases. By physically associating with an inactive cyclin-dependent kinase core (cdk), cyclins lead to the formation of active kinase holoenzymes which recognize and phosphorylate an array of cellular substrate molecules. The phosphorylating activity of these holoenzymes is responsible for regulating the passage of cells through the replication cycle. Cyclins associate with their partners (the cyclin-dependent kinases, CDKs) to be fully active. The KSHV cyclin has highest sequence similarity to the cellular D-type cyclins and is expressed during latency, inferring a possible role in tumorigenesis. It forms active kinase complexes with CDK6 to phosphorylate the retinoblastoma protein (pRb) at authentic sites (Chang *et al.*, 1996). Furthermore, unlike cellular D cyclin/CDK6 complexes, KSHV-cyclin/CDK6 activity is resistant to inhibition by CDK inhibitors (CKI) p16, p21 and p27 (Swanton *et al.*, 1997).

The expression of the KSHV cyclin in latently infected spindle and PEL cells indicates a possible role in either the proliferation or the arrest of differentiation of these cells. Expression of cyclin D1, but not cyclins E or A, inhibits the differentiation of immature myoblasts (Skapek *et al.*, 1995). Human papillomavirus (HPV) E7 has also been shown to uncouple cellular differentiation and proliferation in human keratinocytes (Jones *et al.*, 1997). As KS spindle cells appear to represent undifferentiated endothelial cells (Dupin *et al.*, 1999), this role for the KSHV cyclin is an attractive hypothesis.

Serological assays

There are several KSHV serological assays currently available. The most widely used assays are based on detection of latent or lytic antigens in KSHV-infected PEL cell lines, by either immunofluorescence (IFA) (Gao *et al.*, 1996; Kedes *et al.*, 1996; Simpson *et al.*, 1996) or ELISA (Chatlynne *et al.*, 1998). Assays have also been described which detect antibodies to recombinant KSHV latent and lytic proteins or synthetic peptides. Lytic proteins shown to be immunogenic include ORF 65 (Simpson *et al.*, 1996), ORF 26 (Davis *et al.*, 1997) and ORF K8.1 (Chandran *et al.*, 1998; Raab *et al.*, 1998). The only latent antigen thus far to be used in recombinant assays is ORF 73, which is also the antigen detected in latent IFA (Kedes *et al.*, 1997; Kellam *et al.*, 1997; Rainbow *et al.*, 1997).

No current assay on its own is sensitive enough to detect all KSHV-infected individuals, but the latent IFA and commercially available ELISA tests are considered sensitive and specific enough for large serological surveys (Ablashi *et al.*, 1999; Sitas *et al.*, 1999).

Neoplasms associated with KSHV

Three neoplasms are consistently linked with KSHV infection: KS, a subtype of Castleman's disease and PEL. KSHV sequences have also been described in squamous carcinomas, multiple myeloma and other vascular tumours. Most studies, however, have not confirmed any association between KSHV and these tumours. The detection of KSHV in some of these reports might be due to PCR contamination.

Kaposi's sarcoma. Four observations link KSHV to KS (although none of these findings on their own is sufficient to support a causative role). (1) KSHV DNA is present, by PCR, in all four epidemiological forms of KS and in nearly all KS biopsies tested. However, KSHV DNA is rarely, if at all, detectable in other vascular tumours (International Agency of Cancer Research, 1997). (2) The detection of KSHV DNA by PCR in the peripheral blood of HIV-infected individuals predicts who might subsequently develop KS (Moore *et al.*, 1996; Whitby *et al.*, 1995), indicating that those at risk of KS have a higher viral load than those not at risk. (3) Seroepidemiological surveys show that general populations at risk of developing KS have a higher prevalence of KSHV infection. The incidence of classic KS and AIDS-KS in different populations correlates broadly with the prevalence of the virus in these populations (Boshoff, 1999). (4) In nodular KS lesions, KSHV is latently expressed in nearly all the tumour (spindle) cells (Dupin *et al.*, 1999; Sturzl *et al.*, 1999). This is reminiscent of other virus-driven cancers, e.g. EBV latent infection in PTLD or HPV infection in cervical cancer.

In early KS lesions, only a small proportion (<10 %) of spindle and endothelial cells are positive for KSHV (Dupin *et al.*, 1999), indicating that paracrine mechanisms are probably important in the initiation and progression of KS. In nodular lesions, >90 % of the spindle cells contain latent KSHV, suggesting that KSHV latent proteins provide a growth advantage to the infected cells (Dupin *et al.*, 1999). In culture, KSHV has been shown to infect macro- and microvascular endothelial cells and to provide a growth advantage to these cells (Flore *et al.*, 1998). With other viral oncoproteins like HPV E6 and E7, KSHV can transform human microvascular endothelium (Moses *et al.*, 1999). One current model is that KSHV, like EBV, persists in B cells as an episome expressing only a small number of latent proteins. Reactivation of the virus due to immunosuppression or local cytokine production leads to infection of endothelial precursors. In these precursors, KSHV latent proteins induce proliferation and block differentiation and apoptosis by anti-'anti-viral' cellular immune responses. It is possible that local cytokine production stimulates neo-angiogenesis or neo-lymphangiogenesis and KSHV then infects these immature endothelial cells. There is currently no evidence that KSHV infects endothelial cells in healthy individuals.

Plasmablastic variant multicentric Castleman's disease. Castleman's disease (CD) is a lymphoproliferative disorder (Castleman *et al.*, 1956). CD is more often diagnosed in HIV-infected patients. A systemic variant of CD is associated with multiple organ involvement, especially spleen and lymph nodes, with systemic symptoms such as weight loss and fever (Frizzera *et al.*, 1983). This is called multicentric Castleman's disease (MCD). MCD has been associated with increased circulating IL-6 levels.

KSHV DNA is found in some cases of MCD (Barozzi *et al.*, 1996; Corbellino *et al.*, 1996; Soulier *et al.*, 1995). KSHV is present in plasmablasts in MCD and these plasmablasts are not present in KSHV-negative MCD (Dupin *et al.*, 1999; Parravicini *et al.*, 1998). KSHV-positive plasmablasts belong to the B-cell lineage. KSHV appears also to be present in all tumour cells of plasmablastic lymphomas that develop in patients with the KSHV-positive plasmablastic variant of MCD (Dupin *et al.*, 2000). The development of plasmablastic lymphoma therefore appears to represent a further evolution of this disorder. Unlike KSHV-positive PEL cells, the plasmablasts in MCD are only positive for KSHV, and not for EBV.

Current studies suggest that KSHV-positive MCD has a poorer prognosis than the KSHV-negative cases (Chadburn *et al.*, 1997; Dupin *et al.*, 1999; Parravicini *et al.*, 1998). In contrast to KS lesions that may resolve following partial restoration of the immune system [e.g. by highly active anti-retroviral therapy (HAART) for HIV-positive patients], KSHV-positive MCD often continues to progress.

Primary effusion lymphoma (PEL). PEL is a body-cavity-based lymphoma that usually presents and persists as an effusion: pleural, pericardial or peritoneal. The lymphoma cells in these cases are negative for most lineage-associated antigens, although immunoglobulin gene rearrangement studies indicate a B-cell origin. Cesarman *et al.* (1995) found that KSHV was specifically associated with PEL, but not with other high-grade AIDS-related lymphomas.

These lymphomas occur predominantly in HIV-positive individuals with advanced stages of immunosuppression, but are occasionally seen in HIV-negative patients. Like KS, which can occur in the same patient, PEL occurs primarily in homosexual men and not in other HIV-positive risk groups.

PEL cells contain between 50 and 150 copies each of the KSHV genome. The majority of, but not all, PELs are co-infected with EBV (Cesarman *et al.*, 1995), suggesting that the two viruses may cooperate in neoplastic transformation. Terminal repeat analysis indicates that EBV is monoclonal in most cases (Cesarman *et al.*, 1996; Nador *et al.*, 1996), implying that EBV was present in tumour cells prior to clonal expansion. PEL cells consistently lack molecular defects commonly associated with neoplasia of mature B cells, including activation of the protooncogenes *bcl-2*, *bcl-6*, *n-ras* and *k-ras*, as well as mutations of *p53* (Carbone *et al.*, 1996; Nador *et al.*, 1995).

KS, MCD and PEL have all been described in a single patient, and up to 30 % of HIV-infected patients with KSHV-positive MCD will also have or develop KS.

KSHV and immunity

The introduction of aggressive anti-HIV therapies has led to a decline in the incidence of KS in AIDS patients and also in the resolution of KS in those already affected (Jacobson *et al.*, 1999). This suggests that cellular immune responses, compromised in AIDS, but recovering after HAART, could be important in the control of KSHV infection and in the development of KS. This is further supported by the observation that post-transplant KS lesions can regress when immunosuppressive treatment is stopped.

Like EBV, KSHV probably establishes a persistent infection in B cells which is normally controlled by the immune system and the number of KSHV-infected cells is under immunological control. When this immune control declines due to acquired or iatrogenic immunosuppression, the number of KSHV-infected cells increases. Subsequent unchecked proliferation of virally infected cells leads to the development of KSHV-related tumours.

KSHV, like other herpesviruses, is able to elicit HLA class I restricted CTL responses (Osman *et al.*, 1999). In one pilot study, KSHV-specific CTL responses were not present in most patients with KS, indicating that a decline in cellular immune responses against KSHV may be present in HIV-positive patients with KS and could contribute to KS pathogenesis (Osman *et al.*, 1999). This would be reminiscent of the lack of EBV-specific CTLs seen in immunosuppressed patients, which correlates with the onset of EBV-driven lymphoproliferation.

The rapid resolution of KS in some HIV-positive patients started on HAART suggests that a small improvement in immunity might be important in disease control. $CD4^+$ T helper responses, natural killer (NK) and leukocyte-activated killer cells (LAK) might also control the growth of KSHV-positive cells. If we consider murine herpesvirus-68 induced lymphoproliferation as a paradigm for KSHV pathogenesis, then $CD4^+$ lymphocytes might play a specific important role in the control of virus-driven cellular proliferation (Virgin & Speck, 1999).

REFERENCES

Ablashi, D., Chatlynne, L., Cooper, H. & 21 other authors (1999). Seroprevalence of human herpesvirus-8 (HHV-8) in countries of Southeast Asia compared to the USA, the Caribbean, and Africa. *Br J Cancer* **81**, 893–897.

Adams, A. & Lindahl, T. (1975). Epstein-Barr virus genomes with properties of circular DNA molecules in carrier cells. *Proc Natl Acad Sci U S A* **72**, 1477–1481.

Antinori, A., de Rossi, G., Ammassari, A. & 9 other authors (1999). Value of combined approach with thallium-201 single-photon emission computed tomography and Epstein-Barr virus DNA polymerase chain reaction in CSF for the diagnosis of AIDS-related primary CNS lymphoma. *J Clin Oncol* **17**, 554–560.

Armstrong, A. A., Weiss, L. M. & Gallagher, A. (1992). Criteria for the definition of EBV association in Hodgkin's disease. *Leukaemia* **6**, 869–874.

Ballestas, M. E., Chatis, P. A. & Kaye, K. M. (1999). Efficient persistence of extrachromosomal KSHV DNA mediated by latency-associated nuclear antigen. *Science* **284**, 641–644.

Barozzi, P., Luppi, M., Masini, L., Marasca, R., Savarino, M., Morselli, M., Ferrari, M. G., Bevini, M., Bonacorsi, G. & Torelli, G. (1996). Lymphotropic herpesvirus (EBV, HHV-6, HHV-8) DNA sequences in HIV negative Castleman's disease. *J Clin Pathol Mol Pathol* **49**, M232–M235.

Bayley, A. C. (1984). Aggressive Kaposi's sarcoma in Zambia. *Lancet* **i**, 1318.

Beral, V. (1991). Epidemiology of Kaposi's sarcoma. In *Cancer, HIV and AIDS*, vol. 10, pp. 5–22. Edited by V. Beral, H. W. Jaffe & R. A. Weiss. Cold Spring Harbor, NY: Cold Spring Harbor Laboratory.

Beral, V., Peterman, T. A., Berkelman, R. L. & Jaffe, H. W. (1990). Kaposi's sarcoma among persons with AIDS: a sexually transmitted infection? *Lancet* **335**, 123–128.

Bernheim, A., Berger, R. & Lenoir, G. (1981). Cytogenetic studies on African Burkitt's lymphoma cell lines; t(8;14), t(2;8) and t(8;22) translocations. *Cancer Genet Cytogenet* **3**, 307–315.

Boshoff, C. (1999). Kaposi's sarcoma-associated herpesvirus. In *Cancer Surveys: Infections and Human Cancer*, vol. 33, pp. 157–190. Edited by R. Newton, V. Beral & R. A. Weiss. Cold Spring Harbor, NY: Cold Spring Harbor Laboratory.

Burkitt, D. P. (1958). A sarcoma involving the jaws in African children. *Br J Surg* **197**, 218–223.

de Campos-Lima, P. O., Gavioli, R., Zhang, Q. J., Wallace, L. E., Dolcetti, R., Rowe, M., Rickinson, A. B. & Masucci, M. G. (1993). HLA-A11 epitope loss isolates of Epstein-Barr virus from highly A11+ population. *Science* **260**, 98–100.

Carbone, A., Gloghini, A., Vaccher, E. & 8 other authors (1996). Kaposi's sarcoma-associated herpesvirus DNA sequences in AIDS-related and AIDS-unrelated lymphomatous effusions. *Br J Haematol* **94**, 533–543.

Castleman, B., Iverson, L. & Menendez, V. P. (1956). Localized mediastinal lymph-node hyperplasia resembling thymoma. *Cancer* **9**, 822–830.

Cesarman, E., Chang, Y., Moore, P. S., Said, J. W. & Knowles, D. M. (1995). Kaposi's sarcoma-associated herpesvirus-like DNA sequences in AIDS-related body-cavity-based lymphomas. *N Engl J Med* **332**, 1186–1191.

Cesarman, E., Nador, R. G., Aozasa, K., Delsol, G., Said, J. W. & Knowles, D. M. (1996). Kaposi's sarcoma-associated herpesvirus in non-AIDS-related lymphomas occurring in body cavities. *Am J Pathol* **149**, 53–57.

Chadburn, A., Cesarman, E., Nador, R. G., Liu, Y. F. & Knowles, D. M. (1997). Kaposi's sarcoma-associated herpesvirus sequences in benign lymphoid proliferations not associated with human immunodeficiency virus. *Cancer* **80**, 788–797.

Chandran, B., Smith, M. S., Koelle, D. M., Corey, L., Horvat, R. & Goldstein, E. (1998). Reactivities of human sera with human herpesvirus-8 infected BCBL-1 cells and identification of HHV-8-specific proteins and glycoproteins and the encoding cDNAs. *Virology* **243**, 208–217.

Chang, Y., Cesarman, E., Pessin, M. S., Lee, F., Culpepper, J., Knowles, D. M. & Moore, P. S. (1994). Identification of herpesvirus-like DNA sequences in AIDS-associated Kaposi's sarcoma. *Science* **266**, 1865–1869.

Chang, Y., Moore, P. S., Talbot, S. J., Boshoff, C. H., Zarkowska, T., Godden, K., Paterson, H., Weiss, R. A. & Mittnacht, S. (1996). Cyclin encoded by KS herpesvirus. *Nature* **382**, 410.

Chatlynne, L. G., Lapps, W., Handy, M. & 10 other authors (1998). Detection and titration of human herpesvirus-8-specific antibodies in sera from blood donors, acquired immunodeficiency syndrome patients, and Kaposi's sarcoma patients using a whole virus enzyme-linked immunosorbent assay. *Blood* **92**, 53–58.

Cinque, P., Brytting, M., Vago, L. & 8 other authors (1993). Epstein-Barr virus DNA in cerebrospinal fluid from patients with AIDS-related primary lymphoma of the central nervous system. *Lancet* **342**, 398–401.

Corbellino, M., Poirel, L., Aubin, J. T., Paulli, M., Magrini, U., Bestetti, G., Galli, M. & Parravicini, C. (1996). The role of human herpesvirus 8 and Epstein-Barr virus in the pathogenesis of giant lymph node hyperplasia (Castleman's disease). *Clin Infect Dis* **22**, 1120–1121.

Davis, D. A., Humphrey, R. W., Newcomb, F. M., O'Brien, T. R., Goedert, J. J., Strauss, S. E. & Yarchoan, R. (1997). Detection of serum antibodies to a Kaposi's sarcoma-associated herpesvirus specific peptide. *J Infect Dis* **175**, 1071–1079.

Dittmer, D., Lagunoff, M., Renne, R., Stastus, K., Haase, A. & Ganem, D. (1998). A cluster of latently expressed genes in Kaposi's sarcoma-associated herpesvirus. *J Virol* **72**, 8309–8315.

Dupin, N., Fisher, C., Kellam, P. & 10 other authors (1999). Distribution of HHV-8 positive cells in Kaposi's sarcoma, multicentric Castleman's disease, and primary effusion lymphoma. *Proc Natl Acad Sci U S A* **96**, 4546–4551.

Dupin, N., Diss, T., Kellam, P., Tulliez, M., Du, M.-Q., Weiss, R. A., Isaacson, P. G. & Boshoff, C. (2000). HHV-8 is associated with a plasmablastic variant of Castleman's disease that is linked to HHV-8 positive plasmablastic lymphoma. *Blood* **95**, 1406–1412.

Epstein, M., Achong, B. & Barr, Y. (1964). Virus particles in cultured lymphoblasts from Burkitt's lymphoma. *Lancet* **1**, 702–703.

Epstein, M. A., Henle, G., Achong, B. G. & Barr, Y. M. (1965). Morphological and biological studies on a virus in cultured lymphoblasts from Burkitt's lymphoma. *J Exp Med* **121**, 761–770.

Flore, O., Rafii, S., Ely, S., O'Leary, J. J., Hyjek, E. M. & Cesarman, E. (1998). Transformation of primary human endothelial cells by Kaposi's sarcoma-associated herpesvirus. *Nature* **394**, 588–592.

Franceschi, S. & Geddes, M. (1995). Epidemiology of classic Kaposi's sarcoma, with special reference to Mediterranean population. *Tumori* **81**, 308–314.

Frizzera, G., Massarelli, G., Banks, P. M. & Rosai, J. (1983). A systemic lymphoproliferative disorder with morphologic features of Castleman's disease. *Am J Surg Pathol* **7**, 211–231.

Gao, S. J., Kingsley, L., Li, M. & 10 other authors (1996). KSHV antibodies among Americans, Italians and Ugandans with and without Kaposi's sarcoma. *Nat Med* **2**, 925–928.

Goedert, J. J., Cote, T. R., Virgo, P., Scoppa, S. M., Kingma, D. W., Gail, M. H., Jaffe, E. S. & Biggar, R. J. (1998). Spectrum of AIDS-associated malignant disorders. *Lancet* **351**, 1833–1839.

Hayward, G. S. (1999). KSHV strains: the origins and global spread of the virus. *Semin Cancer Biol* **9**, 187–199.

Heslop, H. E. & Rooney, C. M. (1997). Adoptive cellular immunotherapy for EBV lymphoproliferative diseases. *Immunol Rev* **157**, 217–222.

International Agency of Cancer Research (1997). *Monographs on the Evaluation of Carcinogenic Risks to Humans, Epstein-Barr Virus and Kaposi's Sarcoma Herpesvirus/Human Herpesvirus 8*. Lyon, France: IARC Press.

Jacobson, L. P., Yamashita, T. E., Detels, R., Margolick, J. B., Chmiel, J. S., Kingsley, L. A., Melnick, S. & Munoz, A. (1999). Impact of potent anti-retroviral therapy on the incidence of Kaposi's sarcoma and non-Hodgkin's lymphomas among HIV-1 infected individuals. Multicenter AIDS Cohort Study. *J AIDS* **21** (Suppl. 1), S34–S41.

Jones, D. L., Alani, R. M. & Munger, K. (1997). The human papillomavirus E7 oncoprotein can uncouple cellular differentiation and proliferation pathways in human keratinocytes by abrogating p21Cip1-mediated inhibition of cdk2. *Genes Dev* **11**, 2101–2111.

Kaposi, M. (1872). Idiopathisches multiples pigmentsarcom der haut. *Arch Dermatol Syphillis* **4**, 265–273.

Kedes, D. H., Operskalski, E., Busch, M., Kohn, R., Flood, J. & Ganem, D. (1996). The seroepidemiology of human herpesvirus 8 (Kaposi's sarcoma-associated herpesvirus): distribution of infection in KS risk groups and evidence for sexual transmission. *Nat Med* **2**, 918–924.

Kedes, D. H., Lagunoff, M., Renne, R. & Ganem, D. (1997). Identification of the gene encoding the major latency-associated nuclear antigen of the Kaposi's sarcoma-associated herpesvirus. *J Clin Investig* **100**, 2606–2610.

Kellam, P., Boshoff, C., Whitby, D., Matthews, S., Weiss, R. A. & Talbot, S. J. (1997). Identification of a major latent nuclear antigen (LNA-1) in the human herpesvirus 8 (HHV-8) genome. *J Hum Virol* **1**, 19–29.

Khanna, R., Bell, S., Sherritt, M. & 10 other authors (1999). Activation and adoptive transfer of Epstein-Barr virus-specific cytotoxic T cells in solid organ transplant patients with posttransplant lymphoproliferative disease. *Proc Natl Acad Sci U S A* **96**, 10391–10396.

Klein, G. (1994). Epstein-Barr virus strategy in normal and neoplastic B cells. *Cell* **77**, 791–793.

Kuehnle, I., Huls, M. H., Liu, Z., Semmelmann, K., Krance, R. A., Brenner, M. K., Rooney, C. M. & Heslop, H. E. (2000). CD20 monoclonal antibody (rituximab) for therapy of Epstein-Barr virus lymphoma after hemopoietic stem-cell transplantation. *Blood* **95**, 1502–1505.

Lacoste, V., Mauclere, P., Dubreuil, G., Lewis, J., Courbot-Georges, M.-C. & Gessain, A. (2000). Novel gamma-herpesviruses highly related to human herpesvirus 8 and Epstein-Barr virus in chimpanzees and gorillas. *Nature* **6801**, 151–152.

Lisitsyn, N., Lisitsyn, N. & Wigler, M. (1993). Cloning the differences between two complex genomes. *Science* **259**, 946–951.

Lisitsyn, N. A., Lisitsina, N. M., Dalbagni, G., Barker, P., Sanchez, C. A., Gnarra, J., Linehan, W. M., Reid, B. J. & Wigler, M. H. (1995). Comparative genomic analysis of tumours: detection of DNA losses and amplification. *Proc Natl Acad Sci U S A* **92**, 151–155.

McGeoch, D. J. & Davidson, A. J. (1999). The descent of human herpesvirus 8. *Semin Cancer Biol* **9**, 201–209.

Manolov, G. & Manolova, Y. (1972). Marker band in one chromosome 14 from Burkitt lymphoma. *Nature* **237**, 33–34.

Moore, P. S., Kingsley, L. A., Holmberg, S. D., Spira, T., Gupta, P., Hoover, D. R., Parry, J. P., Conley, L. J., Jaffe, H. W. & Chang, Y. (1996). Kaposi's sarcoma-associated herpesvirus infection prior to onset of Kaposi's sarcoma. *AIDS* **10**, 175–180.

Moses, A. V., Fish, K. N., Ruhl, R., Smith, P. P., Strussenberg, J. G., Zhu, L., Chandran, B. & Nelson, J. A. (1999). Long-term infection and transformation of dermal microvascular endothelial cells by human herpesvirus 8. *J Virol* **73**, 6892–6902.

Nador, R. G., Cesarman, E., Knowles, D. M. & Said, J. W. (1995). Herpesvirus-like DNA sequences in a body-cavity-based lymphoma in an HIV-negative patient. *N Engl J Med* **333**, 943.

Nador, R. G., Cesarman, E., Chadburn, A., Dawson, D. B., Ansari, M. Q., Said, J. & Knowles, D. M. (1996). Primary effusion lymphoma: a distinct clinicopathologic entity associated with the Kaposi's sarcoma-associated herpes virus. *Blood* **88**, 645–656.

Nicholas, J., Jian-Chao, Z., Alcendor, D. J. & 9 other authors (1998). Novel organizational features, captured cellular genes, and strain variability within the genome of KSHV/HHV-8. *J Natl Cancer Inst* **23**, 79–88.

Oettle, A. G. (1962). Geographic and racial differences in the frequency of Kaposi's sarcoma as evidence of environmental or genetic causes. In *Symposium on Kaposi's Sarcoma*. Edited by L. V. Ackerman & J. F. Murray. Basel: Karger.

Osman, M., Kubo, T., Gill, Y., Neipel, F., Smith, J., Weiss, R. A., Gazzard, B., Boshoff, C. & Gotch, F. (1999). Identification of human herpesvirus-8-specific cytotoxic T-cell responses. *J Virol* **73**, 6136–6140.

Parravicini, C., Corbellino, M., Paulli, M., Magrini, U., Lazzarino, M., Moore, P. S. & Chang, Y. (1998). Expression of a virus-derived cytokine, KSHV vIL-6, in HIV seronegative Castleman's disease. *Am J Pathol* **6**, 1517–1522.

Penn, I. (1983). Kaposi's sarcoma in immuno-suppressed patients. *J Clin Lab Immunol* **12**, 1–10.

Raab, M.-S., Albrecht, J.-C., Birkmann, A., Yaguboglu, S., Lang, D., Fleckenstein, B. & Neipel, F. (1998). The immunogenic glycoprotein gp35–37 of human herpesvirus 8 is encoded by open reading frame K8.1. *J Virol* **72**, 6725–6731.

Radkov, S., Kellam, P. & Boshoff, C. (2000). The latent nuclear antigen of Kaposi's sarcoma-associated herpesvirus targets the retinoblastoma-E2F pathway and with the oncogene Hras transforms primary rat cells. *Nat Med* **6**, 1121–1127.

Rainbow, L., Platt, G. M., Simpson, G. R., Sarid, R., Gao, S.-J., Stoiber, H., Herrington, C. S., Moore, P. S. & Schulz, T. F. (1997). The 222–234 kd nuclear protein (LNA) of Kaposi's sarcoma-associated herpesvirus (human herpesvirus 8) is encoded by orf 73 and is a component of the latency-associated nuclear antigen. *J Virol* **71**, 5915–5921.

Rickinson, A. B. & Kieff, E. (1996). Epstein-Barr virus. In *Fields Virology*, vol. 2, pp. 2397–2447. Edited by B. N. Fields, D. M. Knipe & P. M. Howley. Philadelphia: Lippincott–Raven.

Rickinson, A. B. & Moss, D. J. (1997). Human cytotoxic T lymphocyte responses to Epstein-Barr virus infection. *Annu Rev Immunol* **15**, 405–431.

Sarid, R., Flore, O., Bohenzky, R. A., Chang, Y. & Moore, P. S. (1998). Transcription mapping of the Kaposi's sarcoma-associated herpesvirus (human herpesvirus 8) genome in a body cavity-based lymphoma cell line (BC-1). *J Virol* **72**, 1005–1012.

Sarid, R., Wiezorek, J. S., Moore, P. S. & Chang, Y. (1999). Characterization and cell cycle regulation of the major Kaposi's sarcoma-associated herpesvirus (Human herpesvirus 8) latent genes and their promoter. *J Virol* **73**, 1438–1446.

Service, P. H. (1981). Kaposi's sarcoma and *Pneumocystis* pneumonia among homosexual men in New York City and California. *MMWR* **30**, 305–308.

Sharp, T. V. & Boshoff, C. (2000). Kaposi's sarcoma-associated herpesvirus: from cell biology to pathogenesis. *Life* **49**, 97–104.

Simpson, G. R., Schulz, T. F., Whitby, D. & 14 other authors (1996). Prevalence of Kaposi's sarcoma associated herpesvirus infection measured by antibodies to recombinant capsid protein and latent immunofluorescence antigen. *Lancet* **348**, 1133–1138.

Sitas, F., Carrara, H., Beral, V. & 9 other authors (1999). Antibodies against human herpesvirus-8 in black South African patients with cancer. *N Engl J Med* **340**, 1863–1871.

Skapek, S. X., Rhee, J., Spicer, D. B. & Lassar, A. B. (1995). Inhibition of myogenic differentiation in proliferating myoblasts by cyclin D1-dependent kinase. *Science* **267**, 1022–1024.

Soulier, J., Grollet, L., Oksenhendler, E., Cacoub, P., Cazals Hatem, D., Babinet, P., d'Agay, M. F., Clauvel, J. P., Raphael, M. & Degos, L. (1995). Kaposi's sarcoma-associated herpesvirus-like DNA sequences in multicentric Castleman's disease. *Blood* **86**, 1276–1280.

Sturzl, M., Hohendi, C., Zietz, C., Castanos-Velez, E., Wunderlich, A., Ascherl, G., Biberfeld, P., Monini, P., Browning, P. J. & Ensoli, B. (1999). Expression of K13/v-FLIP gene of human herpesvirus 8 and apoptosis in Kaposi's sarcoma spindle cells. *J Natl Cancer Inst* **20**, 1725–1733.

Swanton, C., Mann, D. J., Fleckenstein, B., Neipel, F., Peters, G. & Jones, N. (1997). Herpesviral cyclin/Cdk6 complexes evade inhibition by CDK inhibitor proteins. *Nature* **390**, 184–187.

Talbot, S., Weiss, R. A., Kellam, P. & Boshoff, C. (1999). Transcriptional analysis of human herpesvirus-8 (HHV-8) open reading frames 71, 72, 73, K14 and 74 in a primary effusion lymphoma cell line. *Virology* **257**, 84–94.

Virgin, H. W. & Speck, S. H. (1999). Unraveling immunity to gamma-herpesviruses: a new model for understanding the role of immunity in chronic virus infection. *Curr Opin Immunol* **11**, 371–379.

Whitby, D., Howard, M. R., Tenant Flowers, M. & 12 other authors (1995). Detection of Kaposi sarcoma associated herpesvirus in peripheral blood of HIV-infected individuals and progression to Kaposi's sarcoma. *Lancet* **346**, 799–802.

Yates, J. L., Warren, N. & Sugden, B. (1985). Stable replication of plasmids derived from Epstein-Barr virus in various mammalian cells. *Nature* **313**, 812–815.

Yu, M. C. (1991). Nasopharyngeal carcinoma: epidemiology and dietary factors. *IARC Sci Publ* **105**, 39–47.

Yu, M. C. & Henderson, B. E. (1987). Intake of cantonese-style salted fish as a cause of nasopharyngeal carcinoma. *IARC Sci Publ* **84**, 547–549.

Yu, M. C., Huang, T. B. & Henderson, B. E. (1989). Diet and nasopharyngeal carcinoma: a case control study in Guangzhou, China. *Int J Cancer* **43**, 1088–1094.

Zheng, Y. M., Tuppin, P., Hubert, A., Jeannel, D., Pan, Y. J., Zeng, Y. & de The, G. (1994). Environmental and dietary risk factors for nasopharyngeal carcinoma: a case-control study in Zangwu County, Guangxi, China. *Br J Cancer* **69**, 508–514.

Zompi, S., Tulliez, M., Conti, F. & 9 other authors (1999). Rituximab (anti-CD20 monoclonal antibody) for the treatment of patients with clonal lymphoproliferative disorders after orthotopic liver transplantation: a report of three cases. *J Hepatol* **32**, 521–527.

Zong, J.-C., Ciufo, D. M., Alcendor, D. J. & 14 other authors (1999). High level variability in the ORF-K1 membrane protein gene at the left end of the Kaposi's sarcoma-associated herpesvirus (HHV-8) genome defines four major virus subtypes and multiple clades in different human populations. *J Virol* **73**, 4156–4170.

Structure and function of the proteins of Marburg and Ebola viruses

Hans-Dieter Klenk,[1] Heinz Feldmann,[1] Viktor E. Volchkov,[1] Valentina A. Volchkova[1] and Winfried Weissenhorn[2]

[1]Institut für Virologie, Philipps-Universität Marburg, Postfach 2360, 35011 Marburg, Germany

[2]European Molecular Biology Laboratory (EMBL) Grenoble Outstation, 6 rue Jules Horowitz, 38000, Grenoble, France

INTRODUCTION

Filoviruses cause fulminant haemorrhagic fever in humans and non-human primates, killing up to 90 % of the infected patients. Since the discovery of Marburg virus (MBGV) in 1967 and the emergence of Ebola virus (EBOV), its better known cousin, a few years later, these infections have therefore been a matter of high public and scientific concern. Although it is clear from the recorded history of filovirus outbreaks that all of them have so far been self-limiting and that the total number of human infections hitherto documented scarcely exceeds a thousand cases, EBOV by now ranges among the most ill-famed human viruses. For a long time, research on filoviruses has been impeded by their high pathogenicity, but with the advent of recombinant DNA technology our knowledge of the genome structures and the replication strategies of these agents has increased significantly.

PATHOPHYSIOLOGY OF FILOVIRUS INFECTIONS

The pathophysiological changes that make filovirus infections so devastating are just beginning to be unravelled. Pathogenesis in fatal infections in human and non-human primates is similar, suggesting the primate system is a reasonable model for studying filovirus haemorrhagic fever (Simpson *et al.*, 1968; Murphy *et al.*, 1971; Ellis *et al.*, 1978; Fisher-Hoch *et al.*, 1985; Ryabchikova *et al.*, 1994). Clinical and biochemical findings support the anatomical observations of extensive liver involvement, renal damage, changes in vascular permeability, including endothelial damage, and activation of the clotting cascade. The visceral organ necrosis is a consequence of virus replication in parenchymal cells. However, no organ, not even the liver, shows sufficient damage to account for death.

SGM symposium 60: New challenges to health: the threat of virus infection.
Editors G. L. Smith, W. L. Irving, J. W. McCauley, D. J. Rowlands. Cambridge University Press. ISBN 0 521 80614 3

Fluid distribution problems and platelet abnormalities are dominant clinical manifestations, reflecting damage of endothelial cells and decrease of platelets. Post-mortem there is little monocyte/macrophage infiltration in sites of parenchymal necrosis, suggesting that a dysfunction of white blood cells, such as macrophages, also occurs. Morphological studies on monkeys infected with the Reston subtype of EBOV from the 1989 epizootic (Geisbert *et al.*, 1992) and monkeys infected experimentally with the Zaire subtype (Ryabchikova *et al.*, 1994) showed that monocytes/macrophages and fibroblasts may be the preferred sites of virus replication in early stages, whereas other cell types may become involved as the disease progresses. Human monocytes/macrophages in culture are also sensitive to infection, resulting in massive production of infectious virus and cell lysis (Feldmann *et al.*, 1996). Although the studies on infected non-human primates did not identify endothelial cells as sites of massive virus replication, *in vitro* studies and post-mortem observations of human cases demonstrated clearly that endothelial cells of human origin are suitable targets for virus replication (Schnittler *et al.*, 1993; Zaki & Peters, 1997). Here infection leads to cell lysis, indicating that damage of endothelial cells may be an important pathophysiological parameter during infection.

In addition to evidence for direct vascular involvement in infected hosts, the role of active mediator molecules in the pathogenesis of the disorders must be discussed. It has been demonstrated that supernatants of filovirus-infected monocyte/macrophage cultures are capable of increasing paraendothelial permeability in an *in vitro* model (Feldmann *et al.*, 1996). Searches for mediators in those supernatants revealed increased levels of secreted TNF-α, the prototype cytokine of macrophages. These data support the concept of a mediator-induced vascular instability and thus increased permeability as a key mechanism for the development of the shock syndrome seen in severe and fatal cases. Thus the syndrome may be comparable to shock in response to various endogenous and exogenous mediators (Schnittler & Feldmann, 1999). The bleeding tendency could be due to endothelial damage caused directly by virus replication as well as indirectly by cytokine-mediated processes. The onset of the bleeding tendency is supported by the loss of the integrity of the endothelium, as demonstrated in tissue and organ culture (Schnittler *et al.*, 1993) as well as in infected animals (Fisher-Hoch *et al.*, 1985), and seems to occur later in infection. The combination of viral replication in endothelial cells and virus-induced cytokine release from monocytes/macrophages may also promote a distinct proinflammatory endothelial phenotype that then triggers the coagulation cascade.

GENOME AND GENOME PRODUCTS OF FILOVIRUSES

MBGV and EBOV belong to the family *Filoviridae*, order *Mononegavirales*. They have non-segmented, negative-stranded, RNA genomes that encode the seven structural

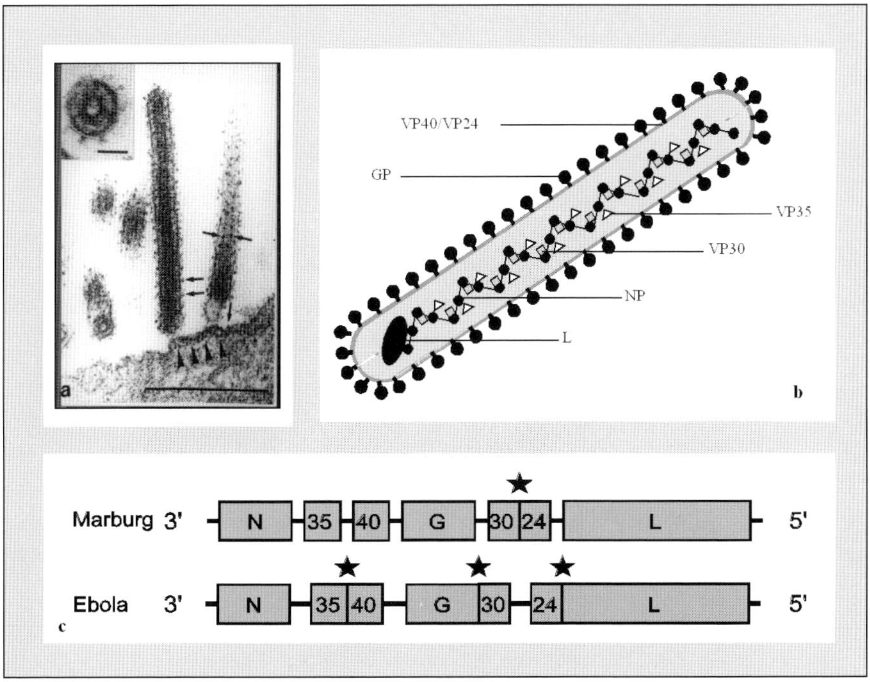

Fig. 1. Structure of filoviruses. (a) Electron micrograph of MBGV particles budding from the surface of human endothelial cells 3 days after infection. Particles consist of a nucleocapsid surrounded by a membrane in which spikes are inserted (arrows). The nucleocapsid contains a central channel (inset). The plasma membrane of infected cells is often thickened at locations where budding occurs (arrowheads) (bar, 0.5 μm; inset bar, 50 nm). (b) Schematic view of a virion. The RNA genome is associated with four viral proteins: the viral polymerase (L), the nucleoprotein (NP) and proteins VP35 and VP30. VP40 and VP24 are matrix proteins. The spikes are formed by trimeric $GP_{1,2}$ complexes. (c) Gene order on the non-segmented negative-strand RNA genome of MBGV and EBOV. Overlapping genes are indicated by asterisks. Taken from Feldmann *et al.* (1997) and reproduced with permission.

proteins in the order 3′-NP-VP35-VP40-GP-VP30-VP24-L-5′ (Feldmann *et al.*, 1992; Sanchez *et al.*, 1993; Volchkov *et al.*, 1993) (Fig. 1). In general, filoviral genes are transcribed into monocistronic subgenomic RNA species (mRNA) (Sanchez & Kiley, 1987; Feldmann *et al.*, 1992).

The ribonucleoprotein complex of filovirus is composed of the viral RNA and four proteins: the major nucleocapsid protein NP, VP35 corresponding to the P protein of other members of the *Mononegavirales*, the polymerase protein L, and VP30. Recently, VP30 of EBOV has been shown to be required for transcription, but not for replication (Mühlberger *et al.*, 1999).

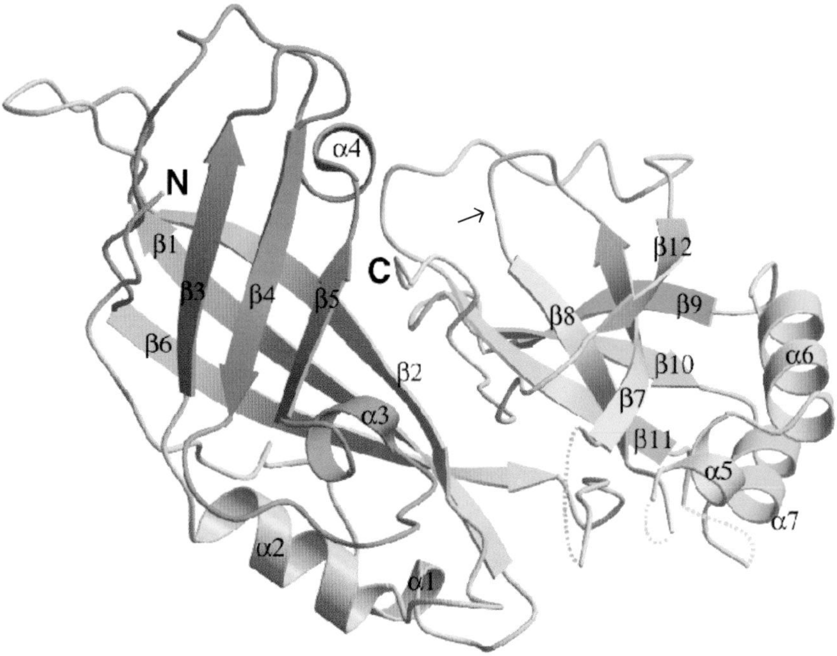

Fig. 2. Structure of the matrix protein VP40 of EBOV as revealed by X-ray crystallography. The ribbon diagram shows the amino-terminal domain (yellow) and the carboxy-terminal domain (blue) are held together by a few contacts and a flexible loop. A trypsin cleavage site after Lys212 is highlighted by an arrow. Trypsinization at this site causes complete disengagement of the carboxy-terminal domain from the amino-terminus and subsequent hexamerization of the amino-terminal domain. From Dessen *et al.* (2000).

Filoviruses have two matrix proteins, VP40 and VP24, of which VP40 is the major one. The structure of VP40 of EBOV has been analysed by X-ray crystallography. The protein crystallizes as a monomer that consists of two domains with unique folds, linked by a flexible dimer (Fig. 2). The structure also reveals that the molecule may be able to switch from a monomeric conformation in solution to a hexameric membrane-bound form in which the amino-terminal domain is responsible for hexamerization whereas the carboxy-terminal domain mediates lipid binding. This membrane-induced conformational change required for hexamerization may also be necessary to achieve lattice formation and may therefore be an important step in virion assembly (Dessen *et al.*, 2000; Ruigrok *et al.*, 2000).

In contrast to all other filoviral genes, including the GP gene of MBGV (Will *et al.*, 1993), the organization and transcription of the fourth gene (GP) of EBOV is unusual, involving transcriptional editing needed to express the envelope glycoprotein (Fig. 3). A

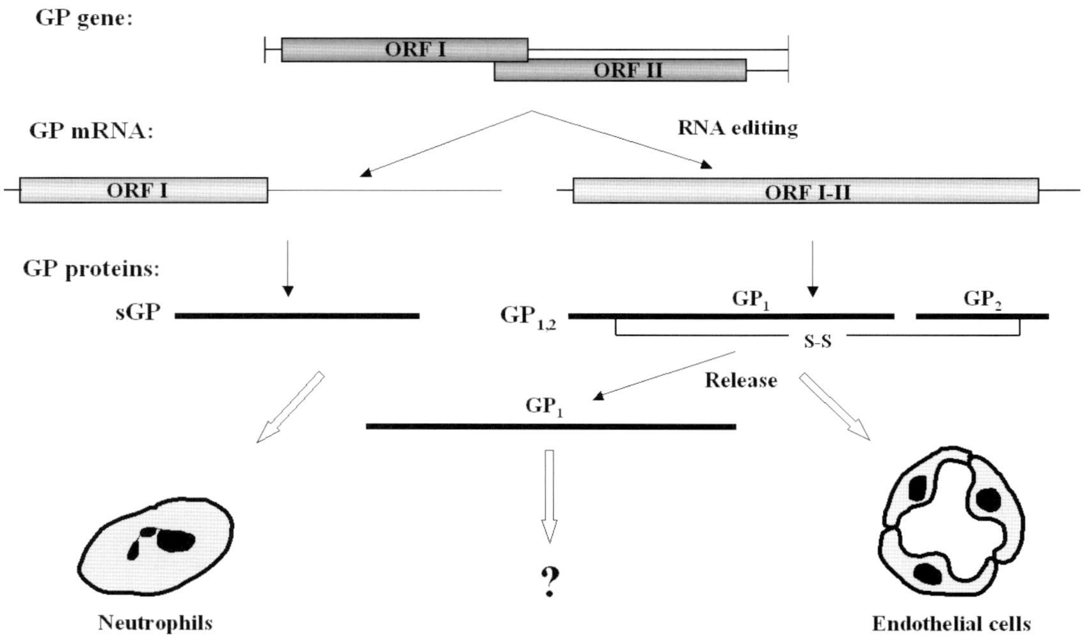

Fig. 3. Biosynthesis of different forms of the EBOV glycoprotein and their target cells. The EBOV surface glycoprotein GP is encoded in two overlapping reading frames (ORF I and II), and expression of GP occurs through transcriptional editing. ORF I encodes a secreted small glycoprotein (sGP) that is expressed from unedited transcripts (Volchkov *et al.*, 1995; Sanchez *et al.*, 1996). The amino-terminal 295 amino acids of sGP are identical with GP, but the last 69 amino acids are different. Mature GP ($GP_{1,2}$) consists of the disulfide-linked (S–S) subunits GP_1 and GP_2 (Volchkov *et al.*, 1998a). Significant amounts of GP_1 are released from expressing cells (Volchkov *et al.*, 1998b). sGP binds to neutrophils and appears to inhibit early activation of these cells (Yang *et al.*, 1998), whereas GP mediates binding of EBOV to endothelial cells and presumably to other cells susceptible to infection. The binding specificity of GP_1 is not known. Taken from Klenk *et al.* (1998).

non-structural, small glycoprotein (sGP) is synthesized from the unedited GP mRNA, which is secreted abundantly from infected cells (Volchkov *et al.*, 1995; Sanchez *et al.*, 1996). sGP is a disulfide-linked dimer, with both subunits aligned in anti-parallel order (Volchkova *et al.*, 1998). It has been reported that by binding to neutrophils sGP may interfere with the activation of these cells and thereby may paralyse the inflammatory defence of the host (Yang *et al.*, 1998). A third mRNA of the GP gene encoding another small glycoprotein (ssGP) has also been identified (Volchkov *et al.*, 1995). The envelope glycoprotein GP of MBGV and EBOV is a trimeric type I transmembrane glycoprotein (Sanchez *et al.*, 1998; Feldmann *et al.*, 1991). The middle region of GP is variable, extremely hydrophilic, and carries the bulk of the glycosylation sites for *N*- and *O*-glycans that account for approximately one-third of the molecular mass (Feldmann *et al.*, 1991; Geyer *et al.*, 1992; Volchkov *et al.*, 1993; Becker *et al.*, 1996). Experimental data on GP function are limited. However, the fact that GP is the only surface protein of virions and that it mediates infection by vesicular stomatitis virus (Takada *et al.*, 1997) and retrovirus pseudotypes (Wool-Lewis & Bates, 1998; Yang *et al.*, 1998) suggest a function in receptor binding and fusion with cellular membranes. There is also evidence that MBGV infects hepatocytes by binding to the asialoglycoprotein receptor of these cells (Becker *et al.*, 1995). Maturation of EBOV GP involves post-translational cleavage of a precursor into the disulfide-linked fragments GP_1 and GP_2 (Volchkov *et al.*, 1998a). GP_1 is not only present in virion spikes, but is also found in soluble form after shedding from the surface of infected cells. Furthermore, membrane vesicles containing the complete envelope glycoprotein are released from infected cells, an action that may be partly responsible for immune modulatory effects associated with EBOV infection (Volchkov *et al.*, 1998b). Thus unlike MBGV glycoprotein, EBOV glycoprotein shows a much higher degree of polymorphism, and there is evidence that the different products of the GP gene have different functions in the infected host (Fig. 3).

By serial passages in guinea pigs, EBOV acquires high pathogenicity for these animals. Sequence comparison of the genomes of wild-type and guinea-pig-adapted variants showed that the increase in pathogenicity was linked to a mutation in VP24, suggesting that this protein may play a role in pathogenicity. Some of the viruses obtained after guinea pig adaptation also showed insertions or deletions of uridine residues at the editing site of the GP gene, resulting in significant changes in the expression of GP and sGP. This observation suggests that the editing site is a hot spot for insertion and deletion of nucleotides, not only at the level of transcription, but also of genome replication (Volchkov *et al.*, 2000).

PROTEOLYTIC PROCESSING OF THE FILOVIRUS GLYCOPROTEIN AS A POTENTIAL DETERMINANT OF PATHOGENICITY

The filovirus GP is cleaved by furin (Volchkov *et al.*, 1998a). This is indicated by the observation that cleavage did not occur when GP was expressed in the furin-defective

LoVo cell line, but that it was restored in these cells by vector-expressed furin. The finding that cleavage was inhibited by a sequence-specific peptidyl chloromethylketone or by mutation of the cleavage site supports this concept. Furin belongs to the proprotein convertases, a family of subtilisin-like, eukaryotic endoproteases that also includes PC1/PC3, PC2, PC4, PACE4, PC5/PC6 and LPC/PC7 (Seidah *et al.*, 1996). These enzymes are expressed differentially in cells and tissues, and they display similar but not identical specificity for basic motifs, such as R-X-K/R-R, at the cleavage site of their substrates. Furin appears to be expressed in most cells. It is a processing enzyme of the constitutive secretory pathway, as seems to be the case with PACE4, PC5/PC6 and LPC/PC7. The expression of PC1/PC3 and PC2 is restricted to the regulated secretory pathway of neuroendocrine cells. Furin is localized predominantly in the *trans*-Golgi network (Molloy *et al.*, 1994; Schäfer *et al.*, 1995), but it is also secreted from cells in a truncated form (Wise *et al.*, 1990; Vey *et al.*, 1995). Proprotein convertases activate numerous cellular proteins (Barr, 1991) and surface proteins of enveloped viruses. Furin appears to be the key enzyme in virus activation (Klenk & Garten, 1994a), but PC5/PC6 (Horimoto *et al.*, 1994) and LPC/PC7 (Hallenberger *et al.*, 1997) are also involved. Thus LPC/PC7 may be responsible for cleavage of the human immunodeficiency virus (HIV) glycoprotein in the furin-deficient LoVo cells. The observation that EBOV GP is not cleaved in these cells is interesting in this context. It is also noteworthy that furin, although ubiquitous, is particularly apparent in hepatocytes and endothelial cells, which are both prime targets of EBOV (Geisbert *et al.*, 1992; Zaki & Peters, 1997). These observations stress the importance of furin as a processing enzyme of GP, but it remains to be seen in future studies if other proprotein convertases can substitute as cleaving enzymes.

Processing by protein convertases is an important control mechanism for the biological activity of viral surface proteins (Klenk & Garten, 1994a, b). Cleavage often occurs next to a protein domain involved in fusion, and it has long been known that in these cases proteolytic cleavage is necessary for fusion activity. Proteolytic cleavage is the first step in the activation of these fusion proteins and is followed by a conformational change resulting in the exposure of the fusion domain (Bullough *et al.*, 1994; Chan *et al.*, 1997; Weissenhorn *et al.*, 1997). The conformational change may be triggered by low pH in endosomes, as is the case with influenza viruses (Skehel *et al.*, 1982), or by the interaction with a secondary receptor protein at the cell surface, as is the case with HIV (Feng *et al.*, 1996). So far we have not been able to demonstrate that cleavage of GP has an effect on fusion activity or on infectivity of EBOV. However, it is interesting that GP_2 contains a sequence of 16 uncharged and hydrophobic amino acids at a short distance (22 amino acids) from the cleavage site which bears some structural similarity to the fusion peptides of retroviruses and has therefore been thought to play a role in EBOV entry (Gallaher, 1996). Furthermore, it appears that cleaved and uncleaved GP differ in folding, as indicated by their different electrophoretic mobilities under non-reducing conditions. Finally, the central structural feature of the GP_2 ectodomain is a long, triple-stranded, coiled coil with three antiparallel helices packed at the surface of this trimer, as has been observed with other viral fusion proteins undergoing proteolytic activation (Fig. 4). These observations are compatible with the view that proteolytic cleavage is a priming mechanism that renders GP susceptible to the conformational change required for fusion.

Finally, it has to be pointed out that proteolytic activation of viral glycoproteins is an important determinant for pathogenicity. Cleavage by furin and other ubiquitous pro-protein convertases has been shown to be responsible for systemic infection caused by highly pathogenic strains of avian influenza and Newcastle disease virus (Klenk & Rott, 1988). It is therefore tempting to speculate that cleavage by furin is also an important factor for the pantropism of EBOV and its rapid dissemination through the organism. Furthermore, variations at the cleavage site of GP may account for differences in the pathogenicity of EBOV (Volchkov *et al.*, 1998a). The pathogenic strains Zaire, Sudan and Ivory Coast, which have the canonical furin motif R-X-K/R-R at the cleavage site, are highly susceptible to cleavage, whereas the Reston strain, which appears to be apathogenic for humans and only moderately pathogenic for at least some monkey species (Fisher-Hoch *et al.*, 1992), has reduced cleavability because of the suboptimal cleavage site sequence K-Q-K-R. That highly pathogenic variants may suddenly emerge from Reston-like strains by mutations restricted to the cleavage site is an intriguing hypothesis. On the other hand, it may be possible to obtain EBOV mutants with even lower cleavability than the Reston strain, and such viruses may have potential as live vaccines. Because furin cleavage can be inhibited not only by peptidyl chloromethylketones as described here but also by less toxic components (Anderson *et al.*, 1993), inhibition of proteolytic cleavage may be a novel concept for treatment of EBOV infections.

The small soluble glycoprotein sGP also undergoes furin cleavage, resulting in the removal of a carboxy-terminal glycopeptide about 40 amino acids in length. The biological significance of sGP cleavage is not known (Volchkova *et al.*, 1999).

POTENTIAL FILOVIRUS MECHANISMS INTERFERING WITH HOST DEFENCE

Recently, immune protection against filovirus infection has been accomplished in animal models employing inactivated virus, recombinant antigens, or cDNA as vaccines (Vanderzanden *et al.*, 1998; Xu *et al.*, 1998). However, fatal filovirus infections usually end with high viraemia and no evidence of an effective humoral immune response. Because circulating monocytes/macrophages are primary target cells in filovirus infections and because the extensive disruption of the parafollicular regions in

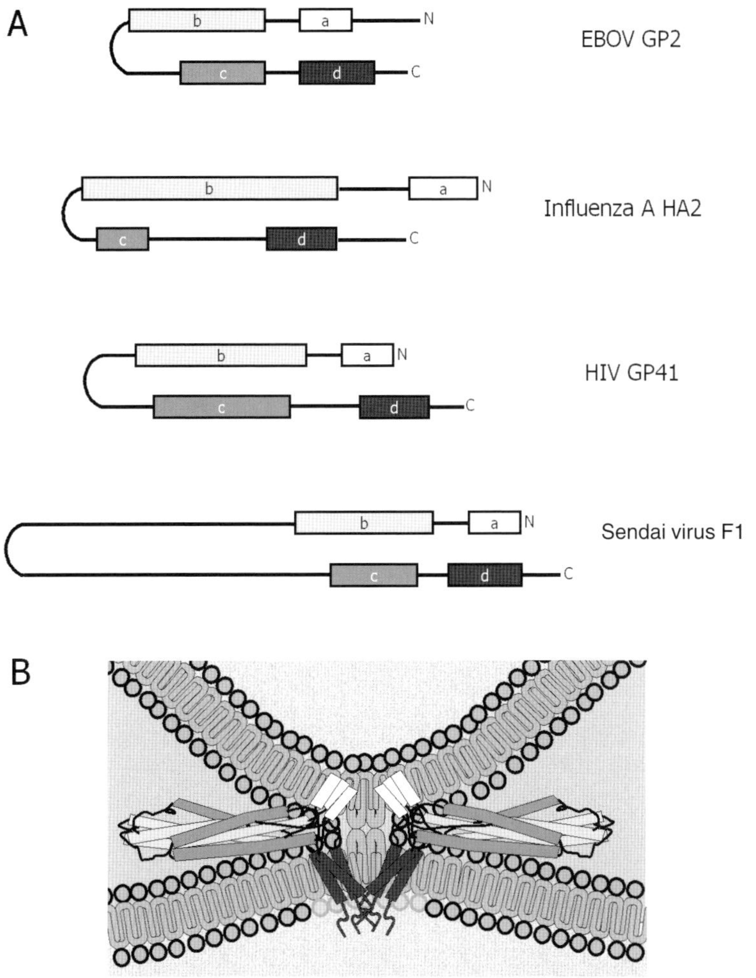

Fig. 4. Filovirus glycoprotein is a potential fusion protein. (A) Structural similarities between EBOV GP_2 (Weissenhorn *et al.*, 1998) and the transmembrane subunits HA2 of the influenza virus haemagglutinin (Bullough *et al.*, 1994; Carr & Kim, 1993), gp41 of the HIV Env protein (Chan *et al.*, 1997; Weissenhorn *et al.*, 1997), and the F1 of the Sendai virus fusion protein (Chambers *et al.*, 1990). Four domains can be discriminated in the fusion active state: the fusion peptide (a), an amino-terminal helix (b), a carboxy-terminal helix (c) and the membrane anchor (d). The transmembrane proteins assemble into trimers in which the large helices form an interior, parallel coiled-coil with the smaller helices packing in an antiparallel fashion at the surface. The fusion peptide and the membrane anchor are therefore located at one end of the rod-like trimers. (B) Fusion model. The close proximity of the fusion peptide and the membrane anchor brings both membranes together and thereby promotes fusion (Weissenhorn *et al.*, 1997). Taken from Feldmann *et al.* (1999) and reproduced with permission.

spleen and lymph nodes results in the destruction of the antigen-presenting dendritic cells, cellular immunity also appears to be affected during filoviral haemorrhagic fever. In addition to these cytolytic effects, the polymorphism of the EBOV glycoprotein suggests a number of other mechanisms regarding virus interference with the host defence. First, the immune reactivity of $GP_{1,2}$ may be modulated by its high carbohydrate content that may cover antigenic epitopes. Second, as already mentioned there is evidence that by binding to neutrophils sGP may block the activation of these cells and thereby interfere with inflammatory reactions. Third, GP_1 released from infected cells by shedding may have a decoy function by binding to GP-specific antibodies. Fourth, a presumably immunosuppressive domain has been identified on GP_2 (Volchkov *et al.*, 1992). Whether any of these mechanisms contributes to pathogenicity remains to be shown in future studies.

REFERENCES

Anderson, E. D., Thomas, L., Hayflick, J. S. & Thomas, G. (1993). Inhibition of HIV-1 gp160-dependent membrane fusion by a furin-directed alpha 1-antitrypsin variant. *J Biol Chem* **268**, 24887–24891.

Barr, P. J. (1991). Mammalian subtilisins: the long-sought dibasic processing endoproteases. *Cell* **66**, 1–3.

Becker, S., Spiess, M. & Klenk, H.-D. (1995). The asialoglycoprotein receptor is a potential liver-specific receptor for Marburg virus. *J Gen Virol* **76**, 393–399.

Becker, S., Klenk, H.-D. & Mühlberger, E. (1996). Intracellular transport and processing of the Marburg virus surface protein in vertebrate and insect cells. *Virology* **225**, 145–155.

Bullough, P. A., Hughson, F. M., Skehel, J. J. & Wiley, D. C. (1994). Structure of influenza haemagglutinin at the pH of membrane fusion. *Nature* **371**, 37–43.

Carr, C. M. & Kim, P. S. (1993). A spring-loaded mechanism for the conformational change of influenza hemagglutinin. *Cell* **21**, 823–832.

Chambers, P., Pringle, C. R. & Easton, A. J. (1990). Heptad repeat sequences are located adjacent to hydrophobic regions in several types of virus fusion glycoproteins. *J Gen Virol* **71**, 3075–3080.

Chan, D. C., Fass, D., Berger, J. M. & Kim, P. S. (1997). Core structure of gp41 from the HIV envelope glycoprotein. *Cell* **89**, 263–273.

Dessen, A., Volchkov, V., Dolnik, O., Klenk, H.-D. & Weissenhorn, W. (2000). Crystal structure of the matrix protein from Ebola virus. *EMBO J* **19**, 4228–4236.

Ellis, D. S., Bowen, E. T., Simpson, D. I. & Stamford, S. (1978). Ebola virus: a comparison, at ultrastructural level, of the behaviour of the Sudan and Zaire strains in monkeys. *Br J Exp Pathol* **59**, 584–593.

Feldmann, H., Will, C., Schikore, M., Slenczka, W. & Klenk, H.-D. (1991). Glycosylation and oligomerization of the spike protein of Marburg virus. *Virology* **182**, 353–356.

Feldmann, H., Mühlberger, E., Randolf, A., Will, C., Kiley, M. P., Sanchez, A. & Klenk, H.-D. (1992). Marburg virus, a filovirus: messenger RNAs, gene order, and regulatory elements of the replication cycle. *Virus Res* **24**, 1–19.

Feldmann, H., Bugany, H., Mahner, F., Klenk, H.-D., Drenckhahn, D. & Schnittler, H.-J. (1996). Filovirus-induced endothelial leakage triggered by infected monocytes/macrophages. *J Virol* **70**, 2208–2214.

Feldmann, H., Volchkov, V. E. & Klenk, H.-D. (1997). Filovirus Marburg et Ebola. *Ann Inst Pasteur/actualités* **8**, 207–222.

Feldmann, H., Volchkov, V. E., Volchkova, V. A. & Klenk, H.-D. (1999). The glycoproteins of Marburg and Ebola virus and their potential roles in pathogenesis. *Arch Virol Suppl* **15**, 159–169.

Feng, Y., Broder, C. C., Kennedy, P. E. & Berger, E. A. (1996). HIV-1 entry cofactor: functional cDNA cloning of a seven-transmembrane, G protein-coupled receptor. *Science* **272**, 872–877.

Fisher-Hoch, S. P., Platt, G. S., Neild, G. H., Southee, T., Baskerville, A., Raymond, R. T., Lloyd, G. & Simpson, D. I. H. (1985). Pathophysiology of shock and hemorrhage in a fulminating viral infection (Ebola). *J Infect Dis* **152**, 887–894.

Fisher-Hoch, S. P., Brammer, T. L., Trappier, S. G. & 9 other authors (1992). Pathogenic potential of filoviruses: role of geographic origin of primate host and virus strain. *J Infect Dis* **166**, 753–763.

Gallaher, W. R. (1996). Similar structural models of the transmembrane proteins of Ebola and avian sarcoma viruses. *Cell* **85**, 477–478.

Geisbert, T. W., Jahrling, P. B., Hanes, M. A. & Zack, P. M. (1992). Association of Ebola-related Reston virus particles and antigen with tissue lesions of monkeys imported to the United States. *J Comp Pathol* **106**, 137–152.

Geyer, H., Will, C., Feldmann, H., Klenk, H.-D. & Geyer, R. (1992). Carbohydrate structure of Marburg virus glycoprotein. *Glycobiology* **2**, 299–312.

Hallenberger, S., Moulard, M., Sordel, M., Klenk, H.-D. & Garten, W. (1997). The role of eukaryotic subtilisin-like endoproteases for the activation of human immunodeficiency virus glycoproteins in natural host cells. *J Virol* **71**, 1036–1045.

Horimoto, T., Nakayama, K., Smeekens, S. P. & Kawaoka, Y. (1994). Proprotein-processing endoproteases PC6 and furin both activate hemagglutinin of virulent avian influenza viruses. *J Virol* **68**, 6074–6078.

Klenk, H.-D. & Garten, W. (1994a). Activation cleavage of viral spike proteins by host proteases. In *Cellular Receptors for Animal Viruses*, pp. 241–280. Edited by E. Wimmer. Cold Spring Harbor, NY: Cold Spring Harbor Laboratory.

Klenk, H.-D. & Garten, W. (1994b). Host cell proteases controlling virus pathogenicity. *Trends Microbiol* **2**, 39–43.

Klenk, H.-D. & Rott, R. (1988). The molecular biology of influenza virus pathogenicity. *Adv Virus Res* **34**, 247–281.

Klenk, H.-D., Volchkov, V. E. & Feldmann, H. (1998). Two strings to the bow of Ebola virus. *Nat Med* **4**, 388–389.

Molloy, S. S., Thomas, L., van Slyke, J. K., Stenberg, P. E. & Thomas, G. (1994). Intracellular trafficking and activation of the furin proprotein convertase: localization to the TGN and recycling from the cell surface. *EMBO J* **13**, 18–33.

Mühlberger, E., Weik, M., Volchkov, V. E., Klenk, H.-D. & Becker, S. (1999). Comparison of the transcription and replication strategies of Marburg virus and Ebola virus using artificial replication systems. *J Virol* **73**, 2333–2342.

Murphy, F. A., Simpson, D. I. H., Whitfield, S. G., Zlotnik, I. & Carter, G. B. (1971). Marburg virus infection in monkeys. *Lab Investig* **24**, 279–291.

Ruigrok, R. W. H., Schon, G., Dessen, A., Forest, E., Volchkov, V., Dolnik, O., Klenk, H.-D. & Weissenhorn, W. (2000). Structural characterization and membrane binding properties of the matrix protein VP40 of Ebola virus. *J Mol Biol* **300**, 103–112.

Ryabchikova, E. I., Kolesnikova, L. V., Tkachev, V. K., Pereboeva, L. A., Baranova, S. G. & Rassadkin, J. N. (1994). Ebola infection in four monkey species. *Ninth International Conference on Negative Strand RNA Viruses*, p. 164. Estoril, Portugal.

Sanchez, A. & Kiley, M. P. (1987). Identification and analysis of Ebola virus messenger RNA. *Virology* **157**, 414–420.

Sanchez, A., Kiley, M. P., Holloway, B. P. & Auperin, D. D. (1993). Sequence analysis of the Ebola virus genome: organization, genetic elements, and comparison with the genome of Marburg virus. *Virus Res* **29**, 215–240.

Sanchez, A., Trappier, S. G., Mahy, B. W., Peters, C. J. & Nichol, S. T. (1996). The virion glycoproteins of Ebola viruses are encoded in two reading frames and are expressed through transcriptional editing. *Proc Natl Acad Sci U S A* **93**, 3602–3607.

Sanchez, A., Yang, Z. Y., Xu, L., Nabel, G. J., Crews, T. & Peters, C. J. (1998). Biochemical analysis of the secreted and virion glycoproteins of Ebola virus. *J Virol* **72**, 6442–6447.

Schäfer, W., Stroh, A., Berghöfer, S., Seiler, J., Vey, M., Kruse, M. L., Kern, H. F., Klenk, H.-D. & Garten, W. (1995). Two independent targeting signals in the cytoplasmic domain determine trans-Golgi network localization and endosomal trafficking of the proprotein convertase furin. *EMBO J* **14**, 2424–2435.

Schnittler, H.-J. & Feldmann, H. (1999). Molecular pathogenesis of filovirus infections: role of macrophages and endothelial cells. *Curr Top Microbiol Immunol* **235**, 175–204.

Schnittler, H. J., Mahner, F., Drenckhahn, D., Klenk, H.-D. & Feldmann, H. (1993). Replication of Marburg virus in human endothelial cells. A possible mechanism for the development of viral hemorrhagic disease. *J Clin Investig* **91**, 1301–1309.

Seidah, N. G., Hamelin, J., Mamarbachi, M., Dong, W., Tadro, H., Mbikay, M., Chretien, M. & Day, R. (1996). cDNA structure, tissue distribution, and chromosomal localization of rat PC7, a novel mammalian proprotein convertase closest to yeast kexin-like proteinases. *Proc Natl Acad Sci U S A* **93**, 3388–3393.

Simpson, D. I. H., Zlotnik, I. & Rutter, D. A. (1968). Vervet monkey disease. Experimental infection of guinea pigs and monkeys with the causative agent. *Br J Exp Pathol* **49**, 458–464.

Skehel, J. J., Bayley, P. M., Brown, E. B., Martin, S. R., Waterfield, M. D., White, J. M., Wilson, I. A. & Wiley, D. C. (1982). Changes in the conformation of influenza virus hemagglutinin at the pH optimum of virus-mediated membrane fusion. *Proc Natl Acad Sci U S A* **79**, 968–972.

Takada, A., Robison, C., Goto, H., Sanchez, A., Murti, K. G., Whittl, M. A. & Kawaoka, Y. (1997). A system for functional analysis of Ebola virus glycoprotein. *Proc Natl Acad Sci U S A* **94**, 14764–14769.

Vanderzanden, L., Bray, M., Fuller, D., Roberts, T., Custer, D., Spik, K., Jahrling, P., Huggins, J., Schmaljohn, A. & Schmaljohn, C. (1998). DNA vaccines expressing either the GP or NP genes of Ebola virus protect mice from lethal challenge. *J Virol* **246**, 134–144.

Vey, M., Schäfer, W., Reis, B., Ohuchi, R., Britt, W., Garten, W., Klenk, H.-D. & Radsak, K. (1995). Proteolytic processing of human cytomegalovirus glycoprotein B (gpUL55) is mediated by the human endoprotease furin. *Virology* **206**, 746–749.

Volchkov, V. E., Blinov, V. M. & Netesov, S. V. (1992). The envelope glycoprotein of Ebola virus contains an immunosuppressive like domain similar to oncogenic retrovirus. *FEBS Lett* **305**, 181–184.

Volchkov, V. E., Blinov, V. M., Kotov, A. N., Chepurnov, A. A. & Netesov, S. V. (1993). The full-length nucleotide sequence of the Ebola virus. *IXth International Congress of Virology*, Glasgow, Scotland, P52–2.

Volchkov, V. E., Becker, S., Volchkova, V. A., Ternovoj, V. A., Kotov, A. N., Netesov, S. V. & Klenk, H.-D. (1995). GP mRNA of Ebola virus is edited by the Ebola virus polymerase and by T7 and vaccinia virus polymerases. *Virology* **214**, 421–430.

Volchkov, V. E., Feldmann, H., Volchkova, V. A. & Klenk, H.-D. (1998a). Processing of the Ebola virus glycoprotein by the proprotein convertase furin. *Proc Natl Acad Sci U S A* **95**, 5762–5767.

Volchkov, V. E., Volchkova, V. A., Slenczka, W., Klenk, H.-D. & Feldmann, H. (1998b). Release of viral glycoproteins during Ebola virus infection. *Virology* **245**, 110–119.

Volchkov, V. E., Chepurnov, A. A., Volchkova, V. A., Ternovoj, V. A. & Klenk, H.-D. (2000). Molecular characterization of guinea pig-adapted variants of Ebola virus. *Virology* **277**, 147–155.

Volchkova, V. A., Feldmann, H., Klenk, H.-D. & Volchkov, V. E. (1998). The nonstructural small glycoprotein sGP of Ebola virus is secreted as an antiparallel-orientated homodimer. *Virology* **250**, 408–441.

Volchkova, V. A., Klenk, H.-D. & Volchkov, V. E. (1999). Delta-peptide is the carboxy-terminal cleavage fragment of the nonstructural small glycoprotein sGP of Ebola virus. *Virology* **265**, 164–171.

Weissenhorn, W., Dessen, A., Harrison, S. C., Skehel, J. J. & Wiley, D. C. (1997). Atomic structure of the ectodomain from HIV-1 gp41. *Nature* **387**, 426–430.

Weissenhorn, W., Calder, L. J., Wharton, S. A., Skehel, J. J. & Wiley, D. (1998). The central structural feature of the membrane fusion protein subunit from the Ebola virus glycoprotein is a long triple-stranded coiled coil. *Proc Natl Acad Sci U S A* **95**, 6032–6036.

Will, C., Mühlberger, E., Linder, D., Slenczka, W., Klenk, H.-D. & Feldmann, H. (1993). Marburg virus gene four encodes the virion membrane protein, a type I transmembrane glycoprotein. *J Virol* **67**, 1203–1210.

Wise, R. J., Barr, P. J., Wong, P. A., Kiefer, M., Brake, A. J. & Kaufman, R. J. (1990). Expression of a human proprotein processing enzyme: correct cleavage of the von Willebrand factor precursor at a paired basic amino acid site. *Proc Natl Acad Sci U S A* **87**, 9378–9382.

Wool-Lewis, R. J. & Bates, P. (1998). Characterization of Ebola virus entry by using pseudotyped viruses: identification of receptor-deficient cell lines. *J Virol* **72**, 3155–3160.

Xu, L., Sanchez, A., Yang, Z., Zaki, S. R., Nabel, E. G., Nichol, S. T. & Nabel, G. J. (1998). Immunization for Ebola virus infection. *Nat Med* **4**, 37–42.

Yang, Z., Delgado, R., Xu, L., Todd, R. F., Nabel, E. G., Sanchez, A. & Nabel, G. J. (1998). Distinct cellular interactions of secreted and transmembrane Ebola virus glycoproteins. *Science* **279**, 1034–1036.

Zaki, S. R. & Peters, C. J. (1997). Viral hemorrhagic fevers. In *The Pathology of Infectious Diseases*, pp. 347–364. Edited by D. H. Connor, F. W. Chandler, D. A. Schwartz, H. J. Manz & E. E. Lack. Norwalk, CT: Appleton and Lange.

Epidemic dengue/dengue haemorrhagic fever as a public health problem in the 21st century

Duane J. Gubler

Division of Vector-Borne Infectious Diseases, National Center for Infectious Diseases, Centers for Disease Control and Prevention, Public Health Service, US Department of Health and Human Services, PO Box 2087, Fort Collins, CO 80522, USA

INTRODUCTION

There was an unprecedented global resurgence of epidemic dengue fever (DF) in the waning years of the 20th century. Both the viruses and the mosquito vectors expanded their geographic distribution, and there was a dramatic increase in the frequency of epidemic dengue, followed by the emergence of the severe form of disease, dengue haemorrhagic fever (DHF), in most tropical regions of the world (Gubler, 1988, 1997, 1998; Halstead, 1992). An estimated 3 billion people currently live in areas at risk for dengue infection (World Health Organization, 1999) (Fig. 1). Each year there are an estimated 50–100 million cases of the mild disease, DF, and several hundred thousand cases of the severe form of disease, DHF, depending on the epidemic activity that particular year. The case fatality rate of DHF cases averages about 5 % (World Health Organization, 1999). In 2001, epidemic DF/DHF is one of the most important emergent tropical infectious diseases in the world, primarily affecting the urban centres of tropical developing countries. This chapter briefly reviews the changing epidemiology of dengue, discusses the factors responsible for the recent resurgence, and reviews options for prevention and control.

NATURAL HISTORY

DF/DHF is caused by infection with dengue viruses. There are four closely related virus serotypes, DEN-1, DEN-2, DEN-3 and DEN-4, which show extensive cross-reactivity in serologic tests, but do not provide cross-protective immunity (World Health Organization, 1997; Gubler, 1998). Thus persons living in an endemic area can be infected with three and probably all four of the dengue serotypes during their lifetime.

SGM symposium 60: New challenges to health: the threat of virus infection.
Editors G. L. Smith, W. L. Irving, J. W. McCauley, D. J. Rowlands. Cambridge University Press. ISBN 0 521 80614 3

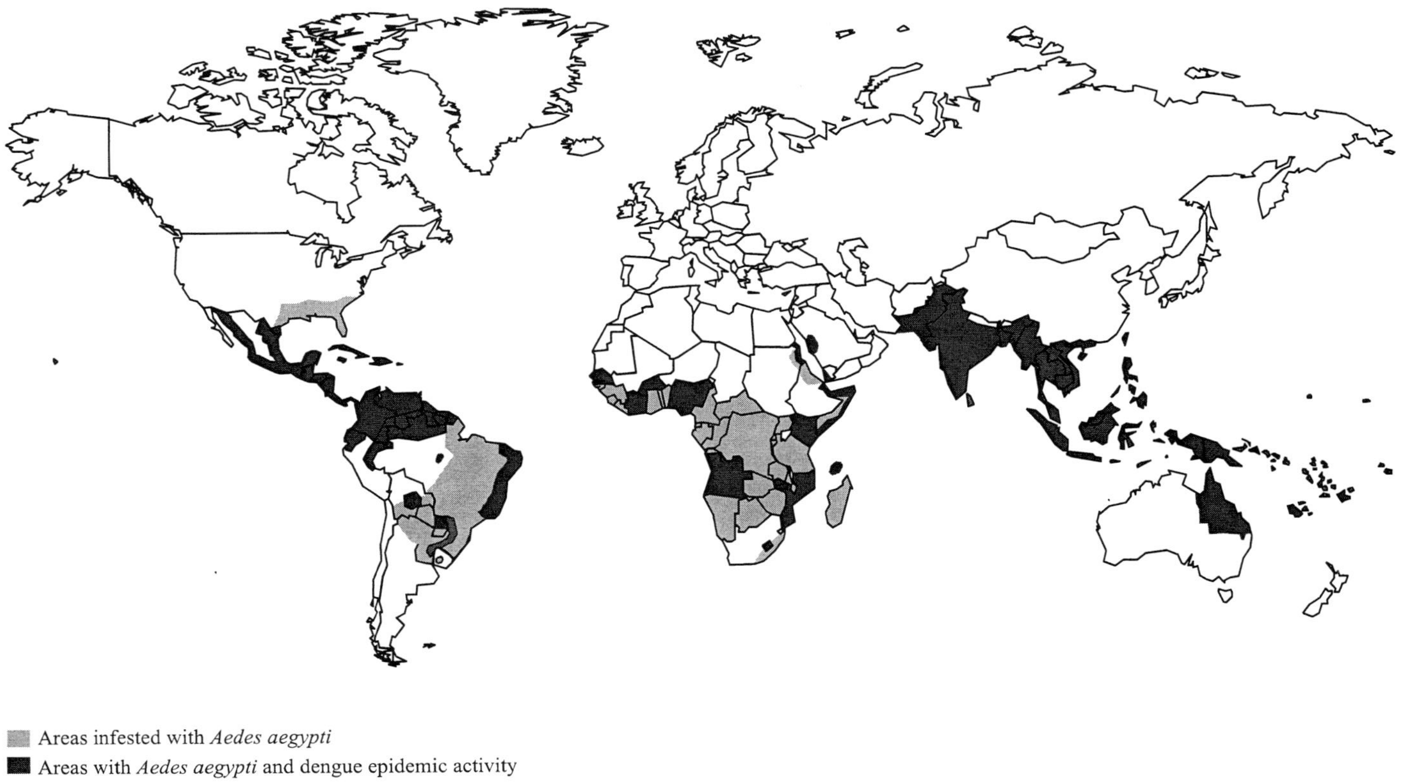

Fig. 1. World distribution of dengue and the principal epidemic vector, *Aedes aegypti*, 2000.

Dengue viruses are transmitted by female *Aedes* (*Stegomyia*) species of mosquitoes (Gubler, 1988, 1997, 1998). *Aedes aegypti*, the principal urban vector, is a small black and white, highly domesticated mosquito that prefers to lay its eggs in artificial water containers commonly found in urban areas of the tropics. Containers found in and around the home, such as those used for water storage, flower vases, old automobile tyres, buckets and various plastic containers, and other junk that collect rainwater, are examples. The adult mosquitoes are highly adapted to living in close association with humans and are rarely noticed, preferring to live indoors and to feed on humans during daylight hours in an unobtrusive and often undetected way (Gubler, 1988, 1997, 1998).

Mosquitoes become infected with dengue viruses by ingesting virus with the blood meal when feeding on persons acutely ill with dengue. From the lumen of the midgut, the virus infects the epithelial cells, passes into the body cavity, and infects most other tissues of the mosquito, including the salivary glands. Once the salivary glands are infected after an extrinsic incubation period of 8–12 days, depending on temperature and other environmental factors, the mosquito becomes infective and can transmit the virus when it takes another blood meal.

Human infection with dengue viruses occurs when an infective mosquito bites a non-immune person. After a period of incubation lasting 3–14 days (mean 4–7 days), the person may experience acute onset of fever accompanied by a variety of nonspecific signs and symptoms (Siler *et al.*, 1926). During this acute febrile period, which may be as short as 2 days and as long as 10 days, a viraemia occurs, which may vary in magnitude and duration (Gubler *et al.*, 1981a, b). If uninfected *Ae. aegypti* mosquitoes bite the ill person during this febrile viraemic stage, those mosquitoes may become infected and subsequently transmit the virus to other, uninfected persons after the extrinsic incubation period of 8–12 days noted above.

CLINICAL PRESENTATION

In humans, dengue virus infection of all four virus serotypes causes a spectrum of illness ranging from inapparent or mild febrile illness to severe and fatal haemorrhagic disease (World Health Organization, 1997, 1999; George & Lum, 1997; Nimmannitya, 1997). The clinical presentation in both children and adults may vary in severity, depending on the strain and serotype of the infecting virus, and the immune status, age and the genetic background of the patient (World Health Organization, 1997; Rosen, 1977; Halstead, 1980; Gubler, 1988, 1998). In dengue endemic areas, acute dengue infections may be clinically nonspecific, especially in children, who often present with signs and symptoms of a nonspecific viral syndrome. The differential diagnosis of dengue during the acute phase of illness includes measles, rubella, influenza, typhoid, leptospirosis, rickettsia, malaria, other viral haemorrhagic fevers, and any other diseases that may present in the acute phase as a nonspecific viral syndrome (World Health Organization, 1997, 1999; George & Lum, 1997; Nimmannitya, 1997; Gubler, 1998).

Classic DF is primarily a disease of older children and adults; it is characterized by sudden onset, fever, and one or more of a number of nonspecific signs and symptoms such as frontal headache, retro-orbital pain, myalgias, arthralgias, nausea and vomiting, weakness and rash (Siler *et al.*, 1926; World Health Organization, 1997, 1999; George & Lum, 1997). Anorexia, altered taste sensation and mild sore throat are not uncommon. Clinical laboratory findings associated with DF include leucopenia and in some patients, thrombocytopenia and elevated liver enzymes (Dietz *et al.*, 1996); haemorrhagic manifestations may occur. DF is generally self-limiting and rarely fatal; the acute illness usually lasts 3–7 days. Convalescence, however, may be prolonged for weeks, with the patient exhibiting weakness and depression. No permanent sequelae are known, and immunity to the infecting virus serotype is lifelong.

DHF is primarily a disease of children under the age of 15 years in Asia, although it may occur in older children and adults in other areas (World Health Organization, 1997, 1999; Dietz *et al.*, 1996; Rigau-Pérez *et al.*, 1998). Like DF, it is characterized by sudden onset of fever and nonspecific signs and symptoms; it is difficult to distinguish DHF from DF and other illnesses during the acute stage. The critical stage in DHF occurs at the time of defervescence when the patient develops a vascular leak syndrome with signs of circulatory failure and haemorrhagic manifestations, primarily skin haemorrhages. Thrombocytopenia ($\leq$100 000 mm^{-3}) and elevated haematocrit are prominent features (World Health Organization, 1997, 1999). DHF can be a very dramatic disease; the patient's condition may deteriorate very rapidly, with onset of shock and death if the vascular leakage is not detected and corrected with fluid replacement therapy. Leucopenia, thrombocytopenia and vascular leakage are constant findings; hepatomegaly and elevated liver enzyme levels are common. Risk factors for developing severe haemorrhagic disease are not fully understood, but as noted above, include the strain and serotype of the infecting dengue virus, and the immune status, age and genetic background of the patient (Rosen, 1977; Halstead, 1980; Gubler, 1988, 1998; World Health Organization, 1997, 1999). The differential diagnosis of DHF includes the same diseases noted above for classic DF.

CHANGING EPIDEMIOLOGY OF DENGUE

DF is an old disease that has had a global distribution in the tropics for over 200 years (Gubler, 1988, 1997, 1998; Halstead, 1980, 1992). Until the 1950s, the disease would occur infrequently as major epidemics in tropical regions of the world. The global epidemiology of dengue viruses first began to change with the ecological disruption that occurred in Southeast Asia during and following World War II (Gubler, 1988, 1997, 1998; Halstead, 1992). During the war, existing water systems were destroyed, and water storage was increased for domestic use as well as for fire control. War equipment was moved frequently between cities and countries, and large amounts were also left behind. This material collected rainwater and, along with water storage containers, made ideal larval habitats for the principal vector mosquito, *Ae. aegypti*. The movement of war materials resulted in the transport of mosquitoes and their eggs to new geographic areas. The result of these ecological changes was a greatly expanded geographic distribution and increased population densities of *Ae. aegypti* in the urban centres of Southeast Asia. In addition, hundreds of thousands of Japanese and Allied soldiers, most of them susceptible to dengue virus infection, were constantly moved between countries in Asia and the Pacific. This provided a mechanism for movement of dengue viruses between cities, countries and other regions, as well for providing susceptible individuals for epidemic transmission. The war years were thus responsible for creating the conditions (the co-circulation of multiple dengue virus serotypes in an area, or hyperendemicity, and high densities of *Ae. aegypti* mosquitoes) that facilitated the emergence of epidemic DHF in Southeast Asia.

In the years following World War II, an unprecedented urbanization of Southeast Asia began, with millions of people moving to the cities of the region. The urban centres in most countries expanded rapidly in an uncontrolled and unplanned fashion. Housing was inadequate for the large numbers of people, and water, sewer and waste management systems deteriorated. The *Ae. aegypti* populations and dengue viruses thrived in this new ecological setting, with increased transmission and increased frequency of dengue epidemics occurring in the indigenous children as the adult populations became immune to the viruses. Moreover, an economic expansion began in the region that continues today. This led to continued urbanization and increased movement of people (and with them, dengue viruses) between cities and countries. Those countries that did not already have multiple virus serotypes co-circulating quickly became hyperendemic. The viruses, often all four serotypes, were (and still are) maintained in a human–*Ae. aegypti*–human transmission cycle in most urban centres of Southeast Asia. The result of these changes was dramatically increased dengue transmission and the emergence of DHF. In every country where the severe disease emerged as a major public health problem, it evolved in a similar manner, first as sporadic cases of DHF occurring for several years, ultimately culminating in a major epidemic. Following the first epidemic, a pattern of epidemic activity was established, with epidemics usually occurring every 3–5 years. Characteristically, successive epidemics became progressively larger as a result of geographic expansion of DHF within the country (World Health Organization, 1997, 1999; Halstead, 1992; Gubler, 1997, 1998).

From the mid-1950s to the 1970s, epidemic DHF was localized in a few Southeast Asian countries. The 1980s and 1990s saw a dramatic geographic expansion of epidemic DHF, west into India, Pakistan, Sri Lanka and the Maldive Islands and east into the Peoples Republic of China (Fig. 1) (Gubler, 1997, 1998). There was also a resurgence of disease in Singapore, which had effectively controlled DHF for nearly 20 years (Fig. 2). Although Singapore still has the most effective prevention and control programme in the region (*Aedes* mosquito vectors are found infesting less than 2 % of premises), there are areas with higher densities (Goh, 1998). Moreover, the human population has a low herd immunity. The critical factor in the resurgence of DF/DHF in Singapore is the thousands of dengue viruses imported to the city by migrant workers and travellers every year. Collectively, these factors ensure continued dengue transmission. It is unlikely that Singapore will be able to effectively prevent outbreaks of DF/DHF until other countries in the region are able to control the disease as well.

Epidemic DF re-emerged as a public health problem in the Americas in the late 1970s after a 40-year quiescence, which was due to an *Ae. aegypti* eradication programme initiated by the Pan American Health Organization (PAHO) in 1946 to prevent urban epidemics of yellow fever (Soper *et al.*, 1943; Schliessman & Calheiros, 1974; Pan American Health Organization, 1979). The programme was successful and eradication was achieved in most countries of the region. Unfortunately, the programme was discontinued in the early 1970s, and failure to eradicate *Ae. aegypti* from all countries of the region resulted in repeated reinfestations by this mosquito of those countries that had achieved eradication. During the 1970s, support for *Ae. aegypti* surveillance and control programmes waned, and they were merged with malaria control programmes in many countries. By the end of the 1980s, *Ae. aegypti* had reinfested much of the area from which it had been eradicated in the 1950s and 1960s (Pan American Health Organization, 1979; Gubler, 1989, 1997). The reinfestation of the region continued during the 1990s, and in 2001, *Ae. aegypti* has a geographic distribution similar to that in the 1930s, before eradication was initiated (Fig. 3), but with higher mosquito densities and in urban centres with much larger human populations.

The expanding geographic distribution of *Ae. aegypti* in the Americas during the 1970s and 1980s coincided with increased movement of dengue viruses both into and within the region (Gubler, 1987, 1989, 1993; Gubler & Trent, 1994). Prior to 1977, only DEN-2 and DEN-3 viruses were known to be present in the Americas, although DEN-1 and DEN-3 were probably present in the early 1940s (Ehrenkranz *et al.*, 1971; Rosen, 1974). A characteristic of dengue in most American countries from the 1950s through the early 1980s was nonendemicity (no viruses present) or hypoendemicity (only a single serotype present) in a country (Ehrenkranz *et al.*, 1971; Pan American Health Organization, 1979; Gubler, 1987, 1989, 1997).

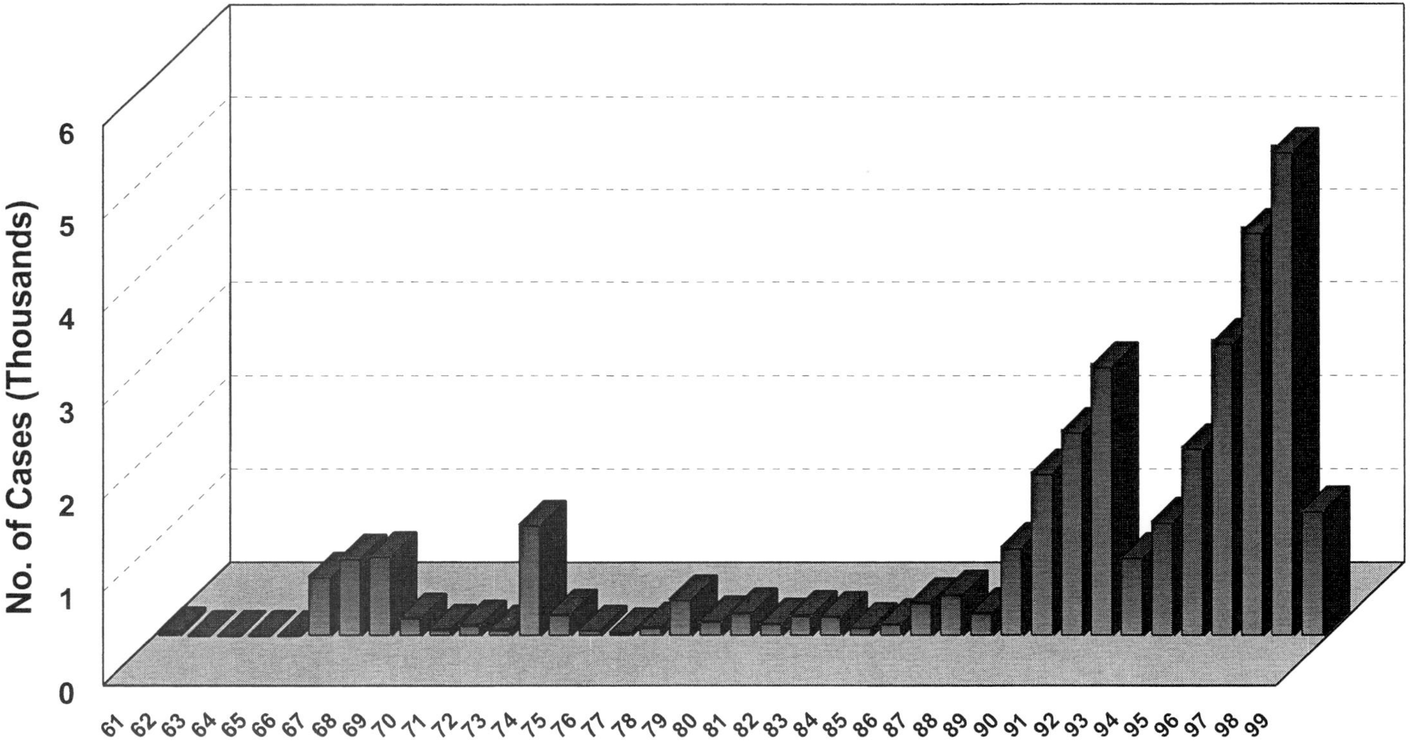

Fig. 2. Re-emergence of epidemic DF/DHF in Singapore.

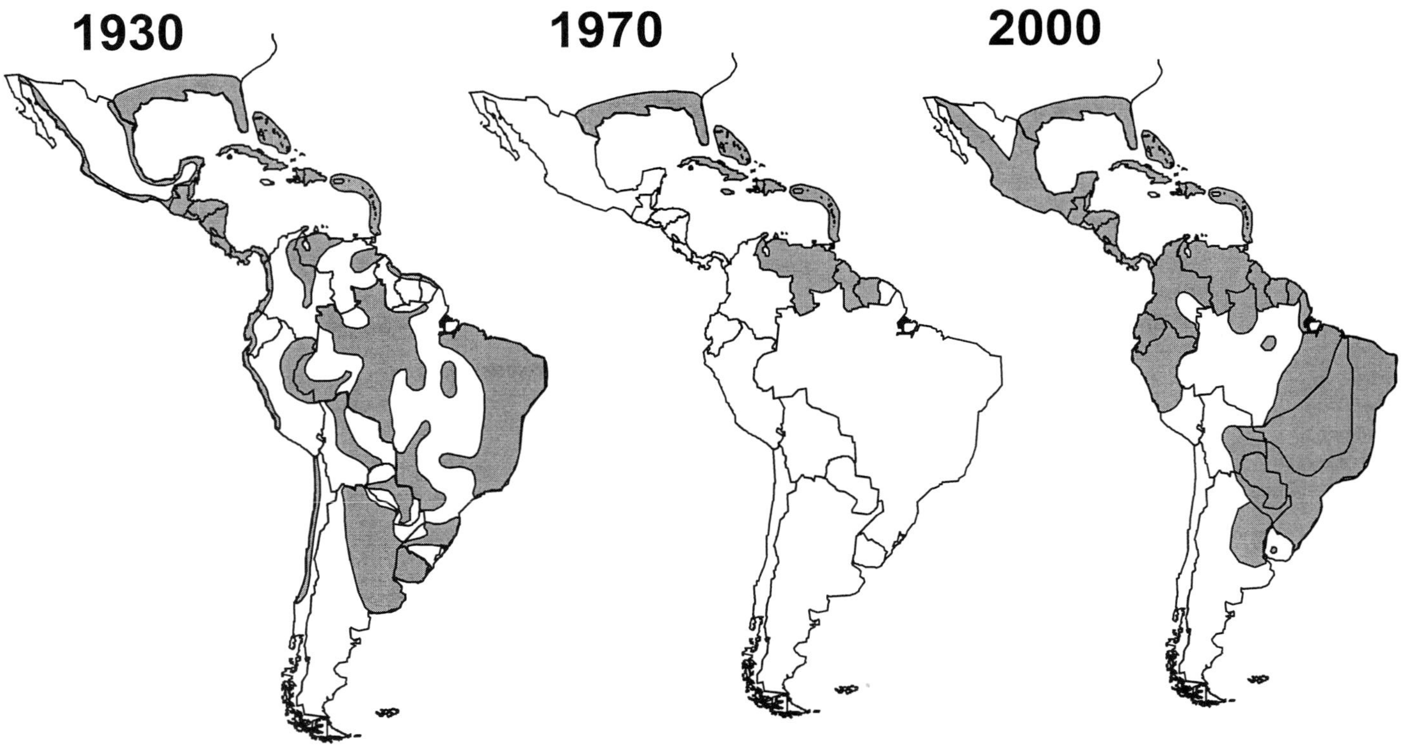

Fig. 3. Geographic distribution of *Aedes aegypti* in the Americas, 1930, 1970 and 2000.

DEN-1 was reintroduced to the American region in 1977, with epidemics in Jamaica and Cuba (Pan American Health Organization, 1979). This serotype subsequently spread throughout the Caribbean Islands, Mexico, Texas, Central America and Northern South America, causing major or minor epidemics over the next 4 years (Pan American Health Organization, 1979; Gubler, 1987, 1993). In 1981, DEN-4 was introduced into the eastern Caribbean Islands, and, like DEN-1, this serotype also spread rapidly to other islands in the Caribbean, and to Mexico, Central America and Northern South America, causing major or minor epidemics, many of them in countries that had experienced recent DEN-1 epidemics (Gubler, 1987, 1993; Gubler & Trent, 1994). Some of these outbreaks (Suriname, 1982; Mexico, 1984; Puerto Rico, 1986; El Salvador, 1987) were associated with the first documented emergence of DHF, occurring in small numbers or sporadically for the most part.

Also in 1981, a new strain of DEN-2 was introduced into Cuba from Southeast Asia (Kouri *et al.*, 1989; Rico-Hesse, 1990; Lewis *et al.*, 1993). Unlike the DEN-1 and DEN-4 epidemics, the 1981 Cuban DEN-2 epidemic was associated with thousands of cases of severe haemorrhagic disease; this was the first major DHF epidemic in the Americas (Kouri *et al.*, 1989). The viruses isolated in Cuba in 1981 were not made available for objective study and controversy arose as to the origin of that strain of DEN-2 (Kouri *et al.*, 1989; Rico-Hesse, 1990; Rico-Hesse *et al.*, 1997; Guzman *et al.*, 1995; Gubler, 1997, 1998). DEN-2 viruses isolated from patients exposed to infection in Jamaica at the time of and shortly after the Cuban epidemic (D. J. Gubler, unpublished data), however, were sequenced (Rico-Hesse, 1990; Lewis *et al.*, 1993). The data suggest that the virus causing the epidemic was a new strain introduced from Asia, most likely from Vietnam, where several thousand Cuban aid personnel were working at the time (Rico-Hesse, 1990; D. J. Gubler, unpublished data).

The second major epidemic of DHF in the Americas, also thought to be caused by this new strain of DEN-2, occurred in Venezuela in 1989–1990 with over 6000 cases and 73 deaths (Pan American Health Organization, 1990). Epidemic DHF of variable intensity caused by this new Southeast Asian genotype of DEN-2 has subsequently occurred in numerous American countries, but none of these epidemics were of the same magnitude and severity as the Cuban epidemic of 1981. Introduction of this genotype of DEN-2 into some countries did not cause epidemic DHF, even though DEN-1 and/or DEN-4 outbreaks had recently preceded its introduction. For example, the Southeast Asian genotype of DEN-2 was introduced into Puerto Rico in 1984, 6 years after this serotype had last been isolated on the island and 2–3 years after outbreaks of DEN-1 and DEN-4 had occurred. The new genotype of DEN-2 circulated on the island for 9 years before causing a major epidemic associated with DHF in 1994 (Rigau-Pérez *et al.*, 2000). An epidemic in Peru in 1995 caused by the old American genotype of

DEN-2 was not associated with DHF (Watts *et al.*, 1999). Currently, there appear to be at least three genotypes of DEN-2 circulating in the American region (Rico-Hesse *et al.*, 1997).

In 1994, a new strain of DEN-3 was introduced into the American region, causing a major epidemic of DF/DHF in Nicaragua and a small outbreak associated with classical DF in Panama (Centers for Disease Control, 1995). This virus, which was shown to be genetically distinct from the DEN-3 virus that previously occurred in the Americas (Lanciotti *et al.*, 1994; R. Lanciotti, I. Quiros, G. G. Clark & D. J. Gubler, unpublished data), apparently was also a recent introduction from Asia. The virus spread throughout Central America and Mexico in 1995 and 1996, causing major epidemics. In January 1998, this new genotype of DEN-3 was detected in Puerto Rico but did not cause an epidemic despite the fact that this serotype had disappeared from the island in 1978. DEN-3 virus has been detected in 17 countries in the Caribbean and Central and South America, but has as yet caused no other major epidemics.

The sequence of events associated with the changing epidemiology of dengue in the Americas in the 1970s, 1980s and 1990s was nearly identical to that which occurred in Southeast Asia in the 1950s, 1960s and 1970s (Gubler, 1987, 1993; Gubler & Trent, 1994). Thus reinvasion of Central and South America by *Ae. aegypti* in the 1970s and 1980s, combined with increased urbanization, increased movement of people and, with them, dengue viruses, resulted in most countries evolving from nonendemicity (no viruses present) or hypoendemicity (one virus present) to hyperendemicity (multiple virus serotypes co-circulating). This resulted in increased frequency of epidemic activity and the emergence of DHF as a major public health problem (Fig. 4). Several countries (Cuba, Venezuela, Brazil, Colombia and Nicaragua) have had major epidemics of DHF. However, outbreaks with sporadic or small numbers of cases of DHF have occurred in many countries throughout the region. In 1980, DHF was not endemic in any American country. Between 1981 and 2000, however, there was a dramatic emergence of DHF, with 28 countries reporting laboratory-confirmed DHF that met the World Health Organization case definition (Fig. 5) (PAHO, unpublished data). This disease is now endemic in most of those countries where multiple dengue virus serotypes co-circulate, and the numbers of cases reported to PAHO have increased dramatically. In 1998, 716 737 cases of DF/DHF were reported by countries in the region, with epidemics in 12 countries. If the disease pattern continues to evolve in the Americas as it did in Southeast Asia, the 21st century will bring more frequent and larger epidemics of DHF.

Surveillance for DF/DHF in Africa has been poor, and epidemics, when they have occurred, were often reported initially as malaria. Although surveillance has not improved, laboratory-confirmed reports of epidemic DF have increased dramatically

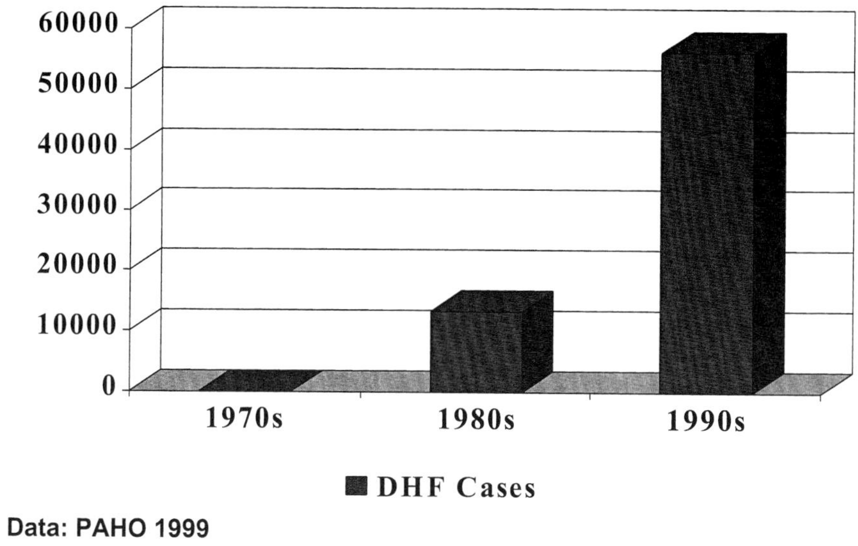

Fig. 4. Emergence of DHF in the Americas by decade, 1970–1999.

since 1980. Limited outbreaks have occurred in West Africa, but the most recent epidemic activity has occurred in East Africa and the Middle East (Gubler, 1997). All four dengue serotypes have been involved, and although sporadic cases of disease clinically compatible with DHF have been reported in some epidemics of DF, DHF in its epidemic form has not yet occurred in Africa or the Middle East.

Surveillance for DF/DHF is passive in most endemic countries, relying on physicians to report suspected disease to the health ministry; only severe cases are typically reported to the World Health Organization. Thus only the tip of the iceberg is reported, making DF/DHF one of the most under-reported tropical infectious diseases in the past 20 years (Gubler, 1997, 1998). As noted above, DF is often reported as malaria. Even so, approximately four times as many DHF cases were reported from 1981 to 1999 than in the previous 30 years (Fig. 6). In addition to its public health importance, DF/DHF has a large economic impact on the communities where it occurs. Recent studies comparing the disability-adjusted life years (DALYs) per million population have shown that DF/DHF has an economic impact of the same order of magnitude as many of the major infectious diseases, such as malaria, hepatitis, sexually transmitted diseases (excluding AIDS), tuberculosis, the childhood cluster and the tropical cluster (Meltzer *et al.*, 1998; Gubler & Meltzer, 1999). In 2000, DF/DHF is the most important arboviral disease affecting humans.

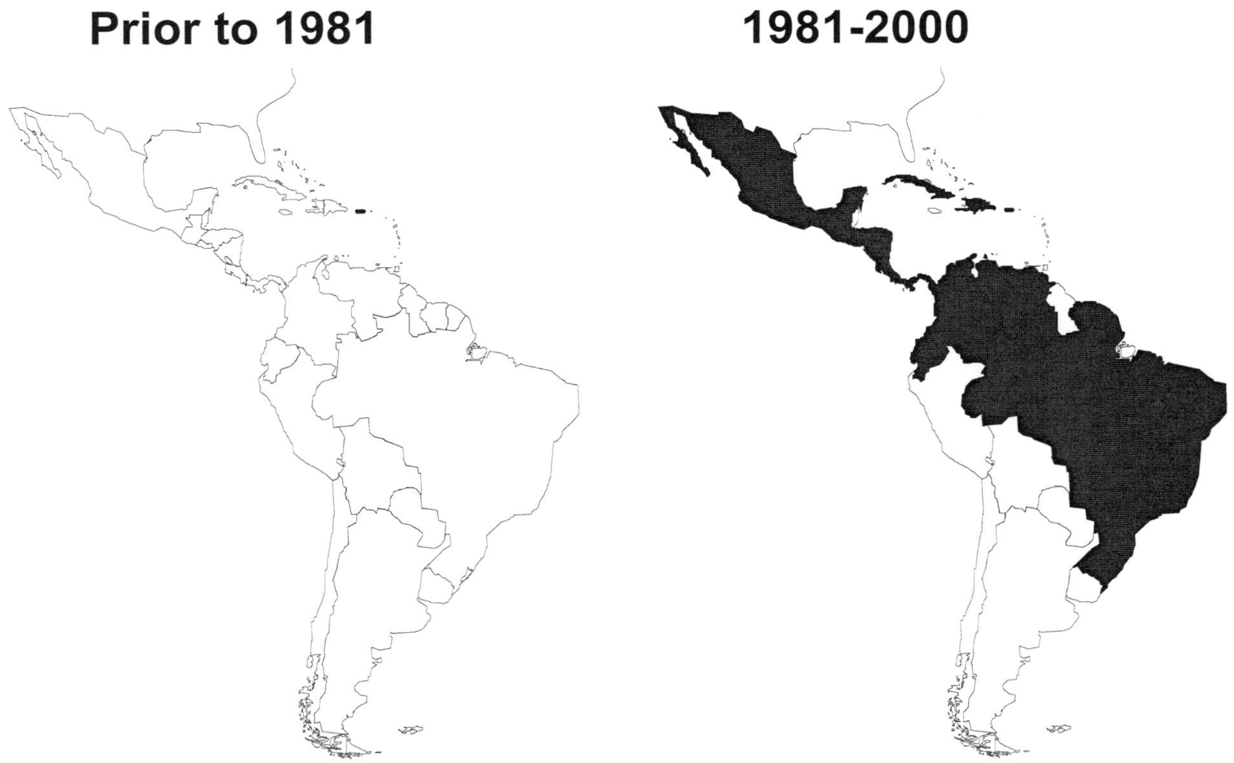

Fig. 5. Geographic spread of DHF in the Americas, prior to 1981, and 1981–2000.

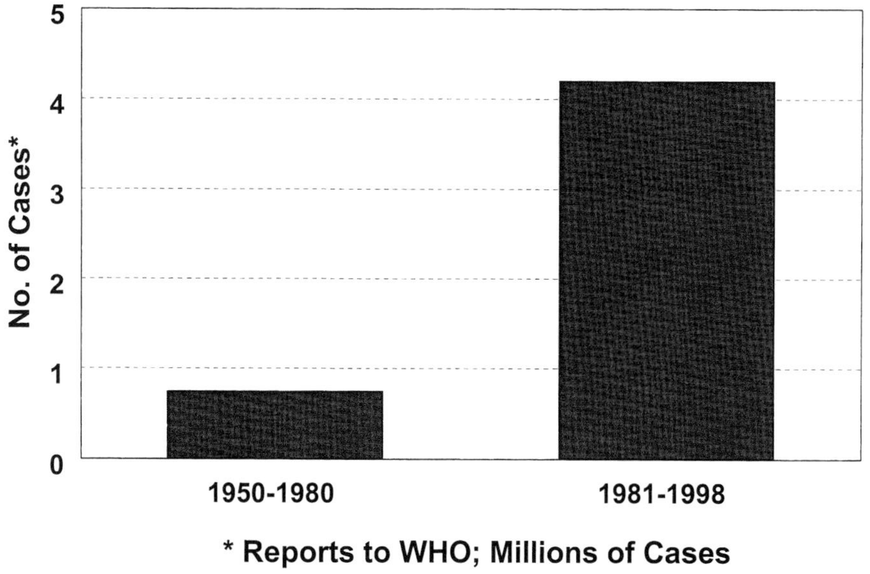

Fig. 6. Cases of DHF reported to the World Health Organization, 1950–1980 and 1981–1998.

FACTORS RESPONSIBLE FOR THE GLOBAL RESURGENCE OF DENGUE

The reasons for the dramatic resurgence of epidemic DF/DHF in the waning years of the 20th century are complex and not fully understood, but are most likely associated with demographic and societal changes (Gubler, 1989, 1997, 1998). The world experienced unprecedented population growth in the last 50 years of the 20th century. It is projected that by the year 2025, the global population will be 8.3 billion people, and by 2050, 10 billion people (Plant, 1996). Recent population increases have occurred primarily in urban centres of developing countries, especially in the tropics. Projections indicate that this trend will continue with 95 % of population growth in the next 25–30 years occurring in developing countries (World Resources Institute, 1996). It is estimated that 39 % of the global population currently live in urban areas, and this will increase to 56 % by the year 2025 (Knudsen & Sloof, 1996).

Coincident with population growth has been uncontrolled and unplanned urbanization in these same countries. These changes have resulted in large, crowded human populations living in urban centres in substandard housing with inadequate water, sewer and waste management systems. Most consumer goods are now packaged in nonbiodegradable plastic or tins, which are discarded into the environment, where they collect rainwater and provide ideal larval habitats for the vector mosquito. Also, used automobile tyres make ideal mosquito larval habitats, and the numbers have increased dramatically in the past 20 years, along with the numbers of automobiles; tyres are very difficult to dispose of from the environment. All of these factors have contributed to the expanded geographic distribution and increased population densities of the principal mosquito vector, *Ae. aegypti*.

Another contributing factor has been the lack of effective *Ae. aegypti* mosquito control in most dengue-endemic countries of the world. Emphasis for the past 30 years has been placed on ultra-low volume space sprays of insecticide for adult mosquito control (Gubler, 1989). This has been shown to be ineffective in controlling *Ae. aegypti* (Gubler, 1989; Newton & Reiter, 1992; Reiter & Gubler, 1997). Thus hundreds of millions of people in urban centres of the tropics are living in intimate association with large populations of an efficient epidemic mosquito vector of dengue viruses, providing ideal conditions for increased transmission of urban mosquito-borne disease such as dengue.

The automobile has also increased the mobility of people within and between population centres in a country, enhancing the ability of dengue viruses to spread locally in infected persons or in mosquitoes, and thus the development and maintenance of multiple virus serotypes (hyperendemicity) in a country or region. The jet airplane has had a similar impact on the rapid and efficient movement of people between regions and continents, and has changed global demographics. It also provides the ideal mechanism for dengue viruses to move with ease in persons incubating the viruses (Gubler, 1989, 1996). The reinfestation of the American tropics by *Ae. aegypti* placed large numbers of susceptible individuals living in permissive urban areas at risk for dengue infection. The numerous epidemics and increased transmission of dengue that subsequently occurred there, in Asia, the Pacific and Africa provided increased opportunity for the viruses to move between countries, both within and between the regions, and has resulted in a constant exchange of dengue viruses and other pathogens, and the development of global hyperendemicity (Fig. 7).

Finally, the public health infrastructure required to deal with epidemic vector-borne infectious diseases has deteriorated during the past 30 years in most countries of the world. Limited financial and human resources, and competing priorities for those resources, have resulted in a 'crisis mentality' among public health officials. The emphasis has thus been on implementing emergency control methods in response to epidemics rather than on developing programmes to prevent epidemic transmission (Gubler, 1989). This approach has been particularly detrimental to dengue prevention and control because in most countries surveillance is very poor; the passive surveillance

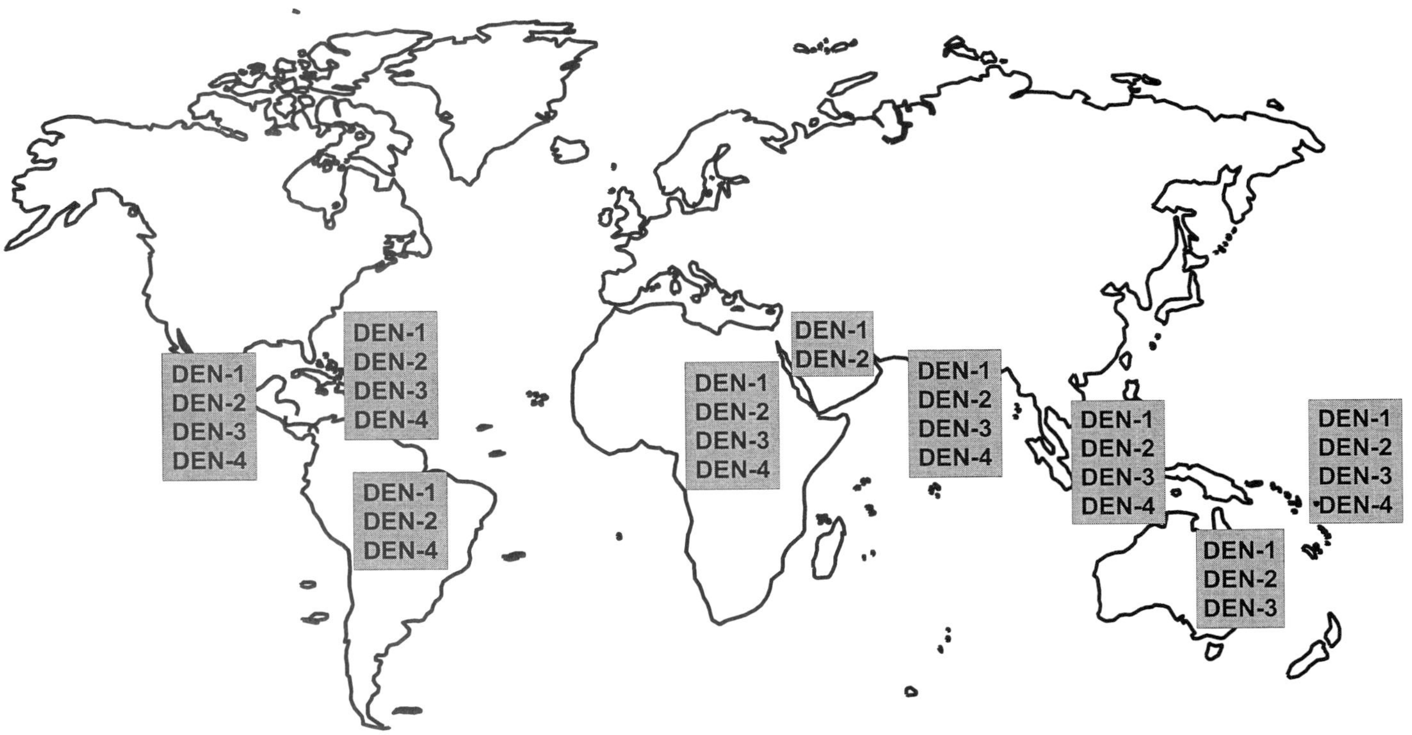

Fig. 7. Known global distribution of dengue viruses in 2000.

systems relied on to detect increased transmission are dependent upon reports by local physicians, who often have a low index of suspicion for dengue and do not consider it in their differential diagnosis of viral syndrome. As a result, epidemics often reach or pass peak transmission before they are detected, and by the time emergency control measures are implemented, it is always too late to have any impact on the course of the epidemic (Gubler, 1989; Newton & Reiter, 1992).

DISEASE PREVENTION AND PROSPECTS FOR THE FUTURE

The prevention of epidemic DF/DHF has become more urgent with the expanding geographic distribution and increased disease incidence in the past 20 years. Unfortunately, tools available to prevent dengue infection are very limited. There is no vaccine currently available (see below) and options for mosquito control are not good.

Effective vaccination to prevent DF/DHF will likely require a tetravalent, live attenuated vaccine. Live, attenuated vaccine candidates for all four dengue virus serotypes have been developed, and have been evaluated in phase I and II trials in Thailand (Bhamarapravati & Yoksan, 1997). Moreover, promising progress has been made on developing second generation, recombinant dengue vaccines using cDNA infectious clone technology (Kinney *et al.*, 1997). Development of new technology such as DNA vaccines also shows promise (Davis & McCluskie, 1999). Despite this progress, however, it is unlikely that an effective, safe and economical dengue vaccine will be available for general use for at least 10 years. Prospects for reversing the trend of increased epidemic DF/DHF in the near future, therefore, are not good. New dengue virus strains and serotypes will likely continue to move in infected air travellers between population centres where *Ae. aegypti* occurs, resulting in continued hyperendemicity, increased frequency of epidemic activity and increased incidence of DHF if effective prevention programmes are not implemented soon. This will require changing the emergency response mentality of government officials, public health professionals and the public to one of disease prevention.

Prevention and control of epidemic DF/DHF currently depends on controlling the mosquito vector, *Ae. aegypti*, in and around the home, where most transmission occurs (Gubler, 1989; Reiter & Gubler, 1997). As noted above, space sprays with insecticides to kill adult mosquitoes are not usually effective unless the insecticide is sprayed indoors every 7–10 days. In large cities, that is logistically impossible with the current resources available. The most effective way to control the mosquitoes that transmit dengue is to control the larvae in the water-holding containers in the domestic environment before they emerge as adults and fly away (Soper *et al.*, 1943; Schliessman & Calheiros, 1974; Gubler, 1989; Reiter & Gubler, 1997).

There are two approaches to larval *Ae. aegypti* control. In the past, the most effective programmes have had a vertical, paramilitary organizational structure, with a large staff and budget (Soper *et al.*, 1943; Schliessman & Calheiros, 1974). The successful American programmes in the 1950s and 1960s were of this type and were also facilitated by the availability of residual insecticides such as DDT that contributed to eliminating the mosquito from the domestic environment. Unfortunately, these vertical programmes have had poor sustainability, because once the mosquito and the disease were controlled, limited health resources were moved to other competing, higher priority public health programmes, and the *Ae. aegypti* population usually rebounded to levels where epidemic transmission occurred (Gubler, 1989). The most recent example of this lack of sustainability is Cuba, where *Ae. aegypti* had been effectively controlled and dengue transmission prevented from 1981 to 1997 (Kouri *et al.*, 1998). The vertically structured Cuban programme failed because of lack of support, resulting in a major DF/DHF epidemic in 1997.

In recent years, emphasis has been placed on community-based, integrated approaches to larval control to provide programme sustainability (Gubler, 1989; Winch *et al.*, 1992; Pan American Health Organization, 1994; World Health Organization, 1999). The rationale is that sustainable *Ae. aegypti* control can only be accomplished by the people who live in the houses where transmission occurs, and who help create the mosquito larval habitats by their lifestyles. Community participation and community ownership of prevention programmes require extensive social marketing of dengue prevention, with health education and community outreach. Unfortunately, this approach is a very slow process, and it will take years for effective disease control to be achieved through community participation alone. It has, therefore, been proposed that a combination top-down and bottom-up approach be used, the former to achieve immediate success and the latter to provide sustainability as the community assumes ownership of the programme (Gubler, 1989). The effectiveness of this approach, however, is not yet known.

In addition to community-based, integrated mosquito control, effective, sustainable prevention programmes for DF/DHF must include several other components (Gubler & Casta-Valez, 1991; Pan American Health Organization, 1994; World Health Organization, 1999). First, an active, laboratory-based surveillance system that can provide early warning for epidemic activity is essential. Health officials in DF/DHF endemic areas must know at any point in time where dengue transmission is occurring, what serotypes are being transmitted, and the severity of illness associated with dengue infection (Gubler & Casta-Valez, 1991; Rigau-Pérez & Gubler, 1997). Moreover, there must be effective use of surveillance data with information exchange and international cooperation. A second component of an effective prevention programme is a contingency plan for rapid response mosquito control to prevent an incipient epidemic when the surveillance system predicts increased dengue transmission. Political and financial support to implement this rapid response plan in a timely manner is critical to preventing major epidemics of DF/DHF. A third component of a sustainable prevention programme is education of the medical community on clinical diagnosis and management of DHF cases. Experience has shown that case fatality rates can be kept acceptably low if physicians and nurses understand the pathophysiologic changes that occur in DHF. Early diagnosis and effective management are the keys to preventing fatalities in this disease (World Health Organization, 1997, 1999; Nimmannitya, 1997).

As noted above, sustainability of prevention programmes will depend on decreasing the reliance on government mosquito control agencies and the transfer of more responsibility for *Ae. aegypti* control to the occupants of houses in urban areas where most dengue transmission occurs. This will require community participation and community ownership of the programme but does not absolve the government of its disease prevention responsibility. The government and community should work together in partnership to achieve effective disease prevention. There is a great need for research to develop more effective disease prevention strategies, including new mosquito control technology and dengue vaccines, and for research on the epidemiology and disease pathogenesis of DF/DHF. Lastly, there is a desperate need to improve the public health infrastructure in disease endemic countries to support community-based, integrated prevention programmes. Only with these changes will we be able to reverse the trend of emergent epidemic DF/DHF in the 21st century.

REFERENCES

Bhamarapravati, N. & Yoksan, S. (1997). Live attenuated tetravalent dengue vaccine. In *Dengue and Dengue Hemorrhagic Fever*, pp. 367–377. Edited by D. J. Gubler & G. Kuno. London: CAB International.

Centers for Disease Control (1995). Imported dengue – United States, 1993–1994. *Morb Mortal Wkly Rep* **44**, 353–356.

Davis, H. L. & McCluskie, M. J. (1999). DNA vaccines for viral diseases. *Microbes Infect* **1**, 7–21.

Dietz, V., Gubler, D. J., Ortiz, S., Kuno, G., Casta-Vélez, A., Sather, G., Gomez, I. & Vergne, E. (1996). The 1986 dengue and dengue hemorrhagic fever epidemic in Puerto Rico: epidemiologic and clinical observations. *P R Health Sci J* **15**, 201–210.

Ehrenkranz, N. J., Ventura, A. K., Cuadrado, R. R., Pond, W. L. & Porter, J. E. (1971). Pandemic dengue in Caribbean countries and the southern United States – past, present and potential problems. *N Engl J Med* **285**, 1460–1469.

George, R. & Lum, L. (1997). Clinical spectrum of dengue infection. In *Dengue and Dengue Hemorrhagic Fever*, pp. 89–113. Edited by D. J. Gubler & G. Kuno. London: CAB International.

Goh, K. T. (1998). Dengue – a reemerging infectious disease in Singapore. In *Dengue in Singapore*, pp. 33–46. Institute of Environmental Epidemiology, Ministry of the

Environment, a World Health Organization Collaborating Center for Environmental Epidemiology, Technical Monograph Series, No. 2, Singapore.

Gubler, D. J. (1987). Dengue and dengue hemorrhagic fever in the Americas. *P R Health Sci J* **6**, 107–111.

Gubler, D. J. (1988). Dengue. In *Epidemiology of Arthropod-Borne Viral Disease*, vol. II, pp. 223–260. Edited by T. P. Monath. Boca Raton, FL: CRC Press.

Gubler, D. J. (1989). *Aedes aegypti* and *Aedes aegypti*-borne disease control in the 1990s: top down or bottom up. *Am J Trop Med Hyg* **40**, 571–578.

Gubler, D. J. (1993). Dengue and dengue hemorrhagic fever in the Americas. In *Dengue Hemorrhagic Fever*, pp. 9–22. Edited by P. Thoncharoen. Regional Publication No. 22. New Delhi, India: World Health Organization Monograph, South East Asia Regional Office.

Gubler, D. J. (1996). Arboviruses as imported disease agents: the need for increased awareness. *Arch Virol* **11**, 21–32.

Gubler, D. J. (1997). Dengue and dengue hemorrhagic fever: its history and resurgence as a global public health problem. In *Dengue and Dengue Hemorrhagic Fever*, pp. 1–22. Edited by D. J. Gubler & G. Kuno. London: CAB International.

Gubler, D. J. (1998). Dengue and dengue hemorrhagic fever. *Clin Microbiol Rev* **11**, 480–496.

Gubler, D. J. & Casta-Valez, A. (1991). A program for prevention and control of epidemic dengue and dengue hemorrhagic fever in Puerto Rico and the U.S. Virgin Islands. *Bull Pan Am Health Organ* **25**, 237–247.

Gubler, D. J. & Meltzer, M. (1999). Impact of dengue/dengue hemorrhagic fever on the developing world. *Adv Virus Res* **53**, 35–70.

Gubler, D. J. & Trent, D. W. (1994). Emergence of epidemic dengue/dengue hemorrhagic fever as a public health problem in the Americas. *Infect Agents Dis* **2**, 383–393.

Gubler, D. J., Suharyono, W., Tan, R., Abidin, M. & Sie, A. (1981a). Viraemia in patients with naturally acquired dengue infection. *Bull WHO* **59**, 623–630.

Gubler, D. J., Suharyono, W., Lubis, I., Eram, S. & Gunarso, S. (1981b). Epidemic dengue 3 in Central Java, associated with low viremia in man. *Am J Trop Med Hyg* **30**, 1094–1099.

Guzman, M. G., Deubel, V., Pelegrino, J. L., Rosario, D., Marrero, M., Sariol, C. & Kouri, G. (1995). Partial nucleotide and amino acid sequences of the envelope and the envelope/nonstructural protein-1 gene junction of four dengue-2 virus strains isolated during the 1981 Cuban epidemic. *Am J Trop Med Hyg* **52**, 241–246.

Halstead, S. B. (1980). Dengue hemorrhagic fever – public health problem and a field for research. *Bull WHO* **58**, 1–21.

Halstead, S. B. (1992). The XXth century dengue pandemic: need for surveillance and research. *World Health Stat Q* **45**, 292–298.

Kinney, R. M., Butrapet, S., Chang, G.-J., Tsuchiya, K. R., Roehrig, J. T., Bhamarapravati, N. & Gubler, D. J. (1997). Construction of infectious cDNA clones for dengue 2 virus: strain 16681 and its attenuated vaccine derivative, strain PDK-53. *Virology* **230**, 300–308.

Knudsen, A. B. & Sloof, R. (1996). Vector-borne disease problems in rapid urbanization: new approaches to vector control. *Bull WHO* **70**, 1–6.

Kouri, G. P., Guzman, M. G., Bravo, J. R. & Triana, C. (1989). Dengue haemorrhagic fever/dengue shock syndrome: lessons from the Cuban epidemic, 1981. *Bull WHO* **67**, 375–380.

Kourí, G., Guzmán, M. G., Valdéz, L., Carbonel, I., del Rosario, D., Vasquez, S., Laferté, J., Delgado, J. & Cabrera, M. V. (1998). Reemergence of dengue in Cuba: a 1997 epidemic in Santiago de Cuba. *Emerg Infect Dis* **4**, 1–4.

Lanciotti, R. S., Lewis, J. G., Gubler, D. J. & Trent, D. W. (1994). Molecular evolution and epidemiology of dengue-3 viruses. *J Gen Virol* **75**, 65–75.

Lewis, J. A., Chang, G. J., Lanciotti, R. S., Kinney, R. M., Mayer, L. W. & Trent, D. W. (1993). Phylogenetic relationships of dengue-2 viruses. *Virology* **197**, 216–224.

Meltzer, M. I., Rigau-Pérez, J. G., Clark, G. G., Reiter, P. & Gubler, D. J. (1998). Using disability-adjusted life years to assess the economic impact of dengue in Puerto Rico: 1984–1994. *Am J Trop Med Hyg* **59**, 265–271.

Newton, E. A. C. & Reiter, P. (1992). A model of the transmission of dengue fever with an evaluation of the impact of ultra-low volume (ULV) insecticide applications on dengue epidemics. *Am J Trop Med Hyg* **47**, 709–720.

Nimmannitya, S. (1997). Dengue hemorrhagic fever: diagnosis and management. In *Dengue and Dengue Hemorrhagic Fever*, pp. 133–145. Edited by D. J. Gubler & G. Kuno. London: CAB International.

Pan American Health Organization (1979). Dengue in the Caribbean, 1977. Proceedings of a Workshop held in Montego Bay, Jamaica, 8–11 May, 1978. Washington, DC: Pan American Health Organization.

Pan American Health Organization (1990). Dengue hemorrhagic fever in Venezuela. *Epidemiol Bull* **11**, 7–9.

Pan American Health Organization (1994). Dengue and dengue haemorrhagic fever in the Americas: guidelines for prevention and control. Washington: World Health Organization/Pan American Health Organization, monograph 548.

Plant, A. E. (1996). Infecting ourselves: how environmental and social disruptions trigger disease. Worldwatch Paper, No. 129. Washington, DC: Worldwatch Institute.

Reiter, P. & Gubler, D. J. (1997). Surveillance and control of urban dengue vectors. In *Dengue and Dengue Hemorrhagic Fever*, pp. 425–462. Edited by D. J. Gubler & G. Kuno. London: CAB International.

Rico-Hesse, R. (1990). Molecular evolution and distribution of dengue viruses type 1 and 2 in Nature. *Virology* **174**, 479–493.

Rico-Hesse, R., Harrison, L. M., Salas, R. A., Tovar, D., Nisalak, A., Ramos, C., Boshell, J., De Mesa, M. T. R., Nogueira, R. M. R. & Travassos Da Rosa, A. (1997). Origins of dengue type 2 viruses associated with increased pathogenicity in the Americas. *Virology* **230**, 244–251.

Rigau-Pérez, J. G. & Gubler, D. J. (1997). Surveillance for dengue and dengue hemorrhagic fever. In *Dengue and Dengue Hemorrhagic Fever*, pp. 405–423. Edited by D. J. Gubler & G. Kuno. London: CAB International.

Rigau-Pérez, J., Clark, G., Gubler, D. J., Reiter, P., Sanders, E. & Vorndam, A. V. (1998). Dengue and dengue haemorrhagic fever. *Lancet* **352**, 971–977.

Rigau-Pérez, J. G., Vordam, V. & Clark, G. G. (2000). The dengue and dengue hemorrhagic fever epidemic in Puerto Rico, 1994–1995. *Am Soc Trop Med Hyg* (in press).

Rosen, L. (1974). Dengue type 3 infection in Panama. *Am J Trop Med Hyg* **23**, 1205–1206.

Rosen, L. (1977). The Emperor's new clothes revisited, or reflections on the pathogenesis of dengue hemorrhagic fever. *Am J Trop Med Hyg* **26**, 337–343.

Schliessman, D. J. & Calheiros, L. B. (1974). A review of the status of yellow fever and *Aedes aegypti* eradication programs in the Americas. *Mosq News* **34**, 1–9.

Siler, J. F., Milton, W. H. & Hitchens, A. P. (1926). *Dengue: its History, Epidemiology, Mechanism of Transmission, Etiology, Clinical Manifestations, Immunity, and Prevention*, pp. 1–302. Manila: Philippine Bureau of Science.

Soper, F. L., Wilson, D. B., Lima, S. & Antunes, W. S. (1943). *The Organization of Permanent Nationwide Anti-Aedes aegypti Measure in Brazil*. New York: The Rockefeller Foundation.

Watts, D. M., Porter, K. R., Putvatana, P., Vasquez, B., Calampa, C., Hayes, C. G. & Halstead, S. B. (1999). Failure of secondary infection with American genotype dengue 2 to cause dengue haemorrhagic fever. *Lancet* **354**, 1431–1434.

Winch, P., Kendall, C. & Gubler, D. J. (1992). Effectiveness of community participation in vector-borne disease control. *Health Policy Planning* **7**, 342–351.

World Health Organization (1997). *Dengue Haemorrhagic Fever, Diagnosis, Treatment, Prevention and Control*. Geneva: World Health Organization.

World Health Organization (1999). *Prevention and Control of Dengue and Dengue Haemorrhagic Fever*. Regional Publication No. 29. New Delhi, India: South East Asia Regional Office, World Health Organization.

World Resources Institute (1996). *World Resources, A Guide to the Global Environment: the Urban Environment*. WRI, Washington, DC: Oxford University Press.

Borna disease virus – a threat for human mental health?

Liv Bode[1] and Hanns Ludwig[2]

[1]Project 23: Bornavirus Infections, Robert Koch-Institut, Nordufer 20, D-13353 Berlin, Germany

[2]Institute of Virology, Free University of Berlin, Königin-Luise-Strasse 49, D-14195 Berlin, Germany

'*Mania* is sickness for one's friends, *depression* for one's self. Both is chemical ... a weakening in the blood.'

From '*Collected Poems*' by Robert Lowell (1917–1977), American poet and Pulitzer prize winner with a documented history of manic-depressive illness.

'Dabei kommt natürlich alles darauf an, dass wir für die betreffenden Seuchen ganz zuverlässige Methoden zum Nachweis der Krankheitserreger besitzen. Insbesondere gilt dies für diejenigen Kranken, welche sich im frühesten Stadium der Krankheit befinden und für die sogenannten Bazillenträger, weil diese beiden Kategorien von Kranken erfahrungsgemäß am meisten zur Verschleppung der Seuchen beitragen. Auf jeden Fall bildet eine sichere Diagnose gewissermaßen den Schlüssel für die moderne Seuchenbekämpfung.' ['Of course it all depends on the availability of reliable methods to detect the causative agents of the pestilence in question. This is particularly true for those patients in very early stages of the disease and for the so-called "carriers", as experience tells us that these two categories of patients are the main contributors when it comes to the spread of the pestilence. In any case, a reliable diagnosis is the key for the modern battle against pestilence.']

From a speech given by Robert Koch 1908 at a convention of most of the MDs of Berlin honouring his return from a successful expedition to East Africa, where he investigated the sleeping disease (published in *Dtsch Med Wochenschr* no. 8).

PROLOGUE

Unlike related viruses infecting both man and animals (for example, influenza virus and rabies virus), in which zoonotic pathways are well established, Borna virus infections of

SGM symposium 60: New challenges to health: the threat of virus infection.
Editors G. L. Smith, W. L. Irving, J. W. McCauley, D. J. Rowlands. Cambridge University Press. ISBN 0 521 80614 3 ©SGM 2001.

man have only been reported recently and their possible zoonotic transmission remains unknown. However, an association of Borna disease virus with mood disorders (Bode, 1995), supported by virus isolates from depressed (Bode *et al.*, 1996) and schizophrenic (Planz *et al.*, 1999; Nakamura *et al.*, 2000) patients, makes this an exciting virus to study to unravel its involvement in human mental health and follow its possible zoonotic routes (Ludwig & Bode, 2000).

A horse disease resulting in behavioural changes and neurological symptoms has been present in Europe for centuries. Its endemic occurrence in the region of Borna, near Leipzig, is crystallized in its name 'Borna-Krankheit', the causative agent of which was finally named Borna disease virus (BDV). A comprehensive review of the historical events in Borna disease and virus research has been published recently (Dürrwald & Ludwig, 1997). During the 1990s, considerable interest in Borna disease arose as a result of the characterization of many biological features, the morphology, and aspects of the molecular biology of the virus. Certainly, the increasing worldwide interest is fuelled by the implication of BDV being involved in psychiatric diseases of humans.

BDV infections are broadly spread in nature. Besides horses, sheep have been found to suffer from Borna disease, causing economic losses. Other farm animals or companion mammals show more or less sporadic disease. Even ostriches occasionally acquire fatal disease. The ability to infect rodents experimentally and study behavioural changes and neurological disease and to infect primates and study their neuropathology has pointed increasingly to the possibility that human infections with analogous patterns and a similar aetiopathogenesis may exist (Ludwig & Bode, 1997). The definition of animal reservoirs and their monitoring, the epidemiology of animal and human infection and antiviral therapy have been summarized and discussed recently (Ludwig & Bode, 2000).

This chapter will pull together the present knowledge on the agent, its biology (Ludwig *et al.*, 1988) and molecular biology (Koprowski & Lipkin, 1995), which led to its classification in its own family (de la Torre *et al.*, 2000), as a member of the order *Mononegavirales* (Pringle, 1995). Furthermore, reviewing clinical aspects of human infections, old and new infection parameters and their interpretation, relevant animal models and the results of antiviral treatment *in vitro* and *in vivo* will lead to strengthening the tenet of BDV's significant influence on human health.

THE AGENT

Biological properties

Neurons of the limbic system are the preferential target cells of BDV under natural conditions, where the infection results in the formation of typical intranuclear inclusion bodies that are pathognomonic for the infection (Joest & Degen, 1911). However, neuropathological studies using naturally infected brains of horses and other animals have shown that in an advanced stage of infection and disease, other cells of the brain, astrocytes and oligodendrocytes, may also become infected (Ludwig *et al.*, 1985; Gosztonyi & Ludwig, 1995; Stitz *et al.*, 1995). *In vitro*, the virus can enter almost every cell type when co-cultivated with explant cells, or when adapted sufficiently to tissue culture (Ludwig *et al.*, 1973; Danner *et al.*, 1978; Herzog & Rott, 1980). Several cell lines persistently infected with strain V (TL) have been established and some of their properties described: Vero/TL, MDBK/TL, MDCK/TL, CRFK/TL, ED/TL, PK15/TL, C6/TL and, from human cells, Oligo/TL. Another line, the C6BV line (propagated by de la Torre's group) which carries the other laboratory strain, abbreviated He/80, has been characterized (for further details, see Ludwig & Bode, 2000). These two strains, V and He/80, are the only two tissue-culture-adapted genetically fully characterized strains (Briese *et al.*, 1994; Cubitt *et al.*, 1994) used worldwide. Strain V is a lapinized horse isolate from the 1920s (see Zwick, 1939), and strain He/80 originates from an experimental animal-adapted horse isolate (Herzog & Rott, 1980). The defined strain V was distributed by the Berlin group; strain He/80 was given to European and Japanese groups by the Giessen group. Recent resequencing indicates that the latter gifts are no longer strain He/80 (unpublished information).

None of the virus strains causes a cytopathic effect, thus assaying this virus required cell-antigen-detection tests. This complicates cell biological and diagnostic investigations. It is known that small to larger antigen deposits (according to the cell type) appear in the nucleus within the infection cycle, a pattern which allows infectivity to be monitored. The growth curve is closely linked to the cell cycle, and virus spreads from cell to cell, forming a focus with a characteristic number of antigen-carrying cells. These observations enabled titration of the virus in a focus assay and the demonstration of neutralization of BDV. Soon thereafter, neutralizing antibodies were detected in sera and cerebrospinal fluid of animals infected naturally or experimentally (Ludwig *et al.*, 1988, 1993).

Primary cells of rabbit origin (e.g. young rabbit brain, spleen and retina cells) are suitable for virus titration. Our group has established a permanent cell line from a young rabbit spleen which is very useful for BDV research, both for studies on cell biology and diagnosis (H. Ludwig & T. Leiskau, unpublished).

In these cells, the virus growth curve in freshly infected cells reaches saturation after 3 or 4 days; the complete replication cycle, however, is not known. However, evidence is accumulating that it is connected with the cell cycle (Ludwig *et al.*, 1988).

Molecular and biochemical properties

The buoyant density of BDV is 1.22 g cm^{-3}. Infectivity is destroyed by heat treatment, acid and alkaline pH, and by treating suspensions with organic solvents, detergents and UV irradiation, as well as with disinfectants containing chlorine or formaldehyde.

Two different BDV strains were isolated originally from horses: strain V and He/80. They have almost identical viral genomes, containing at least six ORFs: (from the 3′ end) N (nucleoprotein), P (phosphoprotein), X (p10 protein), M (matrix protein), G (glycoprotein), L (large/polymerase). The virus genome is a linear, single-stranded, monopartite RNA of 8.915 bases of negative polarity. Generally, the genes map into the scheme of viruses grouped as *Mononegavirales* (Pringle, 1995), except that the X gene has no counterpart (Fig. 1 and Table 1).

The nucleic acid of the genome is not polyadenylated, and extracistronic sequences have been detected at the 3′ (leader) and 5′ (trailer) ends of the genome. The termini of the RNA show inverted partial complementarity. Antigenomic RNAs (positive-stranded full-length) are present in infected cells. In addition to these two full-length 'reference' sequences, which have undergone interspecies passaging, RNA sequences of viruses from natural virus sources have been determined. All these sequences display a high degree of genetic similarity. Particularly in the p40 (ORF I) and p24 (ORF II) genes, the nucleoprotein and the phosphoprotein, very few significant differences were found. This high stability of BDV genomes supports the hypothesis that genomes of currently isolated viruses are the result of a long-term evolutionary process, and that the virus has had a long period of time to adapt to different human and animal hosts and is not evolving rapidly (Ludwig & Bode, 2000; de la Torre *et al.*, 2000).

These properties with the unique feature of nuclear replication and virion morphology led to the creation of a new taxonomic family in the order *Mononegavirales*, named *Bornaviridae* (de la Torre *et al.*, 2000).

Replication, transcription, translation and immunogenicity

Replication occurs in a similar manner to that in other members of the *Mononegavirales* but with the unique property of nuclear replication. Both anti-sense (genomic) and sense (messenger) RNA are present in the nucleus, where transcription and replication take place. *In situ* hybridization has shown that areas of intranuclearly located Joest–Degen bodies are involved in replication of the virus (Gosztonyi & Ludwig, 1995).

Transcription does not use the influenza virus cap-snatching mechanism. The N gene gives rise to monocistronic mRNA, the P and X proteins are encoded by a bicistronic

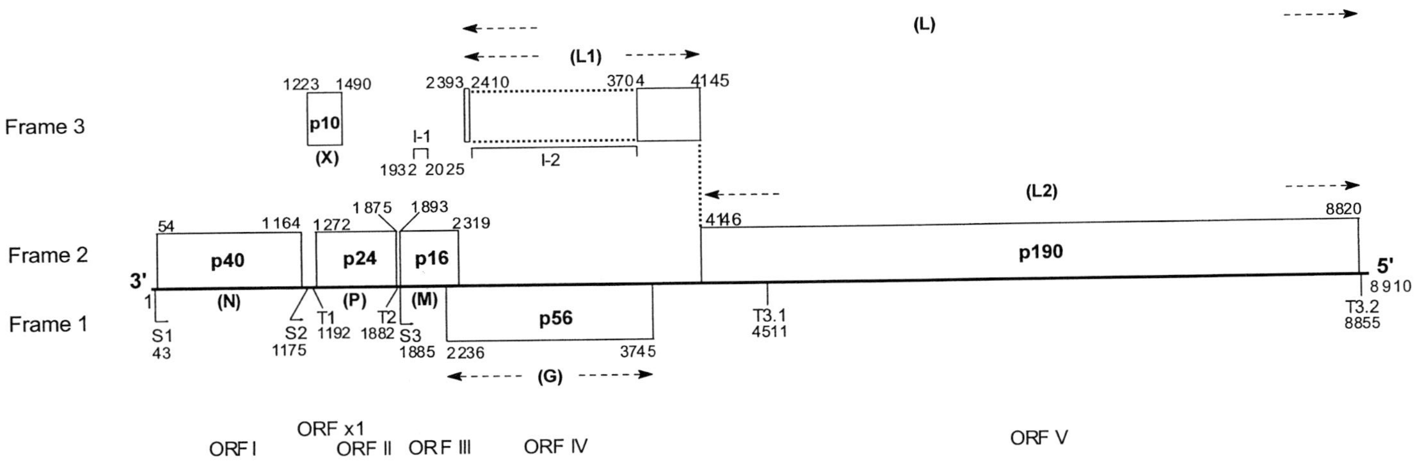

Fig. 1. Organization and physical map of the BDV genome, strain V. The single-stranded RNA (from 3′ to 5′) has six known genes (see also Table 1); their ORFs are shown by boxes. Numbers indicate the nucleotide position of the start codon AUG and stop codons UGA, UAA and UAG in the respective ORFs. Transcription initiation sites (S1, S2, S3) and their nucleotide position on the genome are given by downward lines with small arrows pointing downstream. Transcription termination sites (T1, T2, T3.1, T3.2) and their nucleotide position are indicated by downward lines. The splice sites (intron 1, I-1; and intron 2, I-2) and their nucleotide position are represented by horizontal lines above the main reading frame (Frame 2), which contains the genes for the N, P, M and L2 proteins. Frame 1 with a nucleotide shift of plus 1 contains the gene for the G protein, whereas Frame 3, with a nucleotide shift of plus 2, contains the genes X and L1. These data will be published in a slightly different form by Stoyloff *et al.* (2001), and are based on Briese *et al.* (1994), Schneider *et al.* (1994), de la Torre *et al.* (2000), Walker *et al.* (2000) and our own data.

Table 1. Nucleotides of the BDV genome, strain V (accession no. U04608; total length 8910 nt)

ORF . . .	**I**	**II**	**X**	**III**	**IV**	**V**
Gene product	p40	p24	p10	p16	p56	p190
Final protein	38/40 kDa	24 kDa	10 kDa	gp17 kDa	gp94 kDa	190 kDa
Polymeric forms	No	Dimer (60 kDa)	?	Tetramer (68 kDa)	?	No
Name	Nucleoprotein (N)	Phosphoprotein (P)	X protein	Matrix protein (M)	Glycoprotein (G)	L polymerase
Genome position 3′–5′	54–1164	1272–1875	1223–1490	1893–2319	2236–3745	L1: 2393–2410; 3704–4145 L2: 4146–8820
No. of nucleotides	1110	603	267	426	1509	(18 + 441) + 4674 = 5133
No. of amino acids	370	201	89	142	503	1711

(gene) unit, whereas the M and G proteins and L polymerase gene products originate from polycistronic messages. Two of the BDV transcripts have been shown to be processed by a cellular, post-transcriptional RNA splicing mechanism. The BDV genome carries two introns (I and II), at nucleotide positions 1932–2025 (intron I) and 2410–3703 (intron II). Certainly, RNA splicing (Schneider *et al.*, 1994) contributes to the regulation of expression of gene products, such as the M and G proteins, and probably also the L protein (de la Torre *et al.*, 2000).

The translation products of the N and P mRNAs are most abundant and accumulate in the cell nucleus and later in the cytoplasm. In dissociated cells of the brain or tissue culture, the p24 polypeptide is often present as an aggregate of 60 kDa (Ludwig *et al.*, 1988), in a disulfide-linked dimeric form (de la Torre *et al.*, 2000). These polypeptides constitute the so-called soluble antigen (s-antigen). Their role as major antigenic components in the brain of infected animals and the prominent targets for the humoral immune response were revealed in the 1950s (von Sprockhoff, 1956; Nitzschke, 1963). These two polypeptides have been found in all persistently infected cell lines (Ludwig *et al.*, 1988), and recently in peripheral blood mononuclear cells (PBMCs) and in the plasma (see below).

The glycoproteins of BDV in infected cells as well as the virion deserve special attention, although their relevance in the infection process seems to be less important than in other *Mononegavirales*. The polypeptide backbone of the M protein (ORF III) has a molecular mass of 16 kDa and the post-translationally modified species has an apparent molecular mass of 17 kDa in its glycosylated form. The glycoprotein G (ORF IV) has an apparent molecular mass of 94 kDa, of which 56 kDa make up the polypeptide backbone. Both glycoproteins carry N-linked carbohydrate side chains with terminal 2-acetamido-2-deoxy-β-D-glucopyranose and α-D-mannopyranose residues in a 1–2 linkage. Notably, the M protein has an N-glycosidic modification of the hybrid type with a structure similar to mannan.

MALDI mass spectrometry shows that the M protein exists as a stable tetrameric aggregate with strong hydrophobic interactions with a molecular mass of 69 kDa (Stoyloff *et al.*, 1997). The G protein is metastable in its form. In our studies, breakdown products of molecular masses 46, 48, 57, 74, 82, 84 and 91 kDa were found. The amino acid sequence shows 19 cysteine residues spread over the protein. They could be responsible for defined G protein fragment multimers held together by covalent disulfide bonds. The M and G proteins are membrane-bound with two hydrophobic membrane-spanning domains. Analysis of the subcellular distribution of proteins in Oligo/TL cells showed that the G protein is present mainly in the nuclear envelope, slightly in cytoplasmic membranes and is not present in the nucleoplasm or cytoplasm.

In contrast, the M protein was detected in the nucleoplasm and cytoplasm as well as in the cytoplasmic membrane (Stoyloff *et al.*, 2001; R. Stoyloff, unpublished).

The carbohydrate motifs play an important role in the interaction with the putative receptor for the virion. Possibly, the M protein is involved in virus attachment (Gonzalez-Dunia *et al.*, 1997a). Specific carbohydrate components β-D-GlcNAC and α-D-Man bound to BSA as well as highly purified M protein specifically inhibit BDV infectivity. Additionally, specific anti-carbohydrate antibodies directed against β-D-GlcNAC and α-D-Man–BSA reduce virus infectivity. This indicates that the carbohydrate moieties of BDV participate in neutralization (Stoyloff *et al.*, 1998). Furthermore, enzymic hydrolysis by exoglucosidases of the terminal carbohydrate residues as well as by endoglycosidases, specific for a defined *N*-hybrid linkage, led to a dramatic loss of virus infectivity (Stoyloff *et al.*, 2001).

The ORF V encodes a 180 kDa protein which represents the L polymerase, as in other *Mononegavirales*. This protein shares a high genetic similarity with polymerases of other such viruses (de la Torre *et al.*, 2000).

The N, P, M and G proteins constitute the antigenicity of BDV, with N and P playing the major role (Ludwig & Bode, 2000). A protein complex formed by p40 and p24 represents the principal immunogen in BDV infections, giving rise to a strong non-neutralizing humoral response (Ludwig *et al.*, 1988; Ludwig & Bode, 2000). The N protein is expressed early during the infectious cycle, and exists in two isoforms (38/40 kDa) (Pyper & Gartner, 1997). The protein covers the RNA, probably together with p24 and the L polymerase. It is not clear whether M participates in RNP structures. Infectious RNPs can be isolated from the nucleus (de la Torre *et al.*, 2000).

The virion

Despite earlier attempts to characterize this virus by negative staining or transmission electron microscopy (for further details, see Ludwig & Bode, 1997), it was only recently that Zimmermann *et al.* (1994) visualized virus particles successfully and attributed them to the infectious agent unequivocally. A detailed analysis of released and purified virus (strain V) structures, which banded at 1.22 g ml^{-1} in CsCl, showed not only 90 nm enveloped particles but considerable amounts of 60 nm icosahedra, which may represent defective virus. Additionally, some of the debris-like fibrillar structures probably represent filamentous BDV forms of 100–500 nm (see Fig. 1 in Ludwig & Bode, 1997) comparable to those described for influenza A viruses (R. Compans, personal communication). Convincing evidence for BDV being of icosahedral structure came also from the analyses of strain He/80 in other cells (Kohno *et al.*, 1999). However, the elucidation of the morphogenesis of this unusual virus, which replicates in the nucleus and seems to receive its lipid layer from the outer nuclear membrane (R. Stoyloff, personal communication), awaits further work.

We assume that M and G are in the outer virion envelope since they both are glycosylated and neutralization can be induced by these glycoproteins (Stoyloff *et al.*, 1994, 2001; Kliche *et al.*, 1994; Schneider *et al.*, 1997). Thus the model of the BDV virion (see Fig. 2 in Ludwig & Bode, 1997) carries some of these proteins as spikes (Kohno *et al.*, 1999).

An outstanding biological feature is the low infectivity of Borna virus samples (Ludwig *et al.*, 1988). Whether this is due to a fragile envelope or a low number of particles produced that never reach one infectious unit per cell in the many cell lines we have tested remains unclear. This phenomenon forms the basis for the hypothesis put forward by Gosztonyi & Ludwig (1995, 2001) that, at least *in vivo*, the spread of infectivity occurs via RNPs through the axonal flow. RNPs extractable from tissue culture as well as infectious brains (Nakamura *et al.*, 2000) support the importance of these small infectious structures in aetiopathogenesis.

THE INFECTION

Virus spread in the body

All our studies (Ludwig & Bode, 2000) show that infectious structures spread bimodally in the organism, via nerves (Gosztonyi & Ludwig, 1995) and by blood (Bode *et al.*, 1994). The initial infection is either intranasal or through the nasopharynx or gut. The latter suggestion can be deduced from many reports where horses acquired infection through contaminated water (Dürrwald & Ludwig, 1997). Once the infectious material reaches the brain, replication occurs in preferred limbic areas. How infectivity reaches the reticulo-endothelial system is not known. Excretion of the infectious agent, at least experimentally (Gosztonyi & Ludwig, 1995), occurs through glands or mucosae (nose) having contact to the outside, or through urine. Vertical transmission of the agent from mother to foetus has been shown in both the horse and the rat (L. Bode, P. Thein & H. Ludwig, unpublished data).

The discovery of the remarkable spread of BDV in the body through the nerves came from two elegant experimental designs performed by our group and Narayan's group in rabbits and rats, and is summarized in Fig. 2.

The flow of infectious BDV through blood results in virus isolation from PBMCs (Bode *et al.*, 1996) or granulocyte preparations (where contamination with monocytes is possible; Planz *et al.*, 1999).

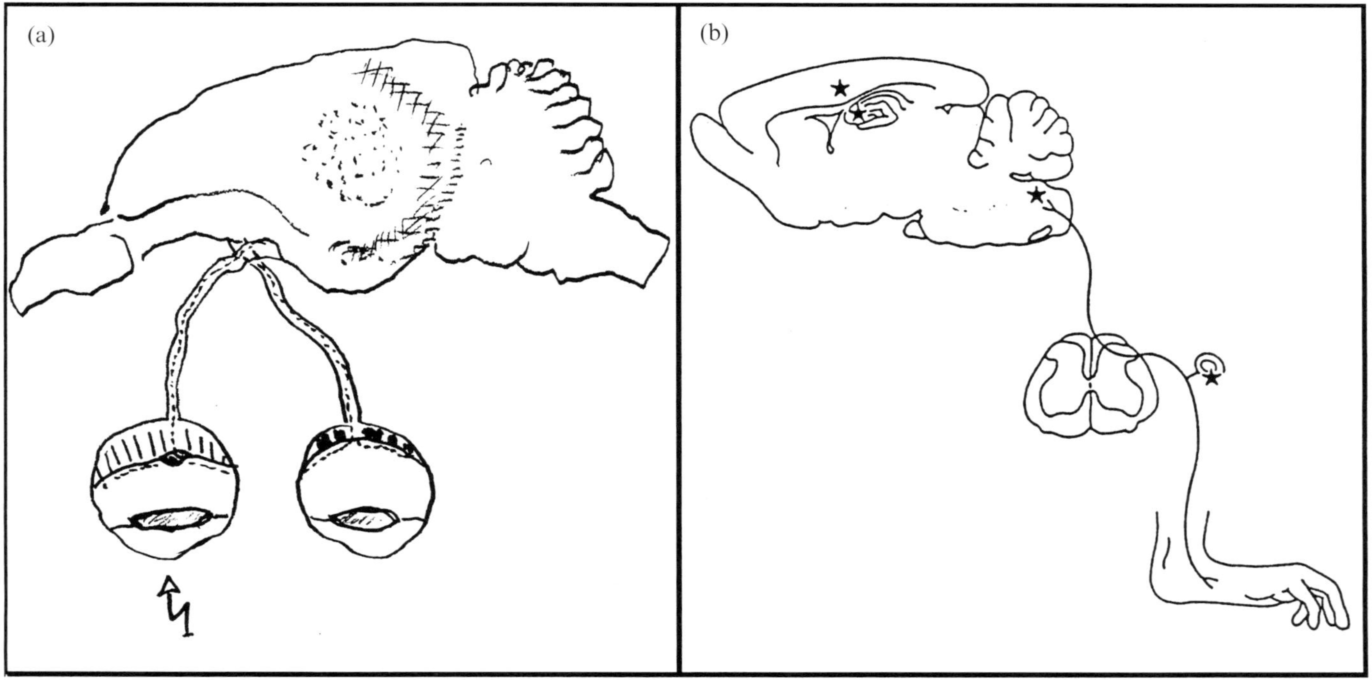

Fig. 2. Spread of BDV in the body. BDV is transported by intra-axonal flow (Gosztonyi & Ludwig, 2001), as demonstrated by experimental infection of rabbits (a) and rats (b). Infection of rabbits (i.c. and i.n.) resulted in a multifocal retinopathy expressed in both eyes, most probably due to focal immunopathological reactions starting from the choroidea (Krey *et al.*, 1979a). The rabbit retina is nourished by diffusion, therefore the optic disc could be lasered (arrow) in one eye (a), whilst the retina remained intact. No virus arrived intra-axonally in the cell layers of the retina, and focal destruction was missing (modified from Krey *et al.*, 1979b). In the rat, footpad infection (b) resulted in an intra-axonal transport of infectivity through nerve pathways via the spinal cord to the brain, reaching the preferred sites in the limbic system. Virus could be monitored along the centripedal pathway (modified from Carbone *et al.*, 1987).

Spread in nature

A broad host spectrum of natural and experimental infections had already been reported decades ago by Zwick (1939) and Nicolau & Galloway (1930). Today, it is generally accepted that humans, horses, cattle, sheep, goats, dogs, cats, rabbits, rats, ostriches and a few zoo animals can carry the virus under natural conditions. Experimentally, rhesus monkeys, tree shrews, horses, sheep, dogs, cats, rabbits, hamsters, guinea pigs, rats, gerbils, mice, ostriches, chicken and quail can be infected. As outlined in great detail, infections without and with symptoms (the latter expressed as behavioural changes or overt disease) diagnosed in species infected naturally or experimentally should be differentiated carefully. All these data and the known epidemiological implications have been extensively reviewed recently (Ludwig & Bode, 2000).

Human infections and disease

Human BDV infections are proven unequivocally (Bode *et al.*, 1995, 1996; Salvatore *et al.*, 1997; Haga *et al.*, 1997b; Nakamura *et al.*, 2000). However, human Borna disease does not exist in the classical sense as coined by the virus-induced horse disease with an occasional fatal outcome (see Ludwig & Bode, 2000).

Monitoring

Human infections are detectable by antibodies, which have been measured by various serological or biochemical methods (see Ludwig & Bode, 2000), but demonstration of virus-specific antigen and nucleic acid is more conclusive (Bode *et al.*, 1995; Deuschle *et al.*, 1998; Kishi *et al.*, 1995; Sauder *et al.*, 1996). However, the isolation of human viruses (see below) leaves no doubt about a natural human infection.

Seroprevalence

Many groups in the world have described BDV-specific antibodies in patients with psychiatric disease (Rott *et al.*, 1985; Bode *et al.*, 1988; Yamaguchi *et al.*, 1999; reviewed by Bode, 1995) with a prevalence of 10–15 %, but often low titres. Antibodies were also found in the healthy controls, but at a significantly lower percentage (1–3 %) than in patients (see Bode, 1995). Long-term observation of patients resulted in up to 30 % seropositive reactants (for a review see Bode, 1995; Ludwig & Bode, 2000). The increased seroprevalence reflects reactivation of the persistent BDV infection over time.

Fig. 3 summarizes our own unpublished data, which correlate with findings of other groups in demonstrating an elevated antibody level in diseased compared with healthy individuals. In addition, our work reflects the fine differences in the humoral immune response in acutely, moderately and severely afflicted manic and depressed patients. In general, however, antibody measurement is of limited diagnostic value, and says

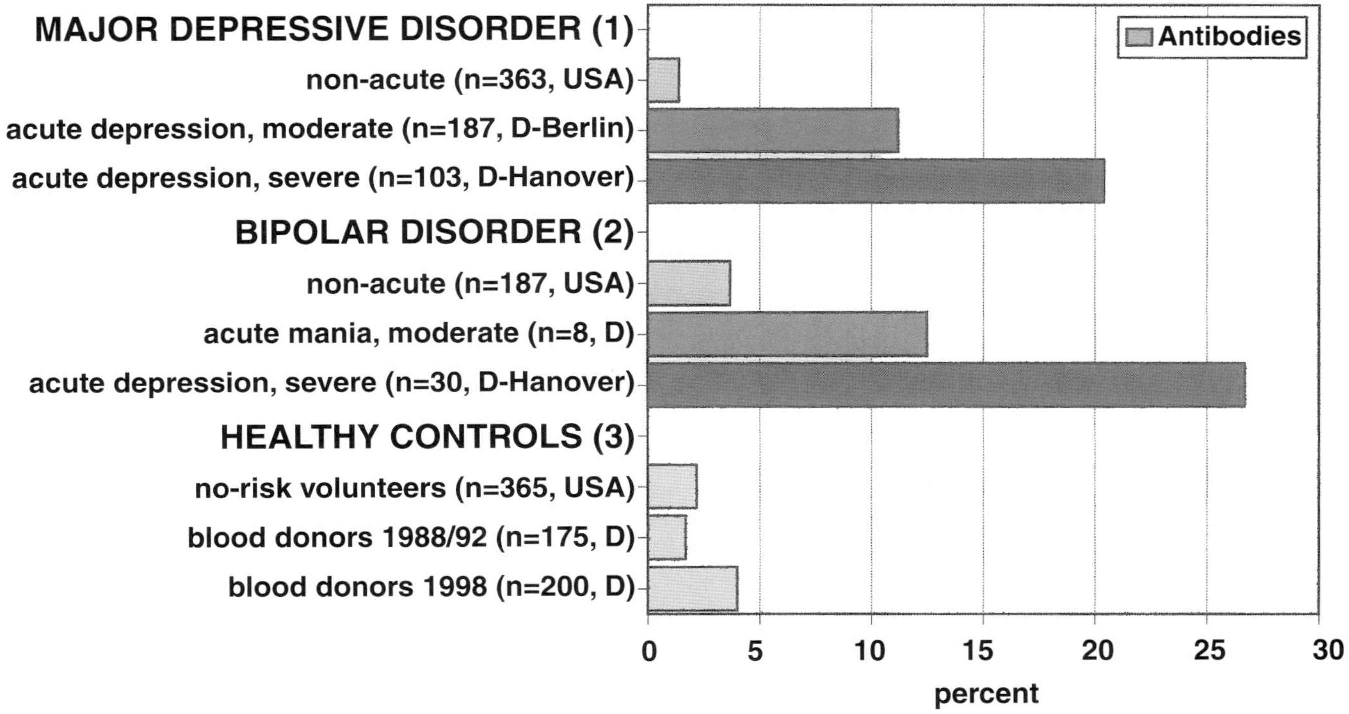

Fig. 3. Seroprevalence in groups of psychiatric patients and healthy controls. Antibodies were measured by double-stain indirect immunofluorescence (Bode *et al.*, 1992, 1995) as point prevalence, and were taken over a time-scale of almost one decade. Besides significant differences between moderately and acutely diseased patients, non-acute patients in their well state showed lower frequencies like the controls.

nothing about the state of infection in terms of viral activity or 'latency'. Neutralizing antibodies have not been detected in human infections.

Antigens

The major complexed BDV antigen was discovered 50 years ago as the so-called s-antigen, which was easily extractable from infected animal brains (von Sprockhoff, 1956). This antigen is known to induce the major humoral immune response also locally in the brain (Ludwig & Thein, 1977; Ludwig *et al.*, 1988). The discovery of antigen (and virus) in PBMCs, the so-called cell antigen (cAG) (Bode *et al.*, 1995, 1996), and the demonstration of RNPs in post-mortem brain (Nakamura *et al.*, 2000), each correlating with disease, strengthened the belief that BDV was involved in human mental disorders.

Although a great body of information on the role of BDV antigens is new (L. Bode and others, unpublished), here we give a short outline of the implications of these findings for epidemiology and prognosis.

Our early knowledge of antigen and possibly virus in the supernatant of infected cells (Ludwig & Becht, 1977) stimulated us to search for BDV antigen in the plasma of infected patients, particularly after we had detected virus antigen in PBMCs by applying our sensitive sandwich ELISA (see Bode *et al.*, 1996). To our great surprise, such plasma antigen (pAG) could be monitored in ELISA by the use of a hyperimmune rabbit-BDV antiserum, and by this became an extraordinary tool to trace BDV infection active in humans and animals.

To demonstrate the potency of the ELISA system in catching free antigen, the immune reactivity of native purified p40 and p24 BDV-specific protein was measured (Fig. 4). In the assay, the monoclonal antibodies could bind up to 7 μg of the native proteins. We have investigated more than 1000 sera. In approximately 20 % of the particularly severely diseased patients, we found the plasma antigen (pAG) in quantities of 7 ng to a maximum of 7 μg. It was of interest that pAG often appeared at the beginning of a depressive disease period, remained constant for some days, and then declined progressively.

Prognostically, this parameter shows that the patient produces excess antigen, which must be 'secreted' from cells. The organ or cellular origin of this antigen remains obscure. Our advanced ELISA system gives a valuable indication of the load of infection. In the plasma of some of the patients, we have found absorbencies higher than 1.0, which are indicative of active replication of the virus and a massive release of antigen. The pAG is composed of the two known major immunogenic components (p40 and p24). In some individual cases, BDV nucleic acid was associated with these antigen

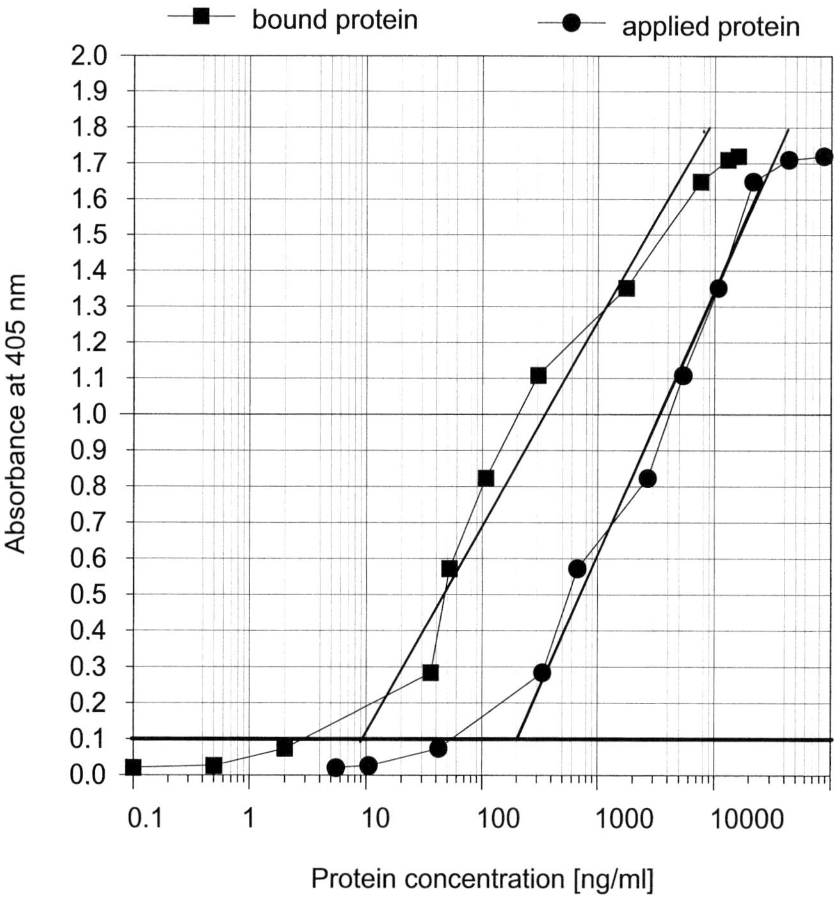

Fig. 4. Binding of native N and P proteins in the sandwich ELISA. Using matrix-coupled anti-p40 and -p24 monoclonal antibodies, native proteins from a suspension of infected horse brain were purified by affinity chromatography. Concentrations were determined by UV spectroscopy at 205 nm. After concentration of the proteins to 100 μg ml^{-1} each, they were mixed in equal proportions and applied to the prepared ELISA plates of the routine sandwich ELISA, where the monoclonals had been bound to trap specific antigen (Bode *et al.*, 1996). After immunospecific binding, the concentration of the N/P mixture was again measured, and the proteins bound were calculated (left line).

complexes, although this conclusion needs confirmation. These observations suggest that RNPs are present in plasma, and currently we are testing whether infectivity is associated with these structures.

Free-floating antigen and the PBMC-bound cAG (see Bode *et al.*, 1995, 1996) are associated with increased replication of virus, and are therefore often found in patients showing severe signs of depression and bipolar disorder. A constant presence of pAG in depressed patients over long periods reflected a rather unfavourable prognosis.

This mass of soluble antigen in the serum initiates a quick immune response and the production of antibodies predicts that immune complexes should be formed. We have, indeed, detected these circulating immune complexes (CICs), occasionally at high levels, in BDV-infected humans and animals, and are on the way to determining their pathogenetic importance (L. Bode and others, unpublished).

Immune complexes

We developed a simple ELISA-based method to trap these complexed structures with p40 and p24-specific monoclonal antibodies, as done in the antigen assay (see above). The antibody component of the trapped complex was then monitored by an anti-species-specific IgG. We submit that this new BDV-EIA system will revolutionize diagnosis of BDV infection in humans and animals (L. Bode and others, unpublished).

From a comparison of antibody detected by indirect fluorescence (IF) analysis, cAG and pAG, as well as CICs detected by ELISA, it became clear that p40/p24 CICs represented the most prominent marker. Surprisingly, CICs showed up in approximately 50 % of patients with more or less clinically well-defined major depression, where approximately 8 % of the patients harboured IF antibodies and approximately 10 % presented with cAG. Higher rates of active infection were prominent in approximately 25 % of CIC-positive healthy control people, where only 2 % antibodies and no cAG were detectable.

Evaluation of these assays and determining their validity over the last 3 years entailed testing more than 3000 human and animal sera. Briefly summarized, the CIC assay turned out to be the optimal marker for monitoring infection. We found that 25 % of individuals were infected in control populations, but in groups with severe depression 100 % were infected. In longitudinal studies, CIC formation represents an excellent measure of the infection status, much more so than monitoring cAG, pAG and antibodies, as illustrated over 6 months in a patient with chronic depression (Fig. 5a).

In another comparative test, CIC and pAG formation were reinvestigated in a group of severely depressed patients with major depression and bipolar disorders. More than 40 % of these patients had shown the activation marker cAG when tested 5 years ago (Ferszt *et al.*, 1999a; Bode *et al.*, 1996). Using our new methods, the patient's sera showed pAG in approximately 40 % of cases, with very high values in 82 % of the samples, and 31 % of these patients were CIC-positive with remarkably high levels.

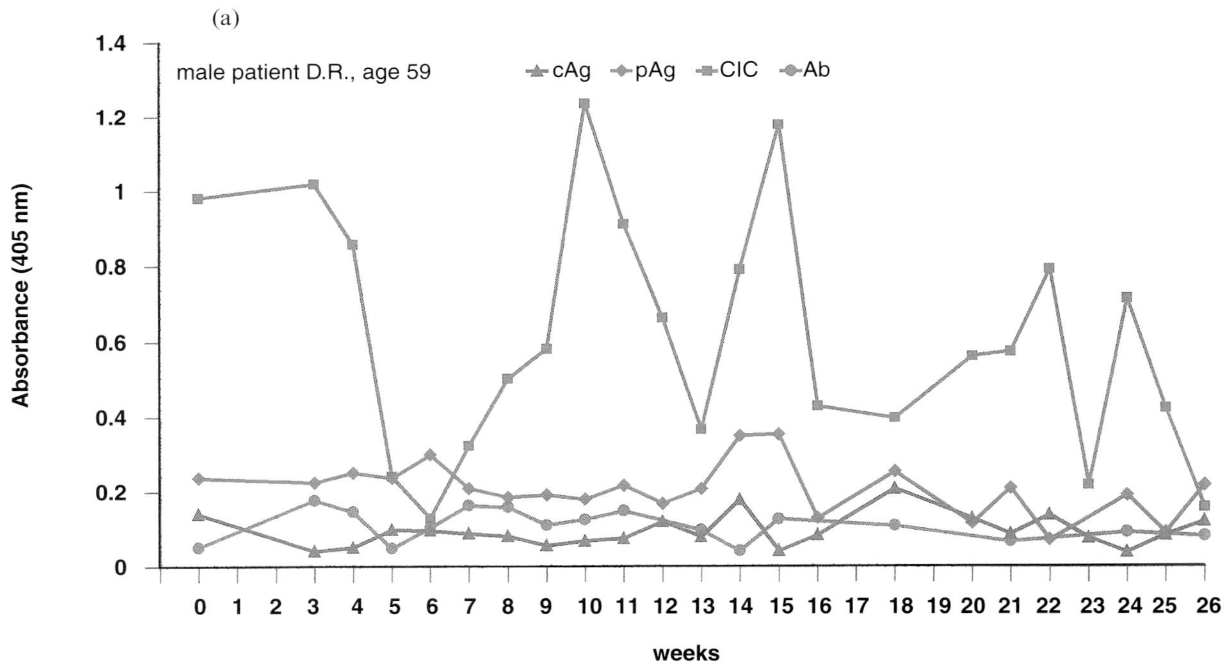

Fig. 5. (a) Long-term follow-up of BDV infection in a chronically depressed patient. Different BDV-specific markers were measured over 26 weeks by ELISA. Absorbance values at the following dilutions are given: 1 : 2 for cAG (cell antigen of PBMCs), 1 : 2 for pAG (plasma antigen), 1 : 20 for CICs (circulating immune complexes) and 1 : 100 for Ab (antibodies). CIC formation peaks approximately bimonthly, paralleling increased disease symptoms. A drop of antibodies with an increase of CICs is clear. Fluctuation in antigen levels appears to be characteristic of chronic patients.

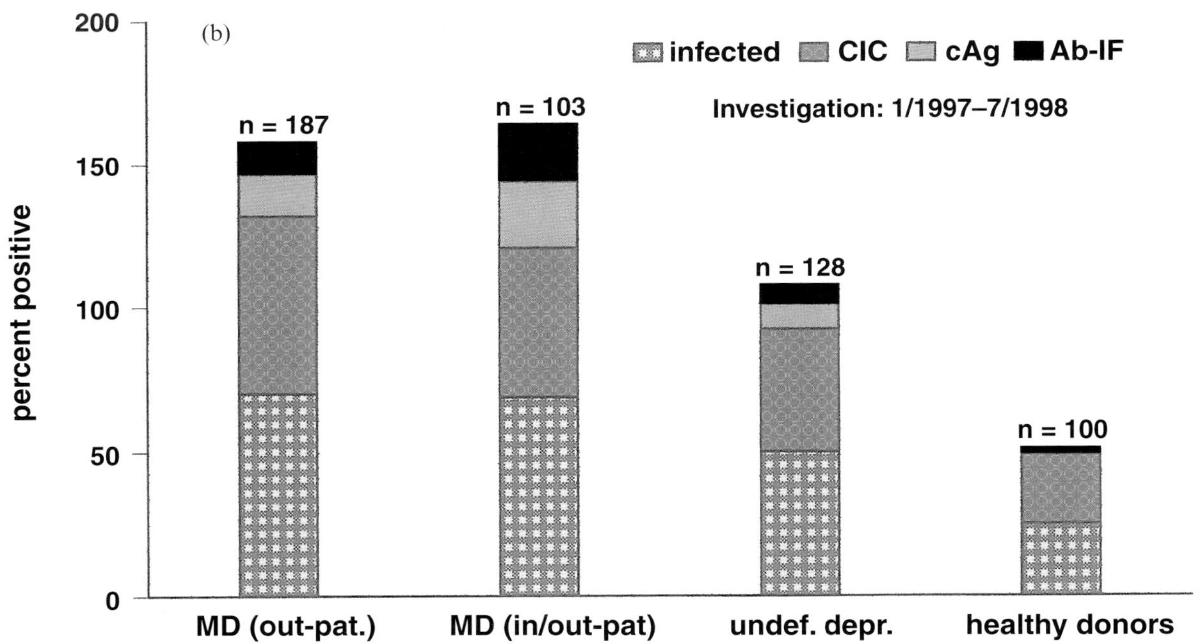

Fig. 5. *(cont.)*
(b) BDV infection monitoring by CICs compared with previous markers. In significant numbers of individuals, the known marker values for cAG (cell antigen) and Ab-IF (antibodies by the indirect fluorescence technique) were compared with the percentage of CICs (circulating immune complexes) in cohorts of mood disorder (MD) and undefined depressive patients, and controls. From all parameters of positive patients, the degree of infected individuals was calculated. Undoubtedly the level of CICs gives the most significant picture for an infection. Note that the point prevalence of 'latent' symptomless infection can reach 25 %, whereas in mood disorder patients this can be threefold higher.

Based on thorough clinical investigations over recent years, patients with recurrent major depression and bipolar disorder reached an infection rate of 100 % when several parameters were taken together with CIC percentages higher than 90 % versus up to 30 % CIC-positive sera in healthy donors. A comparative analysis of a representative study with more than 100 individuals is shown in Fig. 5(b). The CIC marker reflects a most outstanding marker of infection (L. Bode and others, unpublished).

In patients with recurrent mood disorders (major depression and bipolar disorder), and possibly also obsessive compulsive disorder (OCD), our studies over the years suggest a correlation between the amount of CICs and pAG (plasma antigenaemia) and the severity of the symptoms (L. Bode, D. Dietrich & R. Ferszt, unpublished). Similar observations were made recently investigating BDV antigen in cerebrospinal fluid of such patients (Deuschle *et al.*, 1998). The plasma antigenaemia, paralleled with high CIC levels, reflects an extremely high production of viral components that could not be complexed by antibody production. The low level of antibodies found in natural BDV infections and discussed with some surprise by several research groups (Bode, 1995; Sauder *et al.*, 1996; Planz *et al.*, 1999) can now be explained by our discovery of CICs.

Early on, one of the pathological mechanisms of these psychiatric diseases was postulated to be that BDV-specific structural elements are a prerequisite for the interference with neurotransmitter equilibrium (Gosztonyi & Ludwig, 1984b, 2001; Bode *et al.*, 1996). The CICs, measured in animals at least, seem to be such small complexes that deposits in kidneys and other organs are yet undetected in this infection model (G. Gosztonyi, unpublished). In other virus infections, like LCM, Aleutian mink disease or non-microbial diseases (Theofilopoulos & Dixon, 1979; Oldstone, 1984), CIC formation and the immunopathology caused by them are known to be pathogenic.

Nucleic acid

Analysis of BDV RNA from PBMCs of psychiatric patients has been used worldwide following the original discovery by Bode *et al.* (1995). Reports on mRNA or genomic RNA from diseased patient groups, as well as blood donors, have been described in Germany and the US (Bode *et al.*, 1995; de la Torre *et al.*, 1996a; Sauder *et al.*, 1996; Planz *et al.*, 1998), Japan (Iwahashi *et al.*, 1997; Iwata *et al.*, 1998), China (Chen *et al.*, 1999), Austria (Nowotny & Kolodziejek, 2000) and Australia (C. Arunagiri, personal communication), to mention only a few from the many reports (see Ludwig & Bode, 2000). The overwhelming majority detected elevated numbers of positive reactants in psychiatric groups. One German group found no BDV RNA in psychiatric patients (Lieb *et al.*, 1997) and questioned the human origin of BDV simply based on low genetic divergence by postulating that the data obtained were due to contamination with laboratory virus (Schwemmle *et al.*, 1999).

We have demonstrated that these reservations are unjustified (Bode *et al.*, 2000). Comparing 31 natural sequences (18 of human, 13 of animal origin) with those of strain V in a fragment of 440 nucleotides of the p40 gene (Fig. 6), divergence between 0.23 and 4.31 % was found, within a range usually found in different Bornaviruses. One-third of the mutations turned out to be common, whether from human or animal sources. Surprisingly, those found in some human sequences were also present in reference strain He/80. The degree of divergence of the natural isolates from the two laboratory strains depends on the proportion of common versus rare individual mutations. Importantly, unique individual mutations, mainly causing amino acid substitutions compared to strain V, were present in all human strains isolated by our group.

With these data, supported by biological features of the human isolates (see below), the contamination reservation can be rebuffed (Bode, 2000).

Since some groups look for seroprevalence and RNA-positivity in patients, it should be mentioned that a link between seropositivity and presence of viral RNA cannot be made. In support of the involvement of BDV in disease syndromes of humans, although not being an aetiological proof, was the finding of BDV RNA in post-mortem examinations of brains obtained from psychiatric patients (Salvatore *et al.*, 1997; de la Torre *et al.*, 1996b).

Human isolates

Six BDV isolates, all obtained from clinically diseased patients with severe psychiatric symptoms, have been reported. They were sequenced partially and fit into the model that only minor, but significant, mutations differentiate them from the archetype of this virus. Table 2 summarizes the properties of the human viruses.

From a critical review of the data, the conclusion is that human isolates are difficult to obtain. Virus has been recovered only from individuals in longer phases of major mood disorders accompanied by a significant antigen load in blood cells or brain. Virus can be grown easily from fatally diseased horses (Gosztonyi & Ludwig, 1984a; Lüschow, 1999). Prerequisites to isolating the virus from *intra vitam* blood samples of man and animals seem to be that the mononuclear cells carry high enough antigen (and hence replicating viral genomes) to be measurable with our cAG assay, and that co-cultivation has to be carried out, often exceeding 10 passages to allow adaptation to the tissue culture cell system.

Using our published protocols (Bode *et al.*, 1996), recently we have recovered five new isolates from Ficoll-separated citrated blood cell suspensions. They were derived from depressed patients and from malignant human brain tumours, and another virus was

Table 2. Human BDV isolates

No.	Code	Age/sex/time diagnosed (yrs)	Clinical diagnosis	Antigen/cells	Nucleic acid/cells	Isolation (passage pos.)	Amantadine	Exp. infect.	Reference
1	Hu-H1	45/F/12	Bipolar I disorder	Pos./PBMCs	Pos./PBMCs	OL-co-cultivation (p11)	Sensitive	Rabbits, rats	Bode *et al.* (1996)
2	Hu-H2	37/M/14	OCD	Pos./PBMCs	Pos./PBMCs	OL-co-cultivation (p11)	Sensitive	Rabbits, rats	Bode *et al.* (1996)
3	Hu-H3	55/M/36	Bipolar II disorder	Pos./PBMCs	Pos./PBMCs	OL-co-cultivation (p11)	Sensitive	Rabbits, rats	Bode *et al.* (1996)
4	Hu-HUSA1	35/M/4	CFS	Pos./PBMCs	Pos./PBMCs	OL-co-cultivation (p12)	Sensitive	Rabbits	Unpubl.
5	RW98	?/M/?	Schizophrenia	ND/granulocytes	Pos./granulocytes	CRL 1405 cells (p7)	ND	ND	Planz *et al.* (1999)
6	Hu P2Dr	26/M/2	Schizophrenia	Pos./brain	Pos./brain	Gerbil brain (p8)	ND	Gerbils	Nakamura *et al.* (2000)

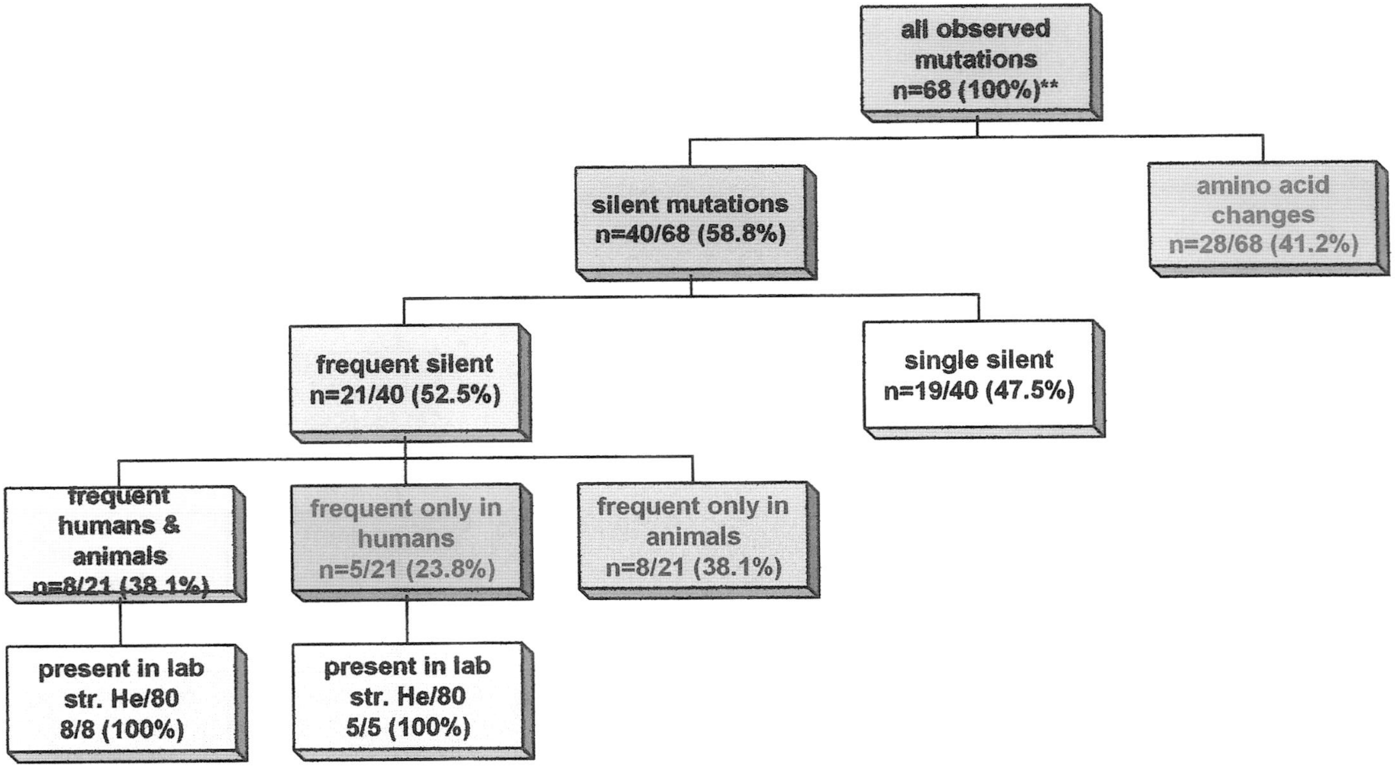

Fig. 6. Sequence comparison of 31 wild-type Borna viruses with reference strain V (441 bp fragment of N protein p40; genome nt 277–717). Sequences are from human PBMCs or brain ($n = 14$), from human isolates derived from PBMCs ($n = 4$) and from brains of different animal species with natural Borna disease ($n = 13$). **, 1–19 mutations per individual sequence (0.23–4.31 % divergence to strain V); divergence of laboratory strain He/80 to strain V: 13 mutations (2.95 %).

from blood samples, kindly supplied by Dr Komaroff, Boston (F. Rantam, R. Ferszt, T. Komaroff & L. Bode, unpublished data). Speculation on the current distribution of defined BDV strains in the world is premature as only relatively few isolates from German (Bode *et al.*, 1996; Planz *et al.*, 1999) and Japanese (Nakamura *et al.*, 2000) groups have been described.

Although the molecular events explaining why infectious BDV isolates could be propagated in some cases but not in others remain obscure, consideration of quasi-species phenomena (Domingo *et al.*, 1995) should be given to the isolation of BDV and BDV-specific nucleic acid sequences. The stability of genome pieces observed worldwide indicates that homologous molecules dominate the swarm of RNA, suggestive of a low mutation rate.

Genomic RNA of true isolates is also surprisingly stable, compared with the swarm RNA picked up from blood cells or organs. This leads to the hypothesis that BDV in its evolution and adaptation to mammals and man has gone through a remarkable evolutionary stasis and has lost virulence by following a trend to more effective persistence.

This also suggests that this neurotropic agent is an evolutionary old pathogen (Bode *et al.*, 1996). Being aware of these considerations, controversies based on the close genetic relationship of BDV strains fade. There are numerous examples in other systems showing that one or a few mutations may cause significant phenotypic variations altering virus properties (as known for human BDV isolates). 'Mutants able to establish persistence rather than killing their host cells, variants with expanded host range, variants resistant to antiviral agents or able to overcome an immune response, may each arise with a finite probability in the ever evolving mutant spectra of quasi-species swarms of RNA viruses' (Domingo *et al.*, 1995). The complicated splicing mechanism seen in BDV leading to a limited number of intact M, G and L-polymerase mRNAs (Stoyloff *et al.*, 2001), together with its complicated intranuclear replication mechanism, may indicate a lower efficiency of replication than known from other viruses, which separates BDV from the rest of the *Mononegavirales* (Pringle, 1995; de la Torre *et al.*, 2000).

Antiviral treatment

Of considerable importance for BDV-associated diseases in man is the possibility of antiviral treatment with amantadine sulfate (AS), which was first introduced successfully by our groups and subsequently was confirmed by others (Bode *et al.*, 1997; Ferszt *et al.*, 1999b; Dietrich *et al.*, 2000).

This therapy started with a female patient suffering from a major mood disorder who additionally acquired Parkinson's disease. Amantadine therapy cured her depression, even clearing the BDV genome below the detection limits of RT-PCR (Bode *et al.*, 1997). Our original data on the amantadine sensitivity of one single human isolate has been confirmed with three other human and three animal isolates *in vitro* (Fig. 7).

It was of considerable interest that a time-dependent reduction of infectivity was clearly related to the initial virus load in culture and was dose-dependent. In the most sensitive US strain of BDV (HU-HUSA1), infection of naïve cells could be inhibited to 50 % (ID_{50}) by a very low dose of 0.6 ng AS ml^{-1} (SD – 0.04). Additionally, we were able to confirm the negative results of our critics (Hallensleben *et al.*, 1997; Cubitt & de la Torre, 1997; Stitz *et al.*, 1998a) when laboratory strains were used. The vaccine strain Dessau was totally resistant to amantadine. We are aware of the potential problem that virus resistant to the drug might appear with time but, based on the genomic stability of BDV, this may be a minor problem – quite different from that seen with influenza virus.

Over the last 2 years, independent clinical trials on moderately depressed out-patients in Berlin and Hanover showed that antiviral therapies are an alternative treatment for patients suffering from activated BDV. In the two open trials, a clinical responder rate of approximately 70 % supported the view that infected patients had a clear benefit from this therapy (Ferszt *et al.*, 1999b; Dietrich *et al.*, 2000), whereas, importantly, many no longer responded to conventional anti-depressants. A virostatic effect was achieved. Recently, a double-blind placebo study with 1.5 years of post-study observation has recently been completed, and this is in the process of statistical evaluation.

In summary, these important amantadine data (*in vitro* and *in vivo*) allow us to conclude that wild-type strains (human, equine) are sensitive to the drug. The drug prevents infection *in vitro* and inhibits virus replication of persistently infected OL cells in a strain-, dose- and time-dependent manner. In contrast, laboratory strains seemed to be more or less insensitive. An anti-depressive effect was observed in up to 70 % of BDV-infected patients. In clinical responders, viral activity tends to drop in the blood, whereas unaltered or increased activity was noticed in non-responders.

Other antiviral substances interact with viral glycoprotein, as described for gp17 (Stoyloff *et al.*, 1996), or interfere with BDV replication (ribavirin) (Mizutani *et al.*, 1998; Jordan *et al.*, 1999). However, treatment with ribavirin is known to cause considerable side effects in humans with other virus infections.

Progress in diagnosis and relevance of human infection

Until the 1970s, Borna disease was always diagnosed by histology post-mortem (Seifried & Spatz, 1930), and later through immune histology (Wagner *et al.*, 1968;

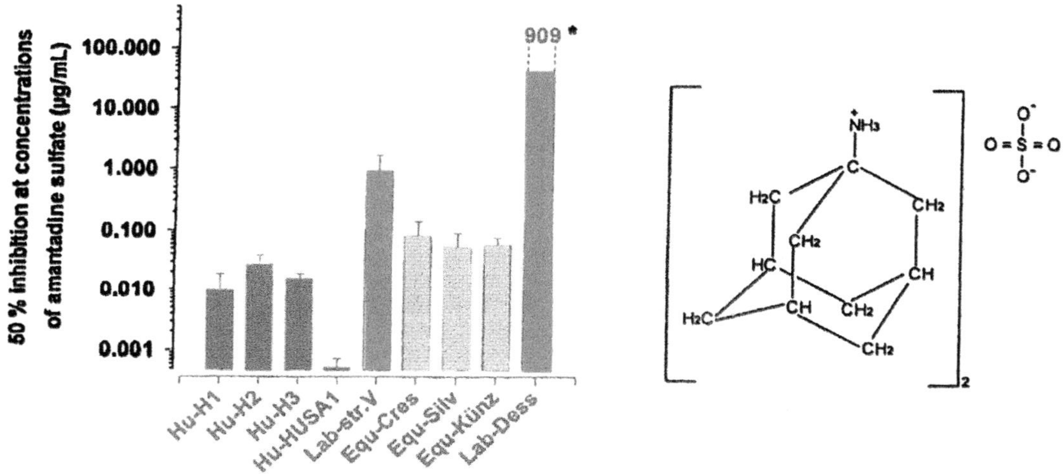

Fig. 7. Influence of amantadine treatment on different BDV isolates *in vitro*. Methods for these experiments follow Bode *et al.* (1997). Briefly, rabbit brain cells or rabbit spleen cells were incubated with a constant dose of amantadine sulfate (see figure of the generic substance at the right side) followed by a defined infection dose. Columns show the 50 % inhibition of infectivity at the indicated concentrations. All wild-type human (Hu-) as well as equine (Equ-) BDV isolates were significantly inhibited. Laboratory strains, animal plus cell culture adapted viruses (vaccine strains V and Dessau), were significantly less sensitive to the drug.

Gosztonyi & Ludwig, 1984a). Since the mid-1980s, the use of serology has heralded an increased understanding of persistent viral infection of man (Rott *et al.*, 1985; Bode *et al.*, 1988, 1992; Ludwig *et al.*, 1988). In the 1990s, work with nucleic acid and the recovery of natural live virus (Bode *et al.*, 1995, 1996; Salvatore *et al.*, 1997; Nakamura *et al.*, 2000) heightened the awareness of human infection, leading to an ongoing debate (Schwemmle *et al.*, 1999; Bode *et al.*, 2000).

However, reliable laboratory indicators, such as antigen quantification and improved antibody testing, have now been introduced. Comparative evaluations performed recently by the Leipzig, Tübingen, Freiburg and Paul Ehrlich groups at the Robert Koch-Institut gave a total consensus using antibody and antigen ELISAs of the Berlin group (unpublished data). Together with the use of immune complexes as a diagnostic tool, almost continuous monitoring of natural infections will now be available (L. Bode and others, unpublished). In vulnerable subjects, positive laboratory markers accompany a disease which is governed by neuronal disturbances (Gosztonyi & Ludwig, 2001; Hornig *et al.*, 2001), thus leaving few reasons to doubt an aetiopathogenetic role of BDV in human health (Bode, 1995; Ludwig *et al.*, 1998; Ferszt *et al.*, 1999a, b; Deuschle *et al.*, 1998; Dietrich *et al.*, 2000).

RELEVANCE OF ANIMAL MODELS

Many animal species can be infected experimentally, but only a few models provide observations that may be relevant to human infection and disease. This may be because most experimental infections have been performed with highly adapted strains, using both an unnatural route of infection and high infection doses. In only a few cases were experimental infections done with isolates close to those found in natural infection (Zwick, 1939; Nitzschke, 1963; Hirano *et al.*, 1983; Bode *et al.*, 1996).

Experimental infection at least revealed that BDV causes a persistent infection of the central nervous system (CNS) of hamster, rabbit, rat and chicken, with neuropathological signs (Ludwig *et al.*, 1985, 1988; Gosztonyi & Ludwig, 1995). Using different passages of the laboratory strain, and animals of differing ages, a variety of clinical pictures and expression of disease symptoms could be seen.

The humoral immune response (non-neutralizing antibodies) was found to be predominantly directed against the s-antigen, and later in infection (onwards from approximately 2 months post-infection) neutralizing antibodies can appear, significantly in mice, but also in obese rats, and with high titres in chicken (Ludwig *et al.*, 1993). Although neutralizing antibodies seemed to inhibit virus spread in the body, and somehow contributed to survival (see below, obese rats) (Gosztonyi & Ludwig,

1995; Stitz *et al.*, 1998b), it is now known that the general humoral response under experimental conditions never matches that of natural infections, and seems to be of low significance (Ludwig & Bode, 2000).

Additionally, intensive studies have been done on the cellular response, with CD4 and CD8 cells being involved in immunopathologic reactions, but with apparently no relevance to the clinical situation in naturally infected horses, sheep or human patients. Immunopathology may contribute to severe, progressing disease in the laboratory animal (Stitz *et al.*, 1995), but it is far away from representing factors which govern initial stages of the infection process. The initial stages are often associated with changes in normal behaviour and life (Ludwig *et al.*, 1988; Bode, 1995) and may from clinical viewpoints return to a healthy state. Certainly experimental infection of rabbits and rats remains valuable for monitoring new BDV isolates and the clinical and neuropathological properties of the infectious non-adapted agent, but the exhaustive animal experimentation during the last 50 years appears to be of doubtful significance to natural BDV infection (Ludwig & Bode, 2000). Nevertheless, we will summarize below points of relevance in animals infected experimentally, the findings and interpretation of which may be of some value in elucidating the BDV infection process and its psychiatric and clinical expression in humans.

Rabbit

This animal still seems to be the most sensitive species with which to study infectivity of BDV isolates. Over the years, it became clear that, contrary to textbooks, after intracerebral inoculation wild-type isolates did not always lead to typical disease and exitus. Obviously virulence differences exist (Ludwig *et al.*, 1985, 1988).

Survival following infection with BDV-positive horse brain suspension can be observed. Early studies carried out by our group used this rodent because of its sensitivity to BDV for monitoring subtle clinical parameters and detection of early neurological disease. Body weight was of greater value to monitor the process of infection. A characteristic multifocal retinopathy obviously due to immune pathological events raised considerable interest (Krey *et al.*, 1979a, b). Immune suppression led to a protracted clinical picture with prolonged stable weight curves, but could not prevent fatal disease (Gierend & Ludwig, 1981), as compared later in rats (Narayan *et al.*, 1983). Early phases of the infectious process were monitored by EEG, where Radermacher complexes and a total disorganization of central brain electric potentials could be measured, even before typical disease occurred. Although this model has not been used exhaustively to study brain physiopathology, it gave an insight into the altered electric signalling in the brain (Gierend, 1982).

Furthermore, early neuropathological studies showed that not only neurons but satellite cells became infected during the clinical process, and a *hydrocephalus internus* was the consequence of prolonged disease (Roggendorf *et al.*, 1983).

Recently, valuable data have been obtained from rabbit infections using human isolates. We could demonstrate, for the first time, that Koch's postulates held true for the human virus, and that using weight measurements and studying the immune response were informative biological parameters. It was of interest that infection of rodents with non-adapted human strains was not fatal, but a long persistent infection (lasting over years) was established, indicated by an active immune response (Bode *et al.*, 1996). There is no doubt that the BDV-infected rabbit is an invaluable experimental model to investigate virus isolates of human and animal origin, but better models, which present easier handling, are available to obtain an insight into subtle changes of behaviour (Ludwig & Bode, 2000).

Rat

Over the last 50 years, this small rodent model of infection has been the best studied. Nitzschke (1963) introduced infection of younger animals that remained more or less asymptomatic, and, by passaging the virus, was able to demonstrate differences in virulence. The general features of the rat model, a persistent and tolerant infection following inoculation of 1-day-old (immature) rats, were established by the Berlin (Hirano *et al.*, 1983) and Giessen (Narayan *et al.*, 1983) groups. Subsequently, it has been used intensively and widely to unravel various aspects of this unique biological situation (Ludwig *et al.*, 1988), and served to get the first viral RNA (Lipkin *et al.*, 1990). First, the immature rat could be infected and carried the virus persistently with phases of activation during its life span, but showed no obvious clinical symptoms. Second, the infected adult rat acquired a subacute disease, depending upon the passage history of the adapted virus, which could also exhibit a chronic stage of infection and could lead to the so-called obesity syndrome. Such 'potato' rats (Ludwig & Bode, 2000) survive for more than 1.5 years and develop unusual accumulations of fat tissue in the brown fat area or around the organs (Ludwig *et al.*, 1985), with a progressing *hydrocephalus internus* and degeneration of structures of the limbic system, visible specifically in hippocampal areas and areas in the infundibular region. Vascular degeneration of neurons in the paraventricular nucleus, as well as severe involution of the hippocampus and piriform cortex, have been discussed previously with regard to the aetiopathogenesis (Gosztonyi & Ludwig, 1995).

The work most relevant to BDV infection and human health are results from asymptomatic, persistently infected rats which had undergone a variety of psychological tests

(Dittrich *et al.*, 1989). These rats, when infected as neonates and tested months later, showed three types of disturbances in learning experiments. First, special learning tasks using a hole board resulted in reduced information processes for spatial and temporal signals. Second, a clear reduced ability to avoid pain stimuli, like negative taste or electric shock, could be measured. Third, open field and neophobia tests indicated emotional alterations in the persistently infected rats. Such sensitive psychological data have not been accrued in any other model of virus infection.

The dysfunction in spatial and recognition memory can be explained by the distribution of BDV antigen in special neural structures in areas of the limbic system, like the hippocampus and the amygdala. Such subtle behavioural alterations can also be correlated with distinct antigen-carrying 'layers' in limbic structures (Gosztonyi & Ludwig, 1995; Ludwig *et al.*, 1988).

This all contrasts with the striking disturbances in global behaviour which have been observed in tree shrews (see below) infected as adults (Sprankel *et al.*, 1978), and equally contrasts with observations made in adult rats (Narayan *et al.*, 1983). Experiments performed with rats infected neonatally seemed to be most relevant for human infections (Gosztonyi & Ludwig, 1995; Hornig *et al.*, 2001).

Whereas disturbance in movement and classical behaviour has been attributed more to, and felt to be associated with, immune-mediated phenomena in adult infections (Gosztonyi & Ludwig, 1995, 2001; Solbrig *et al.*, 1996a), infection of neonates is thought to induce distinct changes in activity and cause cerebellar as well as hippocampal dyskinesias (Bautista *et al.*, 1994; Carbone *et al.*, 1996; Dittrich *et al.*, 1989; Plata-Salaman *et al.*, 1999). The behavioural changes with hyperactivity and dyskinetic dystonias have also been associated with prefrontal cortex dysfunction and changes in dopamine receptors (Solbrig *et al.*, 1995, 1996a, b). The association of changed locomotor activity and greater mean activity in neonatally infected rats most probably is due to altered cytokine gene expression in the prefrontal cortex. Furthermore, a cautious interpretation of differences in behaviour of infected and mock-infected rats has been suggested to be linked to altered apoptotic processes rather than to inflammatory events (Hornig *et al.*, 2001; Gosztonyi & Ludwig, 2001).

A most fascinating hypothesis enhanced through the years has been put forward by Gosztonyi & Ludwig (1984b); it proposes that some of the subtle alterations in neonatally infected rats are due to an interference of BDV antigens with neurotransmitter receptors. This concept has been supported by a characteristic and peculiar distribution pattern of BDV antigen in the hippocampal formations (Gosztonyi & Ludwig, 1984a, 1995; Ludwig *et al.*, 1988): virus-specific antigens mark the stratum oriens as well as the stratum radiatum of the CA1 region, and are lacking in the stratum pyramidale and lacunosum-moleculare. This unique pattern is an image of the termination of various afferent neuronal systems and their specific neurotransmitters. In the case of the strata oriens and radiatum, the synapses are glutamatergic and aspartatergic; the non-antigen-filled other two strata have no affinity to glutamate. BDV-specific RNPs, which are the infectious unit travelling intra-axonally, use anterograde transport mechanisms, finally reaching the dentate gyrus. Here the granule cells transport these infectious units anterogradely along the mossy fibre system to the CA3 neurons (see Fig. 16 in Gosztonyi & Ludwig, 1995). These cells further transport virus RNPs and/or viral antigens to presynaptic terminals in the strata oriens and radiatum of the CA1 area of the hippocampus. This explains why synaptic buttons together with their presynaptic axonal segments are filled with BDV antigens, whereas the CA1 pyramidal neurons, as the termination site of the synapses, are antigen-free. Thus RNPs accumulate in terminal segments of the CA3 axons.

The striking difference between CA3 and CA1 neurons, the former carrying BDV antigen and the latter being empty, and both being of the glutamatergic neurotransmitter type, is that CA3 neurons express the kainate 1 (KA-1) receptor, which is lacking on the CA1 neurons (for further details, see Gosztonyi & Ludwig, 2001). Therefore, we postulate here that the KA-1 receptor represents the BDV receptor in the CNS (for further detailed explanations, see Gosztonyi & Ludwig, 2001). This hypothesis gives a logical explanation for many observations. It could also explain why oligodendrocytes and astrocytes, which are known to carry KA-1 receptors, can express BDV. It has been reported by our group and several others that BDV is not only neuronotropic but is also gliotropic. It has been shown earlier, and elaborated recently (Gosztonyi & Ludwig, 1995, 2001), that BDV antigen is expressed in most of the dentate gyrus and CA3 neurons. The neurons in these brain regions are rich in KA-1 receptors. The presence of Borna antigen, as deduced from these explanations, blocks the normal flow of impulses received at the neuronal glutamate receptors. This easily explains the imbalance between glutamatergic and other neurotransmitter systems in the hippocampus. Based on the importance of hippocampal structures for learning and memory processing, we adduce the reactivity of BDV-infected rats is changed through these mechanisms.

In further support of the hypothesis is the fact that the retina contains formations of glutamate receptors. In all animal models, a severe destruction of the retina has been observed (Krey *et al.*, 1979a, 1982; Narayan *et al.*, 1983). Although in general BDV does not cause cytopathic effects, exceptions from this rule seem to occur in sub-acute/chronic infections where the dentate gyrus (granule cells) and the retina show severe degenerative signs. The CA1 neurons of the hippocampus, however, remain morphologically unchanged as do an overwhelming majority of the CA3 neurons. In all our neuropathologic studies, we have found (and this is supported by other groups) that neurons of the dentate gyrus are destroyed selectively, a peculiar and specific feature in BDV infections. This was obvious in natural infections, but also very prominent in experimentally infected rats (Fig. 8a, b) using laboratory or wild-type strains (Gosztonyi & Ludwig, 1995; Lundgren *et al.*, 1995). Nevertheless, there are still insufficient experimentally proven explanations for this typical destruction of the dentate gyrus. Additional to the 'glutamate-receptor affinity', other groups have emphasized an interplay of BDV proteins preferentially with neurotransmitters of the dopamine system (Hornig *et al.*, 2001).

All these observations support the ability of BDV to cause neurotransmitter abnormalities (Lipkin *et al.*, 1988) and of viral antigens to upset the neurotransmitter equilibrium, providing an explanation for the psychological changes during infection with this virus. In this context, experimental data on the interference of BDV infection of neonates with normal play behaviour or the induction of autistic-like behaviour are of interest (Pletnikov *et al.*, 1999). Furthermore, the interference of BDV infection with plasticity processes of neonatal brain cells could in future be useful as a model (Gonzalez-Dunia *et al.*, 2000).

Recently, Mongolian gerbils were inoculated successfully with human suspected BDV material and a new isolate was recovered (Nakamura *et al.*, 2000). Although this rat-like animal seems to be susceptible to BDV, the lack of knowledge of its genetics and immune response and the difficulty in handling it, diminish its value as an experimental model.

Mouse

Borna virus infection of mice was initiated by our group (Kao *et al.*, 1984). This model shows similarity to that of the rat. After multiple passage in the rat, strain V virus was able to multiply to high titres in the mouse. The infection was persistent, without an obvious cellular immune response. As in the rat, BDV antigens accumulate in brain areas of the limbic system. A characteristic example of the rostro-caudal spreading of BDV along the neuraxis is demonstrated in Fig. 9.

During tolerant infections, mice were seen to become fat. No clinical symptoms were seen using inbred mouse strains. It was of significance that these tolerantly infected mice produced very high titres of neutralizing antibodies, although temporally postponed to the anti-s-antigen antibodies (Kao *et al.*, 1984; Ludwig *et al.*, 1988). Others have used the mouse model with other BDV strains, mainly to study questions of persistence and the immune response in the CNS, and could essentially reproduce earlier data.

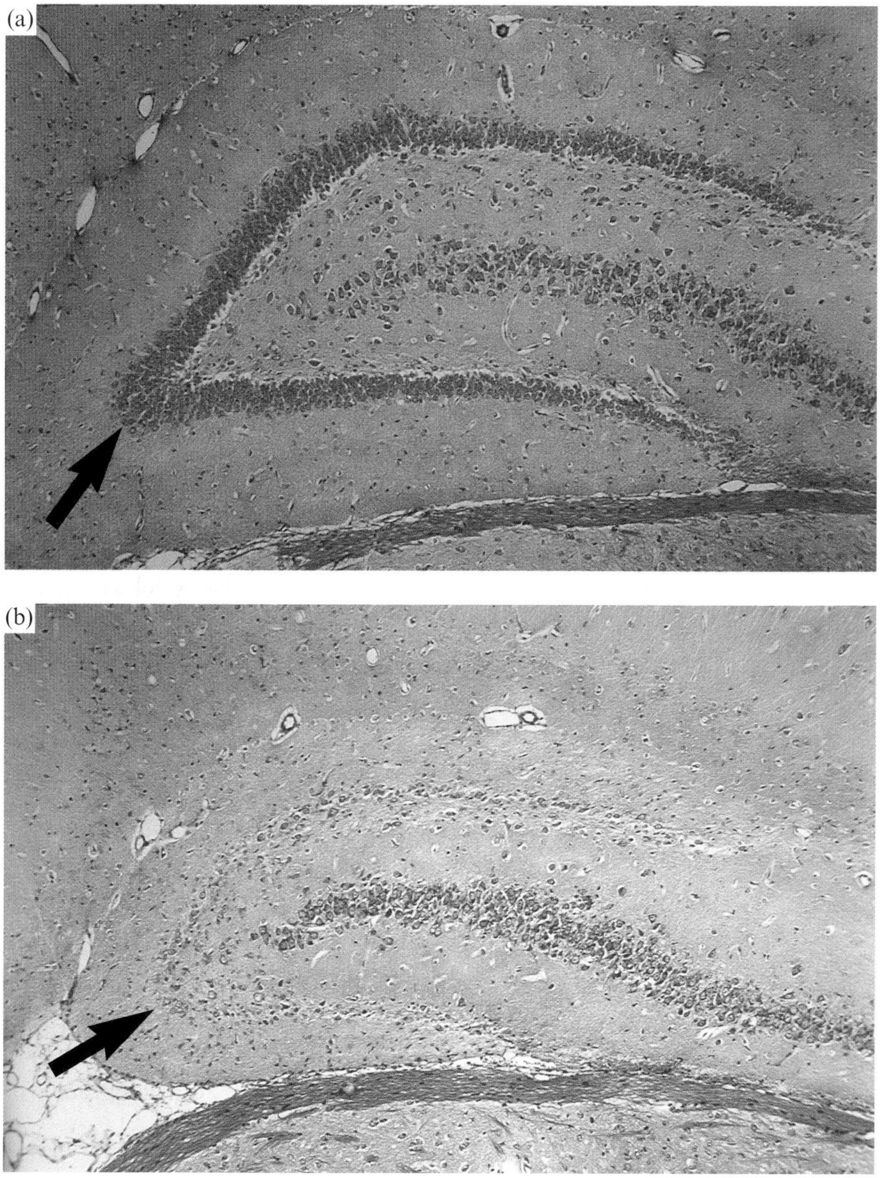

Fig. 8. Hippocampus of the mock-infected (a) and BDV-strain-V-infected (b) rat. Due to long-term infection, the neurons in the preferred CA3b brain region are characteristically destroyed (b); see arrows. For further details see Gosztonyi & Ludwig (1995, 2001).

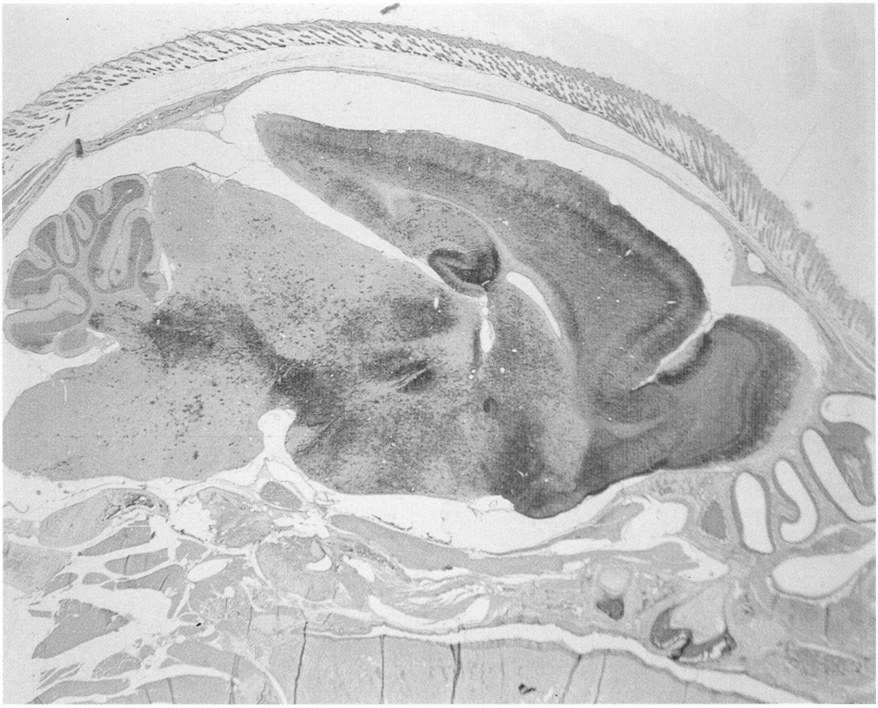

Fig. 9. Distribution of BDV-specific antigen in the mouse brain. Paramedian section of the whole brain of a 4-week-old BALB/c mouse intranasally infected with strain V at day 1 post-partum. The infection has spread rostro-caudally with an antigen expression at preferred sites in brain areas, functionally called the limbic system. Immunohistology is based on the APAAP stain (antigen stains red). Picture kindly provided by Georg Gosztonyi.

From our long-term experience, the mouse model does not portray significant parallels to BDV infections in man. It is less useful than the rat model which, in its tolerant form at least, is a suitable model for understanding some of the fascinating changes in behaviour, interlinked with the possible neurotransmitter alterations, although, unlike in human mood disorders, a return to a healthy state cannot be achieved (Gosztonyi & Ludwig, 2001; Hornig *et al.*, 2001).

Cat

The only relevant experimental infections using adult SPF cats were done by Lundgren *et al.* (1997). The aim was to prove whether 'staggering disease', the Borna disease of the cat (Lundgren *et al.*, 1995; Hübner, 1999), could be induced. Essentially, this model has shown all the problems of artificial infections. Although behavioural alterations and a definite state of infection (silent encephalitis, antibodies) could be modelled, the classical disease did not appear. This again shows that multi-factorial prerequisites are necessary for induction of a full-blown clinical picture. As deduced from other data gathered worldwide, there is no doubt that there is a natural BDV infection of cats with known disease parameters, including subtle changes of normal behaviour (Ludwig & Bode, 2000).

Birds

Experimental infection of chickens, quails and ostriches has been reported to be possible without indicating any important relevance to the features of the human infection. However, it is noteworthy that the chicken produced the highest neutralizing antibody titres ever seen in experimental infection (for further details, see Ludwig & Bode, 2000).

Primate-like animals

Here, rhesus monkeys in two independent settings (Stitz *et al.*, 1980; H. Ludwig & L. Stitz, unpublished) and tree shrews (Sprankel *et al.*, 1978) have been infected experimentally as adults using strain V. Tree shrews were chosen because a colony was kept for study of normal behaviour by Sprankel in Giessen. He concluded that these animals presented a lot of primate-like behavioural patterns, were extremely competent in withstanding injury of all kinds and react sensitively to stress. This animal has turned out to be an exciting model for studying alterations of normal instinctive behaviour caused by BDV infection. However, this model has not yet been explored in wide-ranging studies.

Nevertheless, clear patterns in infected animals were evident. First, some animals passed through hyperactivity followed by decline phases and, after 2 months, showed progressive signs of Borna disease, with paralyses and exitus. Solitary animals expressed temporarily increased hyperkinetic activity compared to pairs of animals. Second, in paired tree shrews significant alterations in the socio-sexual behaviour appeared. It was obvious that communicative behaviour, control of identity of the partners, and the sexual behaviour were reduced significantly. Social activities, like lying in bodily contact, were increased drastically (more than 10-fold). There was an unusual and surprising reversal of the role of sex partners (probably based on increased social behaviour). Third, successful reproduction was reduced due to a change in the polarity of the sexual partners. One offspring of infected parents developed into a dwarf, but was fully integrated in these abnormal (in the widest sense, virus-induced) interactions of the parents (Sprankel *et al.*, 1978). The body of data concerning these psycho-altered features, together with encephalitic reactions specifically in areas of the limbic system, encouraged us to pursue the notion of human infections.

Another interesting model paralleling human infections was the early stages of rhesus monkey (*Macaca mulatta*) infections. In contrast to the shrews, all intracranially and intranasally infected animals finally showed the full-blown Borna disease with clinical and histopathological expression in the eye and brain (Krey *et al.*, 1982).

Of relevance to human health were mood disorders and subtle behavioural changes early in the infection process of this primate. Approximately 3–4 weeks post-infection, the animal keeper reported that some showed subtle changes in their daily behaviour, did not come to the feeding place, were unfriendly to him, appeared aggressive to him, suddenly turned away from him, or unexpectedly attacked him. Furthermore, some of the animals stayed depressed in the corner of the cage. One monkey clearly expressed a depressive mood, and often stared for hours into a corner of its cage. Some weeks later, all the intracranially and intranasally infected monkeys showed increasing locomotor disabilities and ataxias running into the typical neurological disease signs of Borna disease, known in other animals (see Ludwig & Bode, 2000). At these phases, a strong immune response with neutralizing antibodies was measurable (H. Ludwig, L. Bode & L. Stitz, unpublished). These early personality changes and alterations in character of the animals are very reminiscent of BDV infections in man (see above).

EPILOGUE

In this chapter, we have summarized and interpreted data on antibodies, nucleic acid or antigen in human blood and/or brain. A poor correlation of antibodies and direct viral markers exists. New infection markers, like plasma antigen and CICs, have been introduced. Generally, virus markers occur more frequently in psychiatric patients than in healthy controls. This can be observed worldwide. Similar, but not identical, viral sequences are found in humans and animals. Likewise, human and animal virus isolates have similar, but not identical, properties.

With respect to the broad spectrum of mammals which can be infected and show behavioural changes or more severe clinical signs caused by BDV, it is, indeed, not plausible to exclude the possibility that BDV-associated disease can occur in vulnerable humans. Clearly, it is likely, simply due to aetiopathogenetically confirmed disease in so many animal species with which mankind shares a very similar limbic brain structure.

It remains debatable what clinically defined types of disorders may be predominantly associated with human BDV infection, and what primary pathology governs or influences virus infection and disease. Over time, diagnostic tools will certainly improve and help to elucidate these questions.

The source of infection in humans, and likewise the natural reservoir, remain obscure. Whether BDV is a zoonotic agent and its true prevalence rates in the population will be of great interest for the future.

Over time, there have been numerous views and reviews, some more and some less authoritative, focusing on different aspects of BDV infection, and in recent years, including human infections (Joest, 1926; Seifried & Spatz, 1930; Ludwig & Becht, 1977; Bode *et al.*, 1993; Bode & Ludwig, 1997; Haga *et al.*, 1997a; Gonzalez-Dunia *et al.*, 1997b; Hatalski *et al.*, 1997; Bechter, 1998; Staeheli *et al.*, 2000). With our 15 (Bode *et al.*, 1988) and 30 (Wagner *et al.*, 1968) years of experience in the 'Borna-field', we believe that the chapters of Nicolau & Galloway (1928, 1930), broadening the experimental infections and opening up the field to English- and French-speaking readers, Zwick (1939), summarizing clinical and infection aspects of the time, Heinig (1969), summarizing knowledge in endemic areas in Eastern Europe, and Ludwig *et al.* (1988), covering the pathogenesis of the persistent infection, the book of Koprowski & Lipkin (1995), signalling the emerging fields of molecular biology, neuropathology and human infections, and our chapters (Ludwig & Bode, 2000) with the veterinary medical aspects on zoonosis and epidemiology, as well as this article with an emphasis on the growing human public health importance, presenting new methods which strengthen a linkage of infection with mood disorders, and replying also to the demand 'Clearly it is important to develop a reliable test for this virus and sort the matter out once and for ever' (Crawford, 2000), will survive the critics of time. The seminal books on mood disorders and the imbalance of the neurotransmitter equilibrium entering into devastating depression or mania (Goodwin & Jamison, 1990), which might initiate an overflow of creativity (Jamison, 1993), promise a fascinating future for Borna virus research. To retain curiosity and to satisfy long-term sceptics, we leave the question mark in the title: BDV – a threat for human mental health?

DEDICATION

This chapter is dedicated to Rudi Rott (75 years), Giessen, who started post-war Borna virology about 50 years ago at the Zwick-Institut, and introduced one of us (H. L.) to this agent.

ACKNOWLEDGEMENTS

We are grateful to Patricia Reckwald for invaluable help with the immune assays, Tine Leiskau for taking care of the cells, and Christine Bernd and Gülsüm Baykal for technical assistance. We thank Georg Gosztonyi, Detlef Dietrich, Roman Stoyloff, Ron Ferszt, Emanuel Severus, Michael Deuschle and Hinderk Emrich for scientific help and generous support, our American colleagues Ian Lipkin and Juan Carlos de la Torre for previous fruitful joint studies, Toni Komaroff for supply of samples from his CFS study programme, and Fedik Rantam for virus isolation. H. L. is also grateful to the Japanese colleagues Satoshi Sasaki, Masato Furuya, Norio Hirano, Tetsuro Tsukamoto and Koji Furuya who joined our Borna group in the 80s. These long-term studies were supported by the Deutsche Forschungsgemeinschaft and the European Union.

REFERENCES

Bautista, J. R., Schwartz, G. J., de la Torre, J. C., Moran, T. H. & Carbone, K. M. (1994). Early and persistent abnormalities in rats with neonatally acquired Borna disease virus infection. *Brain Res Bull* **34**, 31–40.

Bechter, K. (1998). *Borna Disease Virus, mögliche Ursache neurologischer Störungen des Menschen. Monographien aus dem Gesamtgebiet der Psychiatrie*, 89, pp. 1–177. Steinkopff, Darmstadt.

Bode, L. (1995). Human infections with Borna disease virus and potential pathogenic implications. *Curr Top Microbiol Immunol* **190**, 103–130.

Bode, L. (2000). Human infections with Borna disease virus. In *Proceedings of the 1st UK Workshop on Borna Disease Virus*, pp. 15–16. Rhondda, UK.

Bode, L. & Ludwig, H. (1997). Clinical similarities and close genetic relationship of human and animal Borna disease virus. *Arch Virol Suppl* **13**, 167–182.

Bode, L., Riegel, S., Ludwig, H., Amsterdam, J. D., Lange, W. & Koprowski, H. (1988). Borna disease virus-specific antibodies in patients with HIV infection and with mental disorders. *Lancet* **2**, 689.

Bode, L., Riegel, S., Lange, W. & Ludwig, H. (1992). Human infections with Borna disease virus: seroprevalence in patients with chronic diseases and healthy individuals. *J Med Virol* **36**, 309–315.

Bode, L., Ferszt, R. & Czech, G. (1993). Borna disease virus infection and affective disorders in man. *Arch Virol Suppl* **7**, 159–167.

Bode, L., Steinbach, F. & Ludwig, H. (1994). A novel marker for Borna disease virus infection. *Lancet* **343**, 297–298.

Bode, L., Zimmermann, W., Ferszt, R., Steinbach, F. & Ludwig, H. (1995). Borna disease virus genome transcribed and expressed in psychiatric patients. *Nat Med* **1**, 232–236.

Bode, L., Dürrwald, R., Rantam, F. A., Ferszt, R. & Ludwig, H. (1996). First isolates of infectious human Borna disease virus from patients with mood disorders. *Mol Psychiatry* **1**, 200–212.

Bode, L., Dietrich, D. E., Stoyloff, R., Emrich, H. M. & Ludwig, H. (1997). Amantadine and human Borna disease virus *in vitro* and *in vivo* in an infected patient with bipolar depression. *Lancet* **349**, 178–179.

Bode, L., Stoyloff, R. & Ludwig, H. (2000). Human Bornaviruses and laboratory strains. *Lancet* **355**, 1462.

Briese, T., Schneemann, A., Lewis, A. J., Park, Y.-S., Kim, S., Ludwig, H. & Lipkin, W. I. (1994). Genomic organization of Borna disease virus. *Proc Natl Acad Sci U S A* **91**, 4362–4366.

Carbone, K. M., Duchala, C. S., Griffin, J. W., Kincaid, A. L. & Narayan, O. (1987). Pathogenesis of Borna disease in rats: evidence that intra-axonal spread is the major route for virus dissemination and the determinant for disease incubation. *J Virol* **61**, 3431–3440.

Carbone, K. M., Silvas, P. M., Rubin, S. A., Vogel, M., Moran, T. H. & Schwartz, G. (1996). Quantitative correlation of viral induced damage to the hippocampus and spatial learning and memory deficits. *J Neurovirol* **2**, 195.

Chen, C.-H., Chiu, Y.-L., Shaw, C.-K., Tsai, M. T., Hwang, A. L. & Hsiao, K.-J. (1999). Detection of Borna disease virus RNA from peripheral blood cells in schizophrenic patients and mental health workers. *Mol Psychiatry* **4**, 566–571.

Crawford, D. H. (2000). *The Invisible Enemy, a Natural History of Viruses*, p. 150. New York: Oxford University Press.

Cubitt, B. & de la Torre, J. C. (1997). Amantadine does not have antiviral activity against Borna disease virus. *Arch Virol* **142**, 2035–2042.

Cubitt, B., Oldstone, C. & de la Torre, J. C. (1994). Sequence and genome organization of Borna disease virus. *J Virol* **68**, 1382–1396.

Danner, K., Heubeck, D. & Mayr, A. (1978). *In vitro* studies on Borna virus. I. The use of cell cultures for the demonstration, titration and production of Borna virus. *Arch Virol* **57**, 63–75.

Deuschle, M., Bode, L., Heuser, I., Schmider, J. & Ludwig, H. (1998). Borna disease virus proteins in cerebrospinal fluid of patients with recurrent depression and multiple sclerosis. *Lancet* **352**, 1828–1829.

Dietrich, E. D., Bode, L., Spannhuth, C. W., Lau, T., Huber, T. J., Brodhun, B., Ludwig, H. & Emrich, H. M. (2000). Amantadine in depressive patients with Borna disease virus (BDV) infection: an open trial. *Bipolar Disorder* **2**, 65–70.

Dittrich, W., Bode, L., Ludwig, H., Kao, M. & Schneider, K. (1989). Learning deficiencies in Borna disease virus-infected but clinically healthy rats. *Biol Psychiatry* **26**, 818–828.

Domingo, E., Holland, J., Biebricher, C. & Eigen, M. (1995). Quasi-species: the concept and the word. In *Molecular Basis of Virus Evolution*, pp. 181–191. Edited by A. Gibbs, C. H. Calisher & F. Garcia-Arenal. Cambridge: Cambridge University Press.

Dürrwald, R. & Ludwig, H. (1997). Borna disease virus (BDV), a (zoonotic?) worldwide pathogen. A review of the history of the disease and the virus infection with comprehensive bibliography. *Zentralbl Veterinärmed B* **44**, 147–184.

Ferszt, R., Severus, E., Bode, L., Brehm, M., Kühl, K.-P., Berzewski, H. & Ludwig, H. (1999a). Activated Borna disease virus in affective disorders. *Pharmacopsychiatry* **32**, 93–98.

Ferszt, R., Kühl, K.-P., Bode, L., Severus, E. W., Winzer, B., Berghöfer, A., Beelitz, G., Brodhun, B., Müller-Örlinghausen, B. & Ludwig, H. (1999b). Amantadine revisited: an open trial of amantadine sulfate treatment in chronically depressed patients with Borna disease virus infection. *Pharmacopsychiatry* **32**, 142–147.

Gierend, M. (1982). *Zur Pathogenese der Bornaschen Krankheit beim Kaninchen: Untersuchungen über die zelluläre Imunantwort, die Wirkung von immunsuppressiver Behandlung und die Elektroencephalographie*. DVM thesis, Free University, Berlin.

Gierend, M. & Ludwig, H. (1981). Influence of immunosuppressive treatment of Borna disease in rabbits. *Arch Virol* **67**, 217–228.

Gonzalez-Dunia, D., Cubitt, B., Grässer, F. A. & de la Torre, J. C. (1997a). Characterization of Borna disease virus p56 protein, a surface glycoprotein involved in virus entry. *J Virol* **71**, 3208–3218.

Gonzalez-Dunia, D., Sauder, C. & de la Torre, J. C. (1997b). Borna disease virus and the brain. *Brain Res Bull* **44**, 647–664.

Gonzalez-Dunia, D., Watanabe, M., Syan, S., Mallory, M., Masliah, E. & de la Torre, J. C. (2000). Synaptic pathology in Borna disease virus infection. *J Virol* **74**, 3441–3448.

Goodwin, F. K. & Jamison, K. R. (1990). *Manic-depressive Illness*. New York: Oxford University Press.

Gosztonyi, G. & Ludwig, H. (1984a). Borna disease of horses. An immunohistological and virological study of naturally infected animals. *Acta Neuropathol (Berl)* **64**, 213–221.

Gosztonyi, G. & Ludwig, H. (1984b). Neurotransmitter receptors and viral neurotropism. *Neuropsychiat Clin* **3**, 107–114.

Gosztonyi, G. & Ludwig, H. (1995). Borna disease – neuropathology and pathogenesis. *Curr Top Microbiol Immunol* **190**, 39–73.

Gosztonyi, G. & Ludwig, H. (2001). Interaction of viral proteins with neurotransmitter receptors may protect or destroy neurons. *Curr Top Microbiol Immunol* **253**, 121–144.

Haga, S., Motoi, Y., Ikeda, K. & Japan Borna Study Group (1997a). Borna disease virus and neuropsychiatric disorders. *Lancet* **350**, 592.

Haga, S., Yoshimura, M., Motoi, Y., Arima, K., Aizawa, T., Ikuta, K., Tashiro, M. & Ikeda, K. (1997b). Detection of Borna disease virus genome in normal human brain tissue. *Brain Res* **770**, 307–309.

Hallensleben, W., Zocher, M. & Staeheli, P. (1997). Borna disease virus is not sensitive to amantadine. *Arch Virol* **142**, 2043–2048.

Hatalski, C. G., Lewis, A. J. & Lipkin, W. I. (1997). Borna disease. *Emerg Infect Dis* **3**, 129–135.

Heinig, A. (1969). Die Bornasche Krankheit der Pferde und Schafe. In *Handbuch der Virusinfektionen bei Tieren*, vol. 4, pp. 83–148. Edited by H. Röhrer. Jena: Fischer.

Herzog, S. & Rott, R. (1980). Replication of Borna disease virus in cell cultures. *Med Microbiol Immunol* **168**, 153–158.

Hirano, N., Kao, M. & Ludwig, H. (1983). Persistent, tolerant or subacute infection in Borna disease virus-infected rats. *J Gen Virol* **64**, 1521–1530.

Hornig, M., Solbrig, M., Horscroft, N., Weissenbröck & Lipkin, W. I. (2001). Borna disease virus infection of adult and neonatal rats: models for neuropsychiatric disease. *Curr Top Microbiol Immunol* **253**, 157–178.

Hübner, J. (1999). *Die Borna Disease Virus-Infektion der Katze in Deutschland und ihr Vergleich mit anderen felinen Virusinfektionen*. DVM thesis, Free University, Berlin.

Iwahashi, K., Watanabe, M., Nakamura, K., Suwaki, H., Nakaya, T., Nakamura, Y., Takahashi, H. & Ikuta, K. (1997). Clinical investigation of the relationship between Borna disease virus (BDV) infection and schizophrenia in 67 patients in Japan. *Acta Psychiatr Scand* **96**, 412–415.

Iwata, Y., Takahashi, K., Peng, X., Fukuda, K., Ohno, K., Ogawa, T., Gonda, K., Mori, N., Niwa, S.-I. & Shigeta, S. (1998). Detection and sequence analysis of Borna disease virus p24 RNA from peripheral blood mononuclear cells of patients with mood disorders or schizophrenia and of blood donors. *J Virol* **72**, 10044–10049.

Jamison, K. R. (1993). *Touched with Fire. Manic-depressive Illness and the Artistic Temperament*. New York: Free Press.

Joest, E. (1926). Vergleichend-anatomische Betrachtungen über Encephalitis. *Klin Wochenschr* **5**, 209–211.

Joest, E. & Degen, K. (1911). Untersuchungen über die pathologische Histologie, Pathogenese und postmortale Diagnose der seuchenhaften Gehirn-Rückenmarksentzündung (Bornaschen Krankheit) des Pferdes. *Z Infektionskr Parasitol Krank Hyg Haustiere* **9**, 1–98.

Jordan, I., Briese, T., Averett, D. R. & Lipkin, I. W. (1999). Inhibition of Borna disease virus replication by ribavirin. *J Virol* **73**, 7903–7906.

Kao, M., Ludwig, H. & Gosztonyi, G. (1984). Adaptation of Borna disease virus to the mouse. *J Gen Virol* **65**, 1845–1849.

Kishi, M., Nakaya, T., Nakamura, Y., Kakinuma, M., Takahashi, T. A., Sekiguchi, S., Uchikawa, M., Tadokoro, K., Ikeda, K. & Ikuta, K. (1995). Prevalence of Borna disease virus RNA in peripheral blood mononuclear cells from blood donors. *Med Microbiol Immunol* **184**, 135–138.

Kliche, S., Briese, T., Henschen, A. H., Stitz, L. & Lipkin, W. I. (1994). Characterization of a Borna disease virus glycoprotein, gp18. *J Virol* **68**, 6918–6923.

Kohno, T., Goto, T., Takasaki, T., Morita, C., Nakaya, T., Ikuta, K., Kurane, I., Sano, K. & Nakai, M. (1999). Fine structure and morphogenesis of Borna disease virus. *J Virol* **73**, 760–766.

Koprowski, H. & Lipkin, W. I. (editors) (1995). Borna disease. *Curr Top Microbiol Immunol* **190**, 1–134.

Krey, H. F., Ludwig, H. & Boschek, C. B. (1979a). Multifocal retinopathy in Borna disease virus infected rabbits. *Am J Ophthalmol* **87**, 157–164.

Krey, H. F., Ludwig, H. & Rott, R. (1979b). Spread of infectious virus along the optic nerve into the retina in Borna disease virus-infected rabbits. *Arch Virol* **61**, 283–288.

Krey, H. F., Stitz, L. & Ludwig, H. (1982). Virus-induced pigment epithelitis in rhesus monkeys. Clinical and histological findings. *Ophthalmologica* **185**, 205–213.

Lieb, K., Hallensleben, W., Czygan, M., Stitz, L. & Staeheli, P. (1997). No Borna disease virus-specific RNA detected in blood from psychiatric patients in different regions of Germany. The Bornavirus Study Group. *Lancet* **350**, 1002.

Lipkin, W. I., Carbone, K. M., Wilson, M. C., Duchala, C. S., Narayan, O. & Oldstone, M. B. A. (1988). Neurotransmitter abnormalities in Borna disease. *Brain Res* **475**, 366–370.

Lipkin, W. I., Travis, G. H., Carbone, K. M. & Wilson, M. C. (1990). Isolation and characterization of Borna disease agent cDNA clones. *Proc Natl Acad Sci U S A* **87**, 4184–4188.

Ludwig, H. & Becht, H. (1977). Borna disease – a summary of our present knowledge. In *Slow Virus Infections of the Central Nervous System: Investigational Approaches to Aetiology and Pathogenesis of these Diseases*, pp. 75–83. Edited by V. Ter Meulen & M. Katz. New York: Springer.

Ludwig, H. & Bode, L. (1997). The neuropathogenesis of Borna disease virus infections. *Intervirology* **40**, 185–197.

Ludwig, H. & Bode, L. (2000). Borna disease virus: new aspects on infection, disease, diagnosis and epidemiology. *Rev Sci Tech Off Int Epiz* **19**, 259–288.

Ludwig, H. & Thein, P. (1977). Demonstration of specific antibodies in the central nervous system of horses naturally infected with Borna disease virus. *Med Microbiol Immunol* **163**, 215–226.

Ludwig, H., Becht, H. & Groh, L. (1973). Borna disease (BD), a slow virus infection. Biological properties of the virus. *Med Microbiol Immunol* **158**, 275–289.

Ludwig, H., Kraft, W., Kao, M., Gosztonyi, G., Dahme, E. & Krey, H. (1985). Borna-Virus Infektion (Borna-Krankheit) bei natürlich und experimentell infizierten Tieren: Ihre Bedeutung für Forschung und Praxis. *Tierärztl Praxis* **13**, 421–453.

Ludwig, H., Bode, L. & Gosztonyi, G. (1988). Borna disease: a persistent virus infection of the central nervous system. *Prog Med Virol* **35**, 107–151.

Ludwig, H., Furuya, K., Bode, L., Klein, N., Dürrwald, R. & Lee, D. S. (1993). Biology and neurobiology of Borna disease viruses (BDV), defined by antibodies, neutralizability and their pathogenic potential. *Arch Virol Suppl* **7**, 111–133.

Ludwig, H., Bode, L., Schedlowski, M., Emrich, H. M. & Dietrich, D. E. (1998). Stress and human Borna virus infection. In *Stress and the Nervous System*, pp. 119–128. Edited by C. L. Bolis & J. Licinio. WHO/RPS/98·2. Geneva: World Health Organization.

Lundgren, A.-L., Zimmermann, W., Bode, L., Czech, G., Gosztonyi, G., Lindberg, R. & Ludwig, H. (1995). Staggering disease in cats: isolation and characterization of the feline Borna disease virus. *J Gen Virol* **76**, 2215–2222.

Lundgren, A.-L., Johannisson, A., Zimmermann, W., Bode, L., Rozell, B., Muluneh, A., Lindberg, R. & Ludwig, H. (1997). Neurological disease and encephalitis in cats experimentally infected with Borna disease virus. *Acta Neuropathol (Berl)* **93**, 391–401.

Lüschow, D. (1999). *Borna-Disease-Virus (BDV)-Infektionen und Erkrankungen bei Equiden: serologische und molekularepidemiologische Untersuchungen unter besonderer Berücksichtigung der phylogenetischen Sequenzanalyse*. DVM thesis, Free University, Berlin.

Mizutani, T., Inagaki, H., Araki, K., Kariwa, H., Arikawa, J. & Takashima, I. (1998). Inhibition of Borna disease virus replication by ribavirin in persistently infected cells. *Arch Virol* **143**, 2039–2044.

Nakamura, Y., Takahashi, H., Shoya, Y. & 11 other authors (2000). Isolation of Borna disease virus from human brain tissue. *J Virol* **74**, 4601–4611.

Narayan, O., Herzog, S., Frese, K., Scheefers, H. & Rott, R. (1983). Behavioural disease in rats caused by immunopathological responses to persistent Borna virus in the brain. *Science* **220**, 1401–1403.

Nicolau, S. & Galloway, I. A. (1928). Borna disease and enzootic encephalomyelitis of sheep and cattle. *Medical Research Council Special Reports Series*, vol. 121. London: His Majesty's Stationery Office.

Nicolau, S. & Galloway, I. A. (1930). L'encéphalo-myelite enzootique expérimentale (maladie de Borna). Troisième mémoire. *Ann Inst Pasteur* **45**, 457–523.

Nitzschke, E. (1963). Untersuchungen über die experimentelle Bornavirus-Infektion bei der Ratte. *Zentralbl Veterinärmed B* **10**, 470–527.

Nowotny, N. & Kolodziejek, J. (2000). Demonstration of Borna disease virus nucleic acid in a patient suffering from chronic fatique syndrome. *J Infect Dis* **181**, 1860–1861.

Oldstone, M. B. A. (1984). Virus-induced immune complex formation and disease: definition, regulation and importance. In *Concepts in Viral Pathogenesis*, pp. 201–206. Edited by A. L. Notkins & M. B. A. Oldstone. New York: Springer.

Planz, O., Rentzsch, C., Batra, A., Rziha, H.-J. & Stitz, L. (1998). Persistence of Borna disease virus-specific nucleic acid in blood of a psychiatric patient. *Lancet* **352**, 623.

Planz, O., Rentzsch, C., Batra, A., Batra, A., Winkler, T., Büttner, M., Rziha, H.-J. & Stitz, L. (1999). Pathogenesis of Borna disease virus: granulocyte fractions of psychiatric patients harbour infectious virus in the absence of antiviral antibodies. *J Virol* **73**, 6251–6256.

Plata-Salaman, C. R., Ilysin, S. E., Gayle, D., Romanovitch, A. & Carbone, K. M. (1999). Persistent Borna disease virus infection of neonatal rats causes brain regional changes of mRNAs for cytokines, cytokine receptor components and neuropeptides. *Brain Res Bull* **49**, 441–451.

Pletnikov, M. V., Rubin, S. A., Vasudevan, K., Moran, T. H. & Carbone, K. M. (1999). Developmental brain injury associated with abnormal play behaviour in neonatally Borna disease virus-infected Lewis rats: a model of autism. *Behav Brain Res* **100**, 43–50.

Pringle, C. R. (1995). The order Mononegavirales: evolutionary relationship and mechanisms of variation. In *Molecular Basis of Virus Evolution*, pp. 426–437. Edited by A. Gibbs, C. H. Calisher & F. Garcia-Arenal. Cambridge: Cambridge University Press.

Pyper, J. M. & Gartner, A. E. (1997). Molecular basis for the differential subcellular localization of the 38- and 39-kilodalton structural proteins of Borna disease virus. *J Virol* **71**, 5133–5139.

Roggendorf, W., Sasaki, S. & Ludwig, H. (1983). Light microscope and immunohistological investigations on the brain of Borna disease virus-infected rabbits. *Neuropathol Appl Neurobiol* **9**, 287–296.

Rott, R., Herzog, S., Fleischer, B., Winokur, A., Amsterdam, J., Dyson, W. & Koprowski, H. (1985). Detection of serum antibodies to Borna disease virus in patients with psychiatric disorders. *Science* **228**, 755–756.

Salvatore, M., Morzunov, S., Schwemmle, M. & Lipkin, W. I. (1997). Borna disease virus in brains of North American and European people with schizophrenia and bipolar disorders. Bornavirus Study Group. *Lancet* **349**, 1813–1814.

Sauder, C., Müller, A., Cubitt, B. & 8 other authors (1996). Detection of Borna disease virus (BDV) antibodies and BDV RNA in psychiatric patients: evidence for high sequence conservation of human blood-derived BDV RNA. *J Virol* **70**, 7713–7724.

Schneider, P. A., Schneemann, A. & Lipkin, W. I. (1994). RNA splicing in Borna disease virus, a non-segmented, negative-strand virus. *J Virol* **68**, 5007–5012.

Schneider, P. A., Hatalski, C. G., Lewis, A. J. & Lipkin, W. I. (1997). Biochemical and functional analysis of the Borna disease virus G protein. *J Virol* **71**, 331–336.

Schwemmle, M., Jehle, C., Formella, S. & Staeheli, P. (1999). Sequence similarities between human bornavirus isolates and laboratory strains question human origin. *Lancet* **354**, 1973–1974.

Seifried, O. & Spatz, H. (1930). Die Ausbreitung der enzephalitischen Reaktion bei der Bornaschen Krankheit der Pferde und deren Beziehung zu der Enzephalitis epidemica, der Heine-Medinschen Krankheit und der Lyssa des Menschen. Eine vergleichend-pathologische Studie. *Z ges Neurol Psychiat* **124**, 317–383.

Solbrig, M. V., Fallon, J. H. & Lipkin, W. I. (1995). Behavioral disturbances and pharmacology of Borna disease. *Curr Top Microbiol Immunol* **190**, 93–99.

Solbrig, M. L., Koob, G. F., Fallon, J. H., Reid, S. & Lipkin, W. I. (1996a). Prefrontal cortex dysfunction in Borna disease virus (BDV) infected rats. *Biol Psychiatry* **40**, 629–636.

Solbrig, M. L., Koob, G. F., Joyce, J. N. & Lipkin, W. I. (1996b). A neural substrate of hyperactivity in Borna disease: changes in dopamine receptors. *Virology* **222**, 332–338.

Sprankel, H., Richarz, K., Ludwig, H. & Rott, R. (1978). Behavior alterations in tree shrews (*Tupaia glis*, Diard 1820) induced by Borna disease virus. *Med Microbiol Immunol* **165**, 1–18.

von Sprockhoff, H. (1956). Zur biologischen Charakterisierung des Borna-s-Antigens. *Z Immunitätsforsch* **113**, 379–385.

Staeheli, P., Sauder, C., Hausmann, J., Ehrensperger, F. & Schwemmle, M. (2000). Epidemiology of Borna disease virus. *J Gen Virol* **81**, 2123–2135.

Stitz, L., Krey, H. & Ludwig, H. (1980). Borna disease in rhesus monkeys as a model for uveo-cerebral symptoms. *J Med Virol* **6**, 333–340.

Stitz, L., Dietzschold, B. & Carbone, K. M. (1995). Immunopathogenesis of Borna disease. *Curr Top Microbiol Immunol* **190**, 75–92.

Stitz, L., Planz, O. & Bilzer, T. (1998a). Lack of antiviral effect of amantadine in Borna disease virus infection. *Med Microbiol Immunol* **186**, 195–200.

Stitz, L., Nöske, K., Planz, O., Furrer, E., Lipkin, W. I. & Bilzer, T. (1998b). A functional role for neutralizing antibodies in Borna disease: influence on virus tropism outside the central nervous system. *J Virol* **72**, 8884–8892.

Stoyloff, R., Briese, T., Borchers, K., Zimmermann, W. & Ludwig, H. (1994). N-glycosylated protein(s) are important for the infectivity of Borna disease virus (BDV). *Arch Virol* **137**, 405–409.

Stoyloff, R., Bode, L., Wendt, H., Mulzer, J. & Ludwig, H. (1996). The hydrophobic mannose derivative 1B6TM efficiently inhibits Borna disease virus *in vitro*. *Antivir Chem Chemother* **7**, 197–202.

Stoyloff, R., Strecker, A., Bode, L., Franke, P., Ludwig, H. & Hucho, F. (1997). The glycosylated matrix protein of Borna disease virus is a tetrameric, membrane-bound viral component essential for infection. *Eur J Biochem* **246**, 252–257.

Stoyloff, R., Bode, L., Borchers, K. & Ludwig, H. (1998). Neutralization of Borna disease virus depends upon terminal carbohydrate residues (α-D-Man, β-D-GlcNAc) of glycoproteins gp17 and gp94. *Intervirology* **41**, 135–140.

Stoyloff, R., Bode, L. & Ludwig, H. (2001). Genetic structure and functional features of the glycoproteins M and G of the Borna disease virus. *Virus Genes* (in press).

Theofilopoulos, A. N. & Dixon, F. J. (1979). The biology and detection of immune complexes. *Adv Immunol* **28**, 89–220.

de la Torre, J. C., Bode, L., Dürrwald, R., Cubitt, B. & Ludwig, H. (1996a). Sequence characterization of human Borna disease virus. *Virus Res* **44**, 33–44.

de la Torre, J. C., Gonzalez-Dunia, D., Cubitt, B., Mallory, M., Mueller-Lantzsch, N., Grässer, F. A., Hansen, L. A. & Masliah, E. (1996b). Detection of Borna disease virus antigen and RNA in human autopsy brain samples from neuropsychiatric patients. *Virology* **223**, 272–282.

de la Torre, J. C., Bode, L., Carbone, K. M., Dietzschold, B., Ikuta, K., Lipkin, W. I., Ludwig, H., Richt, J. A., Staeheli, P. & Stitz, L. (2000). Family Bornaviridae. In *Virus Taxonomy*, pp. 531–538. Edited by M. H. V. Van Regenmortel, C. M. Fauquet & D. H. L. Bishop. London: Academic Press.

Wagner, K., Ludwig, H. & Paulsen, J. (1968). Fluoreszenzserologischer Nachweis von Borna-Virus Antigen. *Berl Münch Tierärztl Wochenschr* **81**, 395–396.

Walker, M. P., Jordan, I., Briese, T., Fischer, N. & Lipkin, W. I. (2000). Expression and characterization of Borna disease virus polymerase. *J Virol* **74**, 4425–4428.

Yamaguchi, K., Sawada, T., Naraki, T. & 14 other authors (1999). Detection of Borna disease virus-reactive antibodies from patients with psychiatric disorders and from horses by electrochemiluminescence immunoassay. *Clin Diagn Lab Immunol* **6**, 696–700.

Zimmermann, W., Breter, H., Rudolph, M. & Ludwig, H. (1994). Borna disease virus: immunoelectron microscopic characterization of cell-free virus and further information about the genome. *J Virol* **68**, 6755–6758.

Zwick, W. (1939). Bornasche Krankheit und Enzephalomyelitis der Tiere. In *Handbuch der Viruskrankheiten*, vol. 2, pp. 254–354. Edited by E. Gildenmeister, E. Haagen & O. Waldmann. Jena: Fischer.

Antiviral drug development and the impact of drug resistance

Graham Darby

Glaxo Wellcome Research and Development, Gunnels Wood Road, Stevenage SG1 2NY, UK

INTRODUCTION

Until recently, the clinical management of viral diseases depended largely on palliative measures designed to alleviate symptoms. Infections were to be avoided if at all possible or, if they could not be avoided, they were to be endured until they naturally resolved. There is, however, a long and impressive history of protection against viral diseases. The Chinese introduced inoculation against smallpox almost a thousand years ago, and they established the principle that priming of the immune system could prevent the more serious consequences of viral infection. This principle has been extensively exploited in modern times with the introduction of effective vaccines to prevent many of the more significant human viral diseases (Hilleman, 1998; Nossal, 1998; Ruff, 1999).

The power of vaccination is most dramatically demonstrated by the fact that for the first time man has been able to intervene successfully to eradicate an infectious disease from the human population. A campaign run by the World Health Organization in the late 1960s and throughout the 1970s resulted in the global eradication of smallpox with the last human case recorded in 1978. A similar campaign is under way to attempt the eradication of poliomyelitis and has involved mass vaccination of many millions of children (Hull *et al.*, 1997). Although the disease has now been eradicated from many parts of the world, it remains in several regions that are plagued by warfare and civil disruption, and the final goal may still be several years away.

Despite the impressive achievements of vaccination programmes, especially in the control of childhood diseases, for many viral diseases the development of vaccines has

SGM symposium 60: New challenges to health: the threat of virus infection.
Editors G. L. Smith, W. L. Irving, J. W. McCauley, D. J. Rowlands. Cambridge University Press. ISBN 0 521 80614 3 ©SGM 2001.

proven extremely difficult. This is largely due to the high multiplication rate of viruses and their error-prone genome replication that leads to a capacity for rapid evolution in the face of selective pressures. The consequence is that viral populations are rarely homogeneous and are often rapidly changing complex mixtures of 'quasi-species' (Goodenow *et al.*, 1989). This works against efforts to produce protective vaccines and may explain, in part at least, the failure to develop effective vaccines for human immunodeficiency virus (HIV) and several other important viral diseases (Burton & Moore, 1998).

Over the past 20 years our ability to manage viral diseases has improved markedly with the discovery and introduction of many effective antiviral drugs (see Table 1). The revolution began in the 1980s with the introduction of aciclovir for the treatment of herpes infections and zidovudine for HIV-related disease, and continued into the 1990s with the launch of more than 20 new antiviral drugs. Increasingly, therefore, management of viral diseases is being facilitated by drug treatment (Bean *et al.*, 1982; Carpenter *et al.*, 2000; Fischl *et al.*, 1987; Goodrich *et al.*, 1991; Mertz *et al.*, 1988; Wade *et al.*, 1984). Nowhere has progress been more dramatic than in the area of HIV disease, where drug treatment has had a marked impact on the quality of life and future prognosis of those infected with this virus (Carpenter *et al.*, 2000; Palella *et al.*, 1998).

Perhaps not surprisingly, viruses have responded to the challenge. When they are faced with the powerful selection pressures exerted by antiviral drugs, the capacity for rapid evolution that frequently thwarted attempts to develop vaccines facilitates the evolution of drug-resistant variants (Craig *et al.*, 1993; Larder & Darby, 1985; Tisdale *et al.*, 1993a, b). This is such an important phenomenon, frequently frustrating efforts to treat these diseases effectively (Larder *et al.*, 1989; Hayden *et al.*, 1989; Wade *et al.*, 1983), that drug resistance has now become a major factor to be considered when developing new drugs.

In this chapter, the intention is to discuss drug discovery and development, focusing particularly on aspects of the process that are influenced by issues surrounding drug resistance.

THE DRUG DEVELOPMENT PROCESS

New drug development is a long and protracted process that can take 10 years or more from concept to product on the market. It is also a high-risk enterprise that demands major investment and commitment. The route to a new drug may be well defined but many difficult and challenging hurdles have to be negotiated along the way. These hurdles can be viewed as milestones along the path that signal the completion of each phase of the journey (Fig. 1).

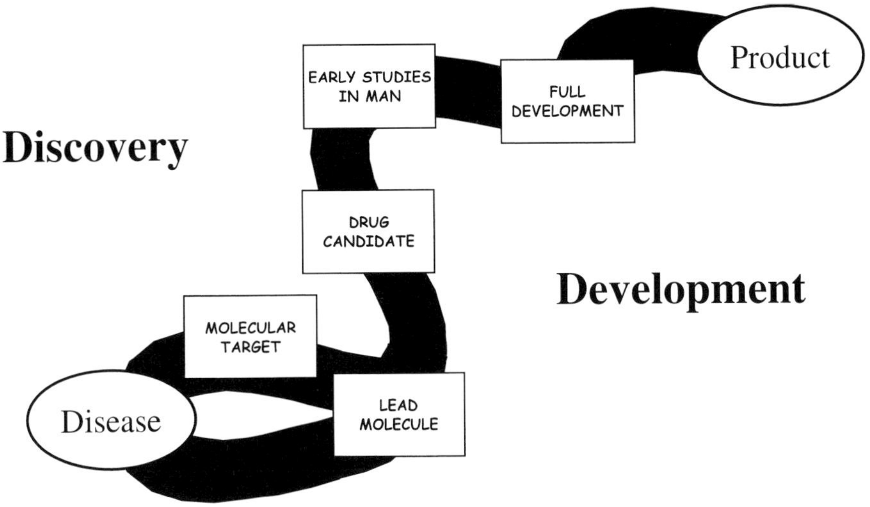

Fig. 1. Major milestones in antiviral drug development. The whole process from start to finish may take more than a decade and there is significant attrition at all stages. Costs escalate significantly in the later stages of the process.

The process begins with disease selection and at this stage the desired characteristics of the product to be developed should also be defined. Specific molecular targets may then be selected, and novel chemical leads identified. These leads are refined by chemical modification into potential drug candidates that are then tested in man to assess their safety. The final part of the journey involves the large-scale clinical testing of the drug to prove its safety and efficacy and this culminates, if successful, in approval of the drug by regulatory authorities.

Considerations relating to drug resistance impact on the drug development process at all its stages. Resistance studies influence our choice of disease and molecular targets, they shed light on the mechanisms of action of drugs and give warnings of potential problems in the clinic. Finally, they may help to guide the use of the drug in disease management.

SELECTING THE DISEASE

Various factors will influence the choice of disease but the major ones to be considered are the degree of morbidity and mortality and the effectiveness of treatment (taken together these represent the unmet medical need). The difficulty is that it is not the current unmet need that needs to be assessed, but the unmet need that is likely to exist when the product becomes available in 10 years or so. It might be foolhardy, for

Table 1. Antiviral drugs launched in the US since 1980

Virus	Drug class	Drug	Molecular target	Disease	Approved (US)
Herpesviruses	Nucleoside analogues	Aciclovir	DNA polymerase	Genital herpes/cold sores/shingles	1982
		Valaciclovir	DNA polymerase	Genital herpes/cold sores/shingles	1995
		Ganciclovir	DNA polymerase	Human cytomegalovirus disease	1996
		Famciclovir	DNA polymerase	Genital herpes/cold sores/shingles	1997
		Penciclovir	DNA polymerase	Genital herpes/cold sores/shingles	1998
	Pyrophosphate analogue	Foscarnet	DNA polymerase	Genital herpes/cold sores	1991
	Nucleotide analogue	Cidofovir	DNA polymerase	Human cytomegalovirus disease	1996
HIV	Nucleoside analogues	Zidovudine	Reverse transcriptase	HIV infection/AIDS	1987
		Didanosine	Reverse transcriptase	HIV infection/AIDS	1991
		Zalcitabine	Reverse transcriptase	HIV infection/AIDS	1992
		Stavudine	Reverse transcriptase	HIV infection/AIDS	1994
		Lamivudine	Reverse transcriptase	HIV infection/AIDS	1996
		Abacavir	Reverse transcriptase	HIV infection/AIDS	1999
	Peptide mimics	Saquinavir	Aspartyl protease	HIV infection/AIDS	1995
		Ritonavir	Aspartyl protease	HIV infection/AIDS	1996
		Indinavir	Aspartyl protease	HIV infection/AIDS	1996
		Nelfinavir	Aspartyl protease	HIV infection/AIDS	1997
		Amprenavir	Aspartyl protease	HIV infection/AIDS	1999

	Non-nucleoside RT inhibitors	Nevirapine	Reverse transcriptase	HIV infection/AIDS	1996
		Delavirdine	Reverse transcriptase	HIV infection/AIDS	1997
		Efavirenz	Reverse transcriptase	HIV infection/AIDS	1998
Influenza	Neuraminidase inhibitors	Zanamivir	Neuraminidase	Influenza	1999
		Oseltamivir	Neuraminidase	Influenza	1999
Hepatitis B	Nucleoside analogue	Lamivudine	Reverse transcriptase	Chronic hepatitis	1999

example, to embark on a drug discovery project for an acute disease if it appears likely that there will be a preventative vaccine shortly available.

In addition, in the corporate world it also has to be asked whether the projected product is likely to generate sufficient return on investment. The high cost of developing each successful drug and the considerable investment in those projects that ultimately fail demand that each drug launched onto the market must generate very significant income.

At this first stage in the drug discovery process, there should be an awareness of drug resistance issues and these may influence disease choices. For example, there are now many drugs available for the treatment of HIV-related disease (Kinchington *et al.*, 1997; Table 1) and so it is therefore reasonable to ask whether new drugs are really needed in this area. Is there likely to be remaining significant unmet need in this disease in 10 years' time?

In fact, there appear to be several compelling reasons why drug development work should continue in HIV and these all centre on issues around resistance. Two important lessons were learned from experience with the early antivirals used in the treatment of HIV. The first was that effective antiviral drugs could have a beneficial clinical effect (Fischl *et al.*, 1987). The second was that drug-resistant variants emerged in patients on chronic treatment and their appearance appears to correlate with clinical failure (Hirsch *et al.*, 1998; Larder *et al.*, 1989).

These observations led to the conclusion that effective control of this virus would be likely to require multiple drugs to be used in combination. Initially the reasoning was that the probability of resistance arising simultaneously to two drugs used in combination would be extremely low and so emergence of clinical resistance would be significantly delayed. This indeed proved to be the case. However, it also became apparent that the more effective the control of virus replication the greater the delay in the appearance of resistance mutations (Carpenter *et al.*, 1998; Balzarini, 1999). The complete suppression of virus replication and the long-term maintenance of that suppression therefore became the goal of antiviral therapy in HIV-related disease (Carpenter *et al.*, 1998; Raboud *et al.*, 1998). Increasingly aggressive drug regimens were introduced with the standard of care being three or four drugs in combination and occasionally as many as eight or nine drugs used for salvage therapy (Carpenter *et al.*, 1998; Katlama *et al.*, 1999).

While such regimens have delivered vastly improved patient outcomes, the resistance issues have not gone away. Even the most potent drug combinations fail to provide complete suppression of replication (Wong *et al.*, 1997) and in many individuals sooner or later drug-resistant virus emerges and virus load rebounds (Hirsch *et al.*, 1998; Monno *et al.*, 1999). There may be a simple explanation for this. Drug regimens that patients are asked to follow are often extremely complex with multiple drugs being taken at different dosing intervals, possibly with different dietary requirements. Additionally, taking anti-HIV drugs may be associated with unpleasant side effects (Carr *et al.*, 1998; Lo *et al.*, 1998). These factors may result in a failure in compliance (Casado *et al.*, 1999), resulting in suboptimal drug concentrations in plasma and tissues and an increase in virus replication rate. Breakdown in control of virus load then leads to the emergence of drug-resistant strains.

The emergence of resistant virus will require a switch in the drugs being used if viral load is to be brought under control once more, but this is becoming difficult as virus variants emerge that show cross-resistance to drugs within classes (Dulioust *et al.*, 1998; Winters *et al.*, 1998a, b). This severely limits the choices available when switching drugs and patients are increasingly being identified who have 'burned through' all available drug classes and who have run out of treatment options (Ledergerber *et al.*, 1999; Montagner *et al.*, 1999).

These experiences tell us that there remains significant unmet need in HIV and that new drugs will be required to keep pace with the emergence of drug resistance.

HIV is a particularly dramatic example where resistance issues are at the forefront in disease management, but this is not the case in all diseases. Two brief examples illustrate this point. Firstly, aciclovir was introduced almost 20 years ago for the treatment of genital herpes. Apart from a single report (Kost *et al.*, 1993), there is no evidence to suggest that resistance is a significant issue in the treatment of this disease in immunocompetent individuals or that the prevalence of resistance is increasing (Christophers *et al.*, 1998; Collins & Ellis, 1993). The picture is somewhat different in the immunocompromised, where around 5 % of treated patients will develop resistant virus often leading to treatment failure (Christophers *et al.*, 1998; Englund *et al.*, 1990; Wade *et al.*, 1983).

Another example is the emerging experience in chronic hepatitis B virus (HBV) infection. Lamivudine is a nucleoside analogue inhibitor of reverse transcriptase that was developed initially for treatment of HIV disease. Interestingly, in HIV-infected individuals, if the drug is used as monotherapy, resistance emerges within a few weeks (Schuurman *et al.*, 1995). This is due to a single base substitution in the reverse transcriptase gene (Schuurman *et al.*, 1995) resulting in an amino acid substitution in the active site of the enzyme (M184V or M184I). Lamivudine is also active against

HBV, and is the first antiviral approved for the treatment of the associated chronic hepatitis. However, the pattern of resistance development is quite distinct.

Lamivudine provides effective control of HBV infection in the majority of individuals with an increasing proportion of treated individuals seroconverting as therapy is maintained over several years (Lai & Yuen, 1999). A small proportion (around 15 % in the first year) will develop virus with mutations in the reverse transcriptase gene that mirror those seen in HIV and that confer a degree of resistance to the drug (Lai *et al.*, 1998; Ling *et al.*, 1996; Tipples *et al.*, 1996), and this may result in a rebound in virus load. However, there is now evidence that the variant virus is less fit than the parent, and that virus load frequently fails to reach pre-treatment levels. Furthermore, rebound is not consistently associated with treatment failure (Lai & Yuen, 1999). Consequently, the urgency for combination therapy is considerably less in chronic hepatitis B than in HIV infection and it may be that a second potent drug used in combination with lamivudine would solve most, if not all, of these problems. Taken along with the fact that there are many drugs in clinical and pre-clinical development for HBV, there appear to be more limited opportunities in this area for new drug development.

DEFINING THE DESIRED PRODUCT

Having decided the diseases to tackle, the next challenge is to define as far as is possible the profile of the drug to be developed. HIV can be used again to illustrate how this profile can be built and how considerations of resistance influence it. It must be borne in mind that the product will not be launched for 10 years and so it is necessary to project into the future and imagine what management of the disease might look like at that time.

The issues in HIV discussed in the previous section suggest that any new HIV drug would be required to have high potency and also be compatible with other HIV drugs so that it could be used in combination. However, consideration of resistance issues requires further characteristics to be built in. The first is self-evident, and that is that the drug should be effective against all circulating viruses – including drug-resistant variants. The second is perhaps less obvious, and that is that characteristics should be built into the product to improve compliance. There are two major factors that will lead to an improvement in compliance: freedom from major side effects and a simple treatment regimen. Taking all these factors into account, the profile illustrated in Fig. 2 might describe the ideal next generation HIV drug. This profile can be used as a guide in subsequent drug discovery and development activities, checking in at various stages along the route to determine if the project is still on course to deliver this profile. As data accumulate on lead series or drug candidates, they can be checked against the profile to ask whether the project is still moving in the right direction. In parallel,

- High antiviral potency
- Compatibility with other HIV drugs
- Activity against all circulating strains
- Freedom from side effects
- Once a day oral dosing
- No dietary constraints

Fig. 2. Required characteristics of the next generation anti-HIV drug. The main requirements in the next generation anti-HIV drug are that it must have efficacy against all the important drug-resistant strains circulating in the population, that it must be in a convenient dosage form, and that it must be free from side effects to encourage compliance.

the profile itself should be checked periodically against the external environment and should be modified in response to significant new developments in the field. The profile provides a clear objective for activities that will continue for perhaps a decade.

While the foregoing discussion relates to HIV, profiles can be built for products in any disease. It might be decided, for example, that in herpes simplex virus (HSV), where resistance is a relatively minor issue and where safe and effective generic antiviral products are available already, the next generation drugs need to be ones that clear infection completely. To achieve this, virus would have to be eliminated from its latent state in local sensory ganglia in the peripheral nervous system. This would be an extremely challenging profile and one that would be difficult to deliver with our current state of knowledge.

SELECTING THE MOLECULAR TARGET

Once the disease is selected and the profile defined, the next problem is to choose an appropriate molecular target.

Viral or cellular?

The first simple criterion that is generally applied to the selection of a molecular target is that it should be a function required for virus multiplication. While it is conceivable that a drug targeted to a non-essential gene function could deliver clinical benefit, there must be greater risks around such a target. Viruses are, of course, intracellular parasites, and their replication is dependent upon functions that are encoded both by the virus genome and by the cell, and so in theory either virus or host cell functions could be targeted. However, there are powerful arguments why it is preferable to target virus rather than host functions.

A goal in developing an antiviral is to create a 'golden bullet', a drug that will block virus replication but will have no adverse effects on the host organism. Clearly, this would be extremely difficult to achieve if targeting a host function. All viruses, for example, depend upon the host cell translation machinery for their protein synthesis and thus for their multiplication. Any product capable of inhibiting translation would be expected to have broad-spectrum antiviral activity. However, any drug that blocked translation would also inhibit protein synthesis in normal cells, leading to cytotoxicity and serious side effects in the host organism. This is an extreme example, but it will generally be true that it would be very difficult to develop a selective drug that targets a host cell function.

The only convincing argument for targeting a host cell function would be that this approach could potentially avoid resistance issues. Blocking a host function required by the virus might preclude mutation of the virus delivering a drug-resistant phenotype. This would certainly be true if the host function were 'free standing', i.e. not dependent upon an interaction with a virus protein for its contribution to virus replication. However, it is conceivable that where there is a direct virus–host interaction, mutation of the virus could lead to resistance. For example, there is considerable interest currently in the possibility of blocking the interaction of HIV with its cell surface receptors (De Santis *et al.*, 1998) to prevent either attachment of the virus to its target cell or penetration of that cell. It is possible that mutation of the viral envelope glycoprotein that interacts with these cell surface receptors, gp120, could enable the virus to use alternative receptors (Ghorpade *et al.*, 1997) and therefore avoid the inhibitory effects of receptor blockade.

Selecting a viral target

As alluded to above, it is safest to select as a molecular target a gene product whose function is essential for virus replication. The genomes of most viruses of medical importance are well understood with complete sequence data available (e.g. Ratner *et al.*, 1985; Davison & Scott, 1986). The more complex viruses have of the order of a hundred or so genes and the least complex, less than 10. In addition, there is usually information relating to gene function available and the products essential for replication have often been identified. The functions required usually include a number implicated in genome replication, including polymerases, structural proteins that are used to build the virions, and frequently regulatory proteins that control virus–cell interactions.

In selecting the target, one of the most important considerations is how the next stage of the process, lead identification, is to be achieved. This is generally facilitated by the availability of a simple assay system that permits measurement of the target function *in vitro* and this is most easily provided where the target is an enzyme. Such systems allow direct screening of potential drug candidates against the target.

Paradoxically, the most common antiviral drug class is that of the nucleoside analogues (Darby, 1995) that are targeted to viral polymerases, and these cannot be identified by *in vitro* screening as they have to be activated to triphosphate in the infected cell. However, viral enzymes have proved to be the most tractable targets and there are drugs available that are targeted to proteases (Roberts *et al.*, 1990), neuraminidase (Von Itzstein *et al.*, 1993) and, in addition, non-nucleosides targeted to polymerases (Merluzzi *et al.*, 1990; Romero *et al.*, 1991). In fact, all antiviral drugs introduced in the past 20 years are targeted to virus enzymes.

Where resistance is a clinical issue with existing drugs, it should be taken into account when selecting the target. So, for example, in HIV where clinical resistance to both reverse transcriptase and protease inhibitors is now relatively common, ideally an alternative target should be selected. The difficulty, of course, is that there are few targets available and the 'easy' ones have already been tackled. The only other well-characterized enzyme functions in HIV are ribonuclease H and integrase, both of which are difficult to work with and neither of which has so far yielded a useful inhibitor. An alternative approach is to target essential regulatory functions, but although these received a lot of attention in the early days of HIV drug discovery, they too have proved to be highly intractable targets.

The situation in hepatitis B, where again the goal is complete control of virus replication through use of a drug combination, is, if anything, more difficult. Lamivudine might be expected to be one component of the drug combination and ideally the second drug should be directed to a different target. Unfortunately, the choice of targets in hepatitis B is even more limited than in HIV. The virus genome (Fig. 3) is extremely compact with multiple overlapping reading frames and, apart from a ribonuclease H function associated with the reverse transcriptase, there is little scope for other targets.

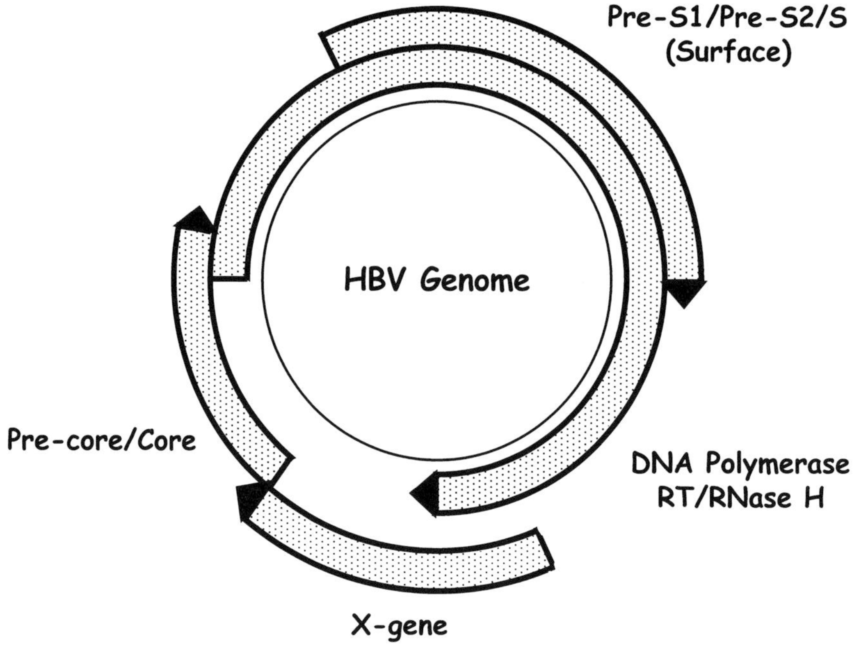

Fig. 3. Open reading frames in the HBV genome. HBV has a compact genome with only one enzymic protein, the polymerase. This gene product has both DNA-dependent and RNA-dependent DNA polymerase activities as well as an RNase activity that acts in concert with the reverse transcriptase. The reading frames of the other genes all overlap the polymerase open reading frame. The overlap is partial in the case of the core protein gene and the regulatory X-gene, and the surface protein gene is completely within the polymerase gene.

IDENTIFICATION OF A LEAD MOLECULE

Armed with a molecular target, the next challenge is to identify a lead molecule. This will be a molecule that interacts appropriately with the target function (e.g. inhibits an enzyme) and one that has chemical characteristics suggesting that it might be refined into a candidate for clinical development. In addition, at this stage it would also be desirable to demonstrate that the molecule has some capability of inhibiting virus growth.

There are two general approaches to lead generation. The first is to screen against the target in an *in vitro* assay, testing all available compounds for their ability to block the function. The second is to adopt a more rational approach, building on knowledge of the target and its natural molecular interactions, e.g. the interactions between an enzyme and its substrate, to direct either screening or the *de novo* synthesis of potential inhibitors.

Screening against a molecular target

The technology around screening has improved dramatically in recent years with the introduction of miniaturization and robotics so that screening which would have taken a year or more in the not-too-distant past can often be accomplished now in a few days. The most serious limitation now appears to be the supply of compounds with sufficient chemical diversity.

Normally screening would be carried out against the wild-type protein and this would certainly be the case if the target were novel. However, if drug resistance at the target is an issue, it may be appropriate to assess compound activities against proteins derived from representative drug-resistant strains. It is unlikely that such a panel would be used in a primary high throughput screen, but any molecules identified in the primary screen could be assessed in a secondary assay for activity against resistant functions. In this way, leads could be identified with the potential for efficacy against resistant strains. As an example, if a next generation non-nucleoside reverse transcriptase inhibitor were to be sought for use in HIV, the primary screen might involve wild-type reverse transcriptase, but any active molecules identified would be tested for their ability to inhibit enzymes derived from the commonly encountered clinically resistant phenotypes.

Rational drug design

It may be that screening is not the method of choice and a more rational approach is taken. Evidence of past success might be a good indicator of the potential for this approach. If, for example, the target is a protease, a good precedent has been set by the work on the aspartyl protease of HIV, a protease involved in cleavage of the polyprotein *gag/pol* precursor. Several groups used knowledge of the cleavage sites recognized by this enzyme to build oligopeptide mimetic inhibitors that were subsequently modified to generate more 'drug-like' molecules (Roberts *et al.*, 1990; Dorsey *et al.*, 1994). Further refinement was guided by detailed structural information generated by X-ray crystallography to develop a second generation of inhibitors (Kaldor *et al.*, 1997). In a more direct approach, the molecular interactions between influenza virus neuraminidase and its substrate, sialic acid, were characterized by X-ray crystallography (Colman *et al.*, 1983), and then used as the starting point to develop inhibitors of the enzyme. This has led to the development of two approved drugs, zanamivir and oseltamivir, with efficacy in the treatment and prophylaxis of influenza (Von Itzstein *et al.*, 1993).

It is possible to combine both rational and screening approaches by using functional knowledge of the protein and its interactions to select compound libraries or sets for screening. Such targeted screening reduces the number of compounds to be screened and at the same time should increase the probability of discovering a lead.

Screening against virus replication

Although today most would use the targeted approach, this is a relatively recent development. Early antiviral drugs were discovered by screening against virus replication in culture. Rather than focusing on one molecular target, a screen against the virus can potentially identify a lead against any essential virus function. The probability of success (per molecule screened) would therefore be significantly higher than if screening were carried out against a single target. The advantages should, however, be balanced against a number of disadvantages such as the difficulty of the screening, which will depend on biological systems (viral growth in cells), and the lack of information generated on the molecular target of active molecules.

With this approach, active molecules are further investigated with an assessment of their effects on uninfected cells. Those molecules that show selective effects against the virus are considered potential leads. If no window exists between the effects on the virus and the cell, it is likely that inhibition is due to an interaction with a cellular function and the chemical series can be discarded. Even if there is an apparent 'therapeutic window' (the ratio of antiviral : cellular activities), the effects could still be due to interaction with a cellular protein, one whose function is not essential for cell growth in culture, but it is more likely that the antiviral effect is due to blockade of a viral function.

Further evidence that the molecule acts at a virus target can be obtained from resistance studies. An attempt should be made to isolate drug-resistant variants following exposure of virus to the molecule in culture. Selecting out drug-resistant variants would indicate that the target is likely to be a virus gene product or at the very least it would implicate a virus function in the mode of action of the drug. Molecular genetics could then be used to identify the specific gene involved.

This approach was used with a series of benzimidazole inhibitors of human cytomegalovirus (HCMV) where studies identified the drug target as the UL89 gene product, a putative terminase – DNase involved in packaging DNA in maturing viral capsids (Krosky *et al.*, 1998). Studies on the replication of virus in the presence and absence of the drug confirmed that the drug induced a defect in particle maturation, resulting in the accumulation of empty capsids (Underwood *et al.*, 1998).

If direct antiviral screening has identified a lead molecule and the target function has been identified, the goal should be to establish an *in vitro* biochemical assay capable of measuring the interaction of the molecule with its target. This assay can then be used in the next phase of the process, lead refinement. It is unlikely that the lead molecule has optimum potency against the target and the availability of the *in vitro* assay permits the building of detailed 'structure activity relationships' (SAR) around the lead series. This essentially defines the features of the molecule that are important in the interaction with the target protein and the nature of the chemical substituents required to optimize those interactions.

DEVELOPMENT OF THE CLINICAL CANDIDATE

The lead molecule must next be refined by chemical modification to optimize its potency and to build in characteristics that will make it a suitable candidate for clinical testing, one that looks to have the potential to deliver the desired product profile.

Benzimidazole inhibitors of HCMV can again be used to illustrate this process. The characteristics of the lead molecule, BDCRB (Fig. 4), were far from those required of a realistic drug candidate. The molecule possessed activity against HCMV, with a degree of selectivity. However, its major problem was its lack of metabolic stability (Townsend *et al.*, 1995). The challenge, therefore, was to increase the selectivity and potency of the molecule while building in the metabolic stability required for oral dosing. At this stage in the development, the mode of action of these molecules had not been identified and so no formal SAR could be developed. The whole-cell viral assay was used to refine the series and this led ultimately to the identification of 1263W94 as the clinical candidate (Chulay *et al.*, 1999).

Comparison of the structures of the lead molecule, BDCRB, and 1263W94 suggests that the modifications required to refine the lead were fairly trivial and that optimization in this case was a simple exercise. This would be far from the truth. The main differences between the lead molecule and the clinical candidate are a change in the stereochemistry around the sugar–base linkage and the substitution of the bromine substituent on the imidazole ring by an aminoalkyl group. It required the synthesis of several hundred novel molecules before this candidate was identified.

Similar challenges were faced in developing inhibitors of the aspartyl protease of HIV. It was not difficult to identify oligopeptide mimetics with a modified scissile bond that were capable of blocking protease function (Roberts *et al.*, 1990). However, such molecules are metabolically unstable, as the peptide bonds are susceptible to hydrolysis, and their molecular masses are generally too high for oral bioavailability. Several groups were able to convert such peptide mimetics to oral drugs with varying degrees of success (Dorsey *et al.*, 1994; Kaldor *et al.*, 1997). Again the difficulties of doing this should not be underestimated and they are reflected in the fact that HIV protease inhibitors used in the clinic are all complex molecules that are synthetically challenging and therefore difficult and expensive to make.

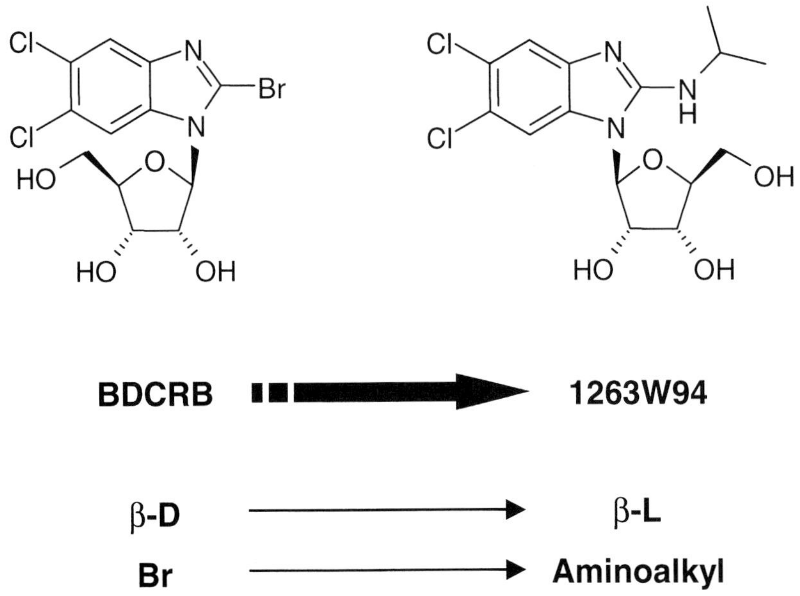

Fig. 4. Lead refinement – benzimidazoles. The lead benzimidazole (BDCRB) differs from the final drug candidate (1263W94) in the nature of the substituent on the imidazole ring and the configuration of the sugar–base linkage.

Before a drug candidate can be tested in man, some further hurdles have to be negotiated. A suitable formulation must be identified to carry out the studies and there must be a minimal package of toxicology data that suggests there will be no major safety issues with the candidate.

In addition, as soon as the clinical candidate has been identified it is prudent to initiate drug-resistance studies on the candidate molecule. Such studies can provide a wealth of information that can help to guide future clinical studies and suggest how the final product might be used in the clinic. In particular, they can flag up at an early stage whether drug resistance is likely to be a major issue.

Resistance studies with the clinical candidate

Isolation of resistant variants. The first question and the simplest to ask is 'how easy is it to isolate resistant variants?' This is generally achieved by passaging the virus in the presence of the drug in culture. It may be that resistant virus emerges very rapidly, perhaps in one or two passages. If this happens – as it does, for example, with the

NNRTIs (non-nucleoside reverse transcriptase inhibitors) and lamivudine with HIV – it should ring warning bells that resistance could be an issue in the clinic (Boucher *et al.*, 1993; Nunberg *et al.*, 1991; Richman *et al.*, 1991; Tisdale *et al.*, 1993a). It is probably indicative that a single mutation in the target gene can result in a dramatic decrease in sensitivity to the drug and that the mutation is compatible with growth in the culture system. With both the NNRTIs and lamivudine, the results in culture were predictive of subsequent experience in the clinic, where clinical resistance was found to emerge within a matter of weeks (Richman *et al.*, 1994; Schuurman *et al.*, 1995).

Rapid isolation of resistant virus in culture does not inevitably correlate with rapid emergence of resistant variants in the clinic. This is presumably because viral growth in culture does not always correlate with growth and pathogenesis in man. This is seen most clearly with the anti-herpes drug aciclovir. Resistance studies in culture showed the rapid emergence of resistant variants (Collins & Darby, 1991; Field *et al.*, 1980; Parris & Harrington, 1982), the variants isolated generally being defective in expression of thymidine kinase (TK: Field *et al.*, 1980; Schnipper & Crumpacker, 1980; Smith *et al.*, 1980). It is now known that loss of TK function is most often a result of frameshift mutations due to insertions or deletions in the TK gene in long runs of Gs (the G-string) or Cs (the C-chord) (Kit *et al.*, 1987). Such variants grow well in culture but they are infrequently seen in the clinic (Burns *et al.*, 1982; Christophers *et al.*, 1998; Englund *et al.*, 1990; Wade *et al.*, 1983).

This example illustrates the need for caution when extrapolating from results in culture to the likely implications in the clinic.

***In vitro* growth characteristics of resistant variants.** It is clear that in addition to characterizing the speed of emergence of resistant variants, other factors need to be taken into account when considering the implications for the clinic. Resistance mutations in fact often have a detrimental effect on the virus, resulting in a less 'fit' phenotype. A relatively simple question to ask is whether the resistant variant is as competent in culture as the wild-type virus. However, standard methods for measuring the rate of viral replication in culture were insufficiently sensitive to measure small differences in replication competency. An elegant solution was to look at the population dynamics of virus mixtures passaged in culture (Harrigan *et al.*, 1998). The drug-resistant variant is mixed with the parent virus and passaged in the absence of drug. The relative concentrations of the two are then monitored as a function of passage number as the fitter of the two variants outgrows the other (Fig. 5).

An analysis of this type has been used to show, for example, that HIV variants isolated in culture by selection with saquinavir are growth-impaired (Martinez-Picado *et al.*,

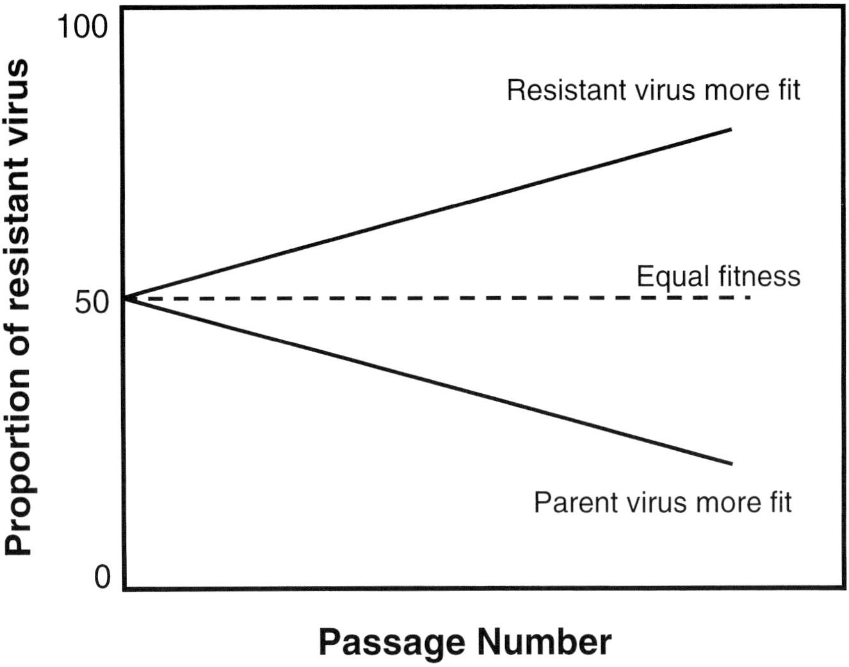

Fig. 5. Viral fitness test. The growth capacity of a virus variant can be compared with that of its parent by taking a combination of the two and passaging them in culture. The relative proportions are monitored at each passage and the fitter of the two outgrows the weaker.

1999). HIV protease is an interesting system because the protease is required to cleave a precursor polyprotein at many non-identical cleavage sites (Debouck, 1992; Henderson *et al.*, 1988). The emergence of resistance through mutation in the protease may alter the active site of the enzyme such that the pattern of recognition of cleavage sites is changed and some become poorly recognized (Gulnik *et al.*, 1995; Mascera *et al.*, 1996). This could account for the impaired growth. However, in time, compensating mutations may appear in the cleavage sites themselves, improving their recognition by the mutant enzyme (Gulnik *et al.*, 1995) and increasing their fitness once more.

Thus when identifying mutations that appear in response to drug pressure, a distinction must be made between the primary resistance mutations and secondary compensatory mutations that restore viral growth potential.

In vivo **growth characteristics.** Data generated in animal models are perhaps even more illuminating than information on *in vitro* characteristics. Where appropriate disease models are available they can yield valuable information. Again this can be illustrated by the aciclovir-resistant mutants of HSV, where animal studies have shown drug-resistant variants to be attenuated with respect to the wild-type virus. TK-deficient variants appear most attenuated but other variants such as DNA polymerase mutants also show some degree of attenuation (Field & Coen, 1986; Field & Wildy, 1978; Larder & Darby, 1985; Larder *et al.*, 1986; Tenser & Edris, 1987). Further *in vivo* studies with TK-deficient variants in the mouse have shown them to be defective in reactivation from latency (Efstathiou *et al.*, 1989). These studies go some way towards explaining the relatively low incidence of aciclovir resistance observed in the population (Christophers *et al.*, 1998).

Genetic characteristics of resistant variants. The genetic lesions responsible for the observed drug-resistant phenotype may also be characterized at this stage. This can be extremely useful, especially if there are a limited number of mutations that can lead to a resistant phenotype. For example, exposure of HIV to lamivudine in culture results in the rapid emergence of variants with mutations resulting in substitutions at residue 184 in the RT gene (Tisdale *et al.*, 1993a), mutations that decrease drug sensitivity by several orders of magnitude. These same mutations also appear rapidly following treatment of individuals with the drug (Schuurman *et al.*, 1995). Sometimes such studies give quite unexpected results. Recently, *in vitro* studies with the new neuraminidase inhibitors of influenza have shown that resistance can arise not only through mutation in the neuraminidase gene (the target protein), but also through mutation in the haemagglutinin gene, the surface glycoprotein that binds the virus to the cell-surface sialic acid receptors (McKimm-Breschkin *et al.*, 1996).

Monitoring a patient's virus for the appearance of mutations known to be associated with decreased drug susceptibility can indicate whether a response to treatment is likely. However, caution is advisable when using this approach. There is the temptation to assume that if virus from a patient has a genetic lesion known to result in reduced drug susceptibility, the patient will not respond to that drug. This is too simplistic a view as there is no absolute correlation between resistant genotype and clinical response (see Fig. 6). Whether or not there is a clinical response to the drug will depend upon the degree of resistance and the drug concentrations achieved during therapy, as well as the fitness of the virus carrying the resistance mutations. So, for example, if the resistant virus is less fit than the drug-sensitive parent, continuous selection pressure from the drug could result in clinical benefit.

The converse may also occur when unexpected (and therefore undetected) lesions appear in the virus. In this situation, the virus would appear to be drug-sensitive but the

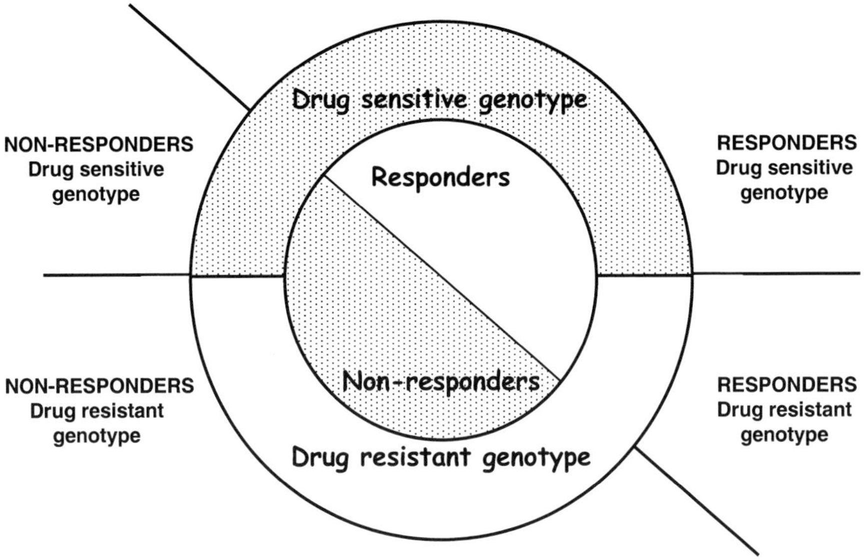

Fig. 6. Relationship between viral genotype and clinical response. Although there is a correlation between genotype and clinical response, this correlation is not perfect. A proportion of individuals whose virus carries mutations associated with decreased drug susceptibility may continue to respond to therapy and some whose virus contains no recognized resistance mutations may fail on therapy.

patient would fail to respond to treatment. Only phenotypic characterization of the virus would reveal this change in drug sensitivity.

One situation where this has occurred is in HIV therapy when patients are treated with drug combinations. If there are two or more drugs of the same class in combination, the selective pressure on the virus gene function may be quite different from that when drugs are used singly in culture, and so novel mutations may appear. This is seen, for example, in the emergence of a multi-resistant phenotype in virus from patients exposed to several nucleoside reverse transcriptase inhibitors simultaneously (Shirsaka *et al.*, 1995).

Despite these 'health warnings', knowledge of potential resistance mutations, if used intelligently, can be a powerful tool in the clinical development of a drug and in the subsequent use of the drug in disease management.

Cross-resistance studies. One other very important aspect of drug resistance that can be assessed at this stage is that of cross-resistance. This can be important if there are, or are likely to be, drugs of the same class in use in the clinic at the same time as the drug under development. Early experience with zidovudine, for example, showed a complex pattern of resistance involving multiple mutations in the RT gene (Larder & Kemp, 1989). As new nucleoside analogues were developed, such as zalcitabine (ddC) and didanosine (ddI), they were assessed to determine their efficacy against zidovudine-resistant viruses (Larder *et al.*, 1989). Little cross-resistance was seen. Similarly, there are now many clinical isolates of HIV showing resistance to protease inhibitors and so any new generation protease inhibitors should ideally be assessed for efficacy against them.

The *in vitro* and *in vivo* drug-resistance studies carried out in the pre-clinical phase help to set the scene for studies in man. They may give strong indications of both what to expect and also what to look for. They may also influence the types of studies done.

EARLY STUDIES IN MAN

The true potential of a drug candidate can only be fully revealed by studies in man. However, initially the primary concern when taking a new drug into man surrounds the safety of the molecule. Nevertheless, as safety data accumulate, attention should turn as early as possible towards a consideration of efficacy.

Studies in healthy volunteers

The first time a drug is taken into man is to look for unwanted side effects in healthy volunteers. The reason for doing these studies in healthy individuals is that if adverse events are seen in significant numbers compared to placebo, they can be ascribed to the drug and not to any underlying clinical condition. Dosing starts at extremely low levels and is gradually escalated towards and beyond the expected clinical dose. Adverse events are monitored carefully and a comparison is made with the pattern seen with placebo.

Once these studies are complete, and assuming the candidate has a clean 'Bill of Health', it can then be taken into infected individuals.

Early studies in infected individuals

The major objective in these studies is once again to look at the safety profile of the drug at the projected clinical doses. These studies will include repeat dosing in which the patients are exposed to drug over an extended time period. Although the primary purpose is to monitor safety, important information can often be collected that will provide insight into whether the drug is likely to have the appropriate efficacy. This may involve direct clinical observations. Challenge studies were done, for example, with the new neuraminidase inhibitor of influenza (zanamivir) in which patients were infected with an influenza A virus (a strain of relatively low virulence) and the clinical outcomes on drug or placebo were compared (Hayden *et al.*, 1996).

More often when looking at new antivirals the first question asked is whether the drug has any impact on virus replication in man. This is relatively straightforward where the virus appears in the bloodstream as is the case with HIV or HBV infection. Simple measurements of virus load, usually by amplification techniques, can be used to monitor the impact of the drug on replication (De Gruttola *et al.*, 1998; Dienstag *et al.*, 1995). Although it is not a clinical effect that is being measured, in HIV there has now been sufficient experience to establish a good correlation between impact on virus load and clinical outcome (Murray *et al.*, 1999).

There may be other surrogates of clinical response that indicate that the drug is working as expected. In chronic hepatitis B, for example, liver enzymes may be monitored to assess the impact on liver inflammation (Dienstag *et al.*, 1995; Lai *et al.*, 1998).

This stage is also the first opportunity to look for the emergence of resistant variants in the clinic. This, at its simplest, involves screening isolates for the presence of resistant genotypes (previously characterized in culture), or better still, it may entail the phenotypic characterization of virus isolates from patients.

Assuming these early studies have a favourable outcome it is likely that the drug candidate will be taken into full clinical development.

FULL CLINICAL DEVELOPMENT

The final, and most costly, phase of drug development is the performance of the definitive clinical trials that will decide whether the clinical candidate can be converted into a significant new medicine. These are likely to involve many hundreds of patients and frequently several thousand. The aim will be to prove a significant impact (>95 % confidence) on a clinically important end point – usually relative to placebo but where treatment already exists, relative to the current standard of care. In HIV disease, the clinical end point in the early studies was often mortality (Fischl *et al.*, 1987), but now treatment has significantly improved, the end point is more often progression to the next stage of disease (Carpenter *et al.*, 1998). Whatever the end point, the trial should demonstrate a clinical benefit to the patient from use of the drug.

During this final phase of development, far larger numbers of patients are exposed to the drug and there is therefore a greater likelihood that resistance will be observed; monitoring should therefore be continued.

The last major hurdle is approval of the drug by regulatory bodies, and once this has been overcome, the drug can be made available to patients who need it.

CONCLUSIONS

Despite the incredible progress that we have seen over the past 20 years with the introduction of more than 20 antiviral products, and dramatic improvements in disease management, there remain significant challenges ahead. The products currently approved are for the treatment of diseases caused by viruses from only four major families (Table 1) and there are still important diseases (e.g. chronic hepatitis caused by hepatitis C virus) where there remains significant unmet need but no available antivirals.

It also remains difficult to achieve total control of chronic viral infections such as HIV and hepatitis B, and where there is viral replication there is the potential for development of resistance. Even if the drugs themselves are capable of exerting control, there are serious issues around compliance with complex and inconvenient drug regimens. It is therefore likely that resistance will continue to be a problem in the future and that in developing new drugs serious attention must be paid to both resistance and compliance issues.

The progress made in the past 20 years has been remarkable and there is no doubt that antiviral drug treatment will remain the cornerstone in the management of virus diseases for many years to come. Although there are considerable challenges ahead, the next 10 years promise to be exciting ones in this fast-moving and important field.

REFERENCES

Balzarini, J. (1999). Suppression of resistance to drugs targeted to human immunodeficiency virus reverse transcriptase by combination therapy. *Biochem Pharmacol* **58**, 1–27.

Bean, B., Braun, C. & Balfour, H. H., Jr (1982). Acyclovir therapy for acute herpes zoster. *Lancet* ii, 118–121.

Boucher, C. A. B., Cammack, N., Schipper, P., Schuurman, R., Rouse, P. L. & Cameron, J. M. (1993). High level resistance to (−) enantiomer of 2′-deoxy-3′-thiacytidine (3TC) in vitro is due to amino acid substitution in the catalytic site of HIV reverse transcriptase. *Antimicrob Agents Chemother* **37**, 2231–2234.

Burns, W. H., Saral, R., Santos, G. W., Laskin, O. L., Leitman, P. S., McLaren, C. & Barry, D. W. (1982). Isolation and characterisation of resistant herpes simplex virus after acyclovir therapy. *Lancet* i, 421–423.

Burton, D. & Moore, J. P. (1998). Why do we not have an HIV vaccine and how can we make one? *Nat Med* **4** (suppl. 5), 495–498.

Carpenter, C. C. J., Fischl, M. A., Hammer, S. M. & 12 other authors (1998). Antiretroviral therapy for HIV infection in 1998: updated recommendations of the International AIDS Society – USA panel. *J Am Med Assoc* **280**, 78–86.

Carpenter, C. C. J., Cooper, D. A., Fischl, M. A. & 15 other authors (2000). Antiretroviral therapy in adults. Updated recommendations of the International AIDS Society – USA panel. *J Am Med Assoc* **283**, 381–390.

Carr, A., Samaras, K., Burton, S., Law, M., Freund, J., Chisholm, D. J. & Cooper, D. A. (1998). A syndrome of peripheral lipodystrophy, hyperlipidaemia and insulin resistance in patients receiving HIV protease inhibitors. *AIDS* **12**, F51–F58.

Casado, J. L., Sabido, R., Perez-Elias, M. J., Antela, A., Oliva, J., Dronda, F., Mejia, B. & Fortun, J. (1999). Percentage of adherence correlates with the risk of protease inhibitor (PI) treatment failure in HIV-infected patients. *Antivir Ther* **4**, 157–161.

Christophers, J., Clayton, J., Craske, J., Ward, R., Collins, P., Trowbridge, M. & Darby, G. (1998). Survey of resistant herpes simplex virus to acyclovir in Northwest England. *Antimicrob Agents Chemother* **42**, 869–874.

Chulay, J., Biron, K. K., Wang, L. & 9 other authors (1999). Development of novel benzimidazole riboside compounds for treatment of cytomegalovirus disease. *Adv Exp Med Biol* **458**, 129–134.

Collins, P. & Darby, G. (1991). Laboratory studies of herpes simplex virus strains resistant to acyclovir. *Rev Med Virol* **1**, 19–28.

Collins, P. & Ellis, N. M. (1993). Sensitivity monitoring of clinical isolates of herpes simplex virus to acyclovir. *J Med Virol* **1**, 58–66.

Colman, P. M., Varhese, J. N. & Laver, W. G. (1983). Structure of the catalytic and antigenic sites in influenza virus neuraminidase. *Nature* **303**, 41–44.

Craig, J. C., Whittaker, L., Duncan, I. B. & Roberts, N. A. (1993). *In vitro* resistance to an inhibitor of HIV protease (Ro 31-8959) relative to inhibitors of reverse transcriptase (AZT and TIBO). *Antivir Chem Chemother* **4**, 335–339.

Darby, G. (1995). In search of the perfect antiviral. *Antivir Chem Chemother* **6** (suppl. 1), 1–10.

Davison, A. J. & Scott, J. E. (1986). The complete DNA sequence of varicella-zoster virus. *J Gen Virol* **67**, 1759–1816.

Debouck, C. (1992). The HIV protease as a therapeutic target for AIDS. *AIDS Res Hum Retrovir* **8**, 153–164.

De Gruttola, V., Hughes, M., Gilbert, P. & Phillips, A. (1998). Trial design in the era of highly effective antiviral drug combinations for HIV infection. *AIDS* **12** (suppl. A), 5149–5156.

De Santis, C., Rizzi, M., Bolognesi, D. & Lusso, P. (1998). Generation of recombinant analogues of the C-C chemokine Rantes with enhanced HIV suppressive and reduced proinflammatory activity. *XII International Conference on AIDS, Geneva*, abstract 11101.

Dienstag, J. L., Perillo, R. P., Schiff, E. R., Bartholemew, M., Vicary, C. & Rubin, M. (1995). A preliminary trial of lamivudine for chronic hepatitis B infection. *N Engl J Med* **333**, 1657–1661.

Dorsey, B. D., Levin, R. B., McDaniel, S. L. & 15 other authors (1994). L-735,524: the design of a potent and orally bioavailable HIV protease inhibitor. *J Med Chem* **37**, 3443–3451.

Dulioust, A., Paulos, S., Gillemot, L., Delavalle, A.-M., Boue, F. & Clavel, F. (1998). Constrained evolution of human immunodeficiency virus type 1 protease during sequential therapy with two distinct protease inhibitors. *J Virol* **73**, 850–854.

Efstathiou, S., Kemp, S., Darby, G. & Minson, A. C. (1989). The role of herpes simplex virus type 1 thymidine kinase in pathogenesis. *J Gen Virol* **70**, 869–879.

Englund, J. A., Zimmerman, E. M., Swierkosz, J. L., Goodman, D. R. & Balfour, H. H. (1990). Herpes simplex virus resistance to acyclovir. A study in a tertiary care center. *Ann Intern Med* **112**, 416–422.

Field, H. J. & Coen, D. M. (1986). Pathogenicity of herpes simplex virus mutants containing drug resistance mutations in the DNA polymerase gene. *J Virol* **60**, 286–289.

Field, H. J. & Wildy, P. (1978). The pathogenicity of thymidine kinase-deficient mutants of herpes simplex virus in mice. *J Hyg* **81**, 267–277.

Field, H. J., Darby, G. & Wildy, P. (1980). Isolation and characterisation of acyclovir-resistant mutants of herpes simplex virus. *J Gen Virol* **49**, 115–124.

Fischl, M., Richman, D. D., Grieco, M. H., Gootlieb, M. S., Volberding, P. A., Laskin, O. L., Groopman, J. E., Mildvan, D., Schooley, R. T., Jackson, G. G., Dirack, D. T., King, D. & the AZT Collaborative Working Group (1987). The efficacy of azidothymidine (AZT) in the treatment of patients with AIDS and AIDS-related complex: a double blind, placebo-controlled trial. *N Engl J Med* **317**, 185–191.

Ghorpade, A., Xia, M., Hyman, B. T., Persidsky, Y., Nukuna, A., Bock, P., Che, M., Limoges, J., Gendelman, H. E. & Mackay, C. R. (1997). Role of the beta-chemokine receptors CCR3 and CCR5 in human immunodeficiency virus type 1 infection of monocytes and microglia. *J Virol* **72**, 3351–3361.

Goodenow, M., Huet, T., Sawin, W., Kwok, S., Sninsky, J. & Wain Hobson, S. (1989). HIV-1 isolates are rapidly evolving quasispecies: evidence for viral mixtures and preferred nucleotide substitutions. *J Acquir Immune Defic Syndr* **2**, 344–352.

Goodrich, J. M., Mori, M., Gleaves, C. A., Du Mond, C., Cays, M., Ebeling, D., Buhles, W. C., Armond, B. & Meyers, J. D. (1991). Early treatment with ganciclovir to prevent cytomegalovirus disease after allogeneic bone marrow transplantation. *N Engl J Med* **325**, 1601–1607.

Gulnik, S. V., Suvorov, L. I., Liu, B., Yu, B., Anderson, B., Mitsuya, H. & Erickson, J. W. (1995). Kinetic characterization and cross-resistance patterns of HIV-1 protease mutants selected under pressure. *Biochemistry* **34**, 9282–9287.

Harrigan, R. P., Bloor, S. & Larder, B. A. (1998). Relative replication fitness of zidovudine resistant human immunodeficiency virus type 1 isolates in vitro. *J Virol* **72**, 3773–3778.

Hayden, F. G., Belshe, R. B., Clover, R. D., Hay, A. J., Oakes, M. G. & Soo, W. (1989). Emergence and apparent transmission of rimantadine-resistant influenza A virus in families. *N Engl J Med* **321**, 1696–1702.

Hayden, F. G., Treanor, J. J., Betts, R. F., Lobo, M., Esinhart, J. D. & Hussey, R. K. (1996). Safety and efficacy of the neuraminidase inhibitor GG167 in experimental influenza. *J Am Med Assoc* **275**, 295–299.

Henderson, L. E., Copeland, T. D., Sowder, R. C., Schulz, A. M. & Oroszlan, S. (1988). Analysis of proteins and peptides purified from sucrose gradient banded HTLV-III. In *Human Retroviruses, Cancer and AIDS: Approaches to Prevention and Therapy*, pp. 135–147. Edited by D. Bolognesi. New York: Alan R. Liss.

Hilleman, M. R. (1998). Six decades of vaccine development – a personal history. *Nat Med* **4** (suppl. 5), 507–514.

Hirsch, M. S., Conway, B., D'Aquila, R. T. & 11 other authors (1998). Antiretroviral drug resistance testing in adults with HIV infection: implications for clinical management. *J Am Med Assoc* **279**, 1984–1991.

Hull, H. F., Birmingham, M. E., Melgaard, B. & Lee, J. W. (1997). Progress towards global polio eradication. *J Infect Dis* **175** (suppl. 1), 34–59.

Kaldor, S. W., Kalish, V. J., Davies, J. F., II & 17 other authors (1997). Viracept (nelfinavir mesylate, AG1343): a potent, orally bioavailable inhibitor of HIV-1 protease. *J Med Chem* **40**, 3979–3985.

Katlama, C., Duvivier, C., Mouroux, M. & 10 other authors (1999). GIGHAART; a rescue therapy for HIV patients with multiple HAART failures. *Antivir Ther* **4** (suppl. 1), 77.

Kinchington, D., Balzarini, J. & Field, H. J. (1997). Current antiviral agents factfile, 3rd edition: part II – human immunodeficiency viruses. *Int Antivir News* **5**, 161–174.

Kit, S., Sheppard, M., Ichimura, H., Nusinoff-Lehrman, S., Ellis, N. M., Fyfe, J. A. & Otsuka, H. (1987). Nucleotide sequence changes in the thymidine kinase gene of herpes simplex virus type 2 clones from an isolate of a patient treated with acyclovir. *Antimicrob Agents Chemother* **31**, 1483–1490.

Kost, R. G., Hill, E. L., Tigges, M. & Straus, S. E. (1993). Brief report: recurrent acyclovir-resistant genital herpes in an immunocompetent patient. *N Engl J Med* **328**, 1777–1781.

Krosky, P. M., Underwood, M. R., Turk, S. R., Feng, K. W.-H., Jain, R. K., Ptak, R. G., Westerman, A. C., Biron, K. K., Townsend, L. B. & Drach, J. C. (1998). Resistance of human cytomegalovirus to benzimidazole ribonucleosides maps to two open reading frames UL89 and UL56. *J Virol* **72**, 4721–4728.

Lai, C. L. & Yuen, M. F. (1999). Clinical experience and follow-up with lamivudine in the Asian population. *Antivir Ther* **4** (suppl. 4), 7.

Lai, C.-L., Chien, R.-N., Leung, N. W. Y. & 9 other authors (1998). A one-year trial of lamivudine for chronic hepatitis B. *N Engl J Med* **339**, 61–68.

Larder, B. A. & Darby, G. (1985). Selection and characterisation of aciclovir-resistant herpes simplex type 1 mutants inducing altered DNA polymerase activities. *Virology* **146**, 262–271.

Larder, B. A. & Kemp, S. D. (1989). Multiple mutations in HIV-1 reverse transcriptase confer high-level resistance to zidovudine (AZT). *Science* **246**, 1155–1158.

Larder, B. A., Lisle, J. J. & Darby, G. (1986). Restoration of wild type pathogenicity to an attenuated DNA polymerase mutant of herpes simplex virus type 1. *J Gen Virol* **67**, 2501–2506.

Larder, B. A., Darby, G. & Richman, D. D. (1989). HIV with reduced sensitivity to zidovudine (AZT) isolated during prolonged therapy. *Science* **243**, 1731–1734.

Ledergerber, B., Egger, M., Opravil, M. & 9 other authors (1999). Clinical progression and virological failure on highly active anti-retroviral therapy. *Lancet* **353**, 863–868.

Ling, R., Mutimer, D., Ahmed, M., Boxall, E. H., Elias, E., Dusheiko, G. M. & Harrison, T. J. (1996). Selection of mutations in hepatitis B virus polymerase during therapy with lamivudine. *Hepatology* **24**, 711–713.

Lo, J. C., Mulligan, K., Tai, V. W., Algren, H. & Schambelan, M. (1998). 'Buffalo hump' in men with HIV infection. *Lancet* **351**, 867–870.

McKimm-Breschkin, J. L., Blick, T. J., Sahasrabudhe, A., Tiong, T., Marshall, D., Hart, G. J., Bethell, R. C. & Penn, C. R. (1996). Generation and characterisation of variants of NWS/G70C influenza virus after in vitro passage in 4-amino-Neu5Ac2en and 4-guanidino-Neu5Ac2en. *Antimicrob Agents Chemother* **40**, 40–46.

Martinez-Picado, J., Savara, A., Sutton, L. & D'Aquila, R. T. (1999). Replicative fitness of protease-resistant mutants of human immunodeficiency virus type 1. *J Virol* **73**, 3744–3752.

Mascera, B., Darby, G., Palu, G., Wright, L. L., Tisdale, M., Myers, R., Blair, E. D. & Furfine, E. S. (1996). Mutations in the viral protease that confer resistance to saquinavir increase the dissociation rate constant of the protease-saquinavir complex. *J Biol Chem* **271**, 33231–33235.

Merluzzi, M., Hargrave, K. D., Labadia, M. & 12 other authors (1990). Inhibition of HIV-1 replication by a nonnucleoside reverse transcriptase inhibitor. *Science* **250**, 1411–1413.

Mertz, G. J., Jones, C. C., Mills, J. & Lemon, S. M. (1988). Long-term acyclovir suppression of frequently recurring genital herpes simplex virus infection. *J Am Med Assoc* **260**, 201–206.

Monno, L., Appice, A., Cavaliere, R., Scarabaggio, T. & Angarano, G. (1999). Highly active antiretroviral failure and protease and reverse transcriptase human immunodeficiency virus type 1 gene mutations. *J Infect Dis* **180**, 568–571.

Montagner, J. S. G., Harrigan, P. R., Jahnke, N. A., Hogg, R. S., Yip, B., Harris, M., Montessori, V. & O'Shaughnessy, M. V. (1999). Multidrug rescue therapy for HIV-infected individuals with prior virological failure to multiple regimens: results from a second cohort. *Antivir Ther* **4**, 241.

Murray, J. S., Elashoff, M. R., Iacono-Connors, L. C., Cvetkovich, T. A. & Struble, K. A. (1999). The use of plasma HIV RNA as a study end-point in efficacy trials of antiretroviral drugs. *AIDS* **13**, 797–804.

Nossal, G. (1998). Living up to the legacy. *Nat Med* **4** (suppl. 5), 475–476.

Nunberg, J. H., Schleif, W. A., Boots, E. G., O'Brien, J. A., Quintero, J. C., Hoffman, J. M., Emini, E. A. & Goldman, M. E. (1991). Resistance to human immunodeficiency virus type-1 specific pyridone reverse transcriptase inhibitors. *J Virol* **65**, 4887–4892.

Palella, F. J., Jr, Delaney, K. M., Moorman, A. C., Loveless, M. O., Fuhrer, J., Satten, G. A., Aschman, D. J. & Holmberg, S. D. (1998). Declining morbidity and mortality among patients with advanced human immunodeficiency virus infection. *N Engl J Med* **338**, 853–860.

Parris, D. S. & Harrington, J. E. (1982). Herpes simplex virus variants resistant to high concentrations of acyclovir exist in clinical isolates. *Antimicrob Agents Chemother* **22**, 71–77.

Raboud, J. M., Montaner, J. S., Conway, B. & 10 other authors (1998). Suppression of plasma viral load below 20 copies/ml is required to achieve a long-term response to therapy. *AIDS* **12**, 1619–1624.

Ratner, L., Haseltine, W. A., Patarea, R. & 16 other authors (1985). Complete nucleotide sequence of the AIDS virus, HTLV-III. *Nature* **313**, 277–284.

Richman, D. D., Shih, C.-K., Lowy, I., Rose, J., Prodanovich, P., Goff, S. & Griffin, J. (1991). Human immunodeficiency virus type 1 mutants resistant to non-nucleoside inhibitors of reverse transcriptase arise in tissue culture. *Proc Natl Acad Sci U S A* **88**, 11241–11245.

Richman, D. D., Havlir, D., Corbeil, J. & 9 other authors (1994). Nevarapine resistance mutations of human immunodeficiency virus type 1 selected during therapy. *J Virol* **68**, 1660–1666.

Roberts, N. A., Martin, J. A., Kinchington, D. & 16 other authors (1990). Rational design of peptide-based proteinase inhibitors. *Science* **248**, 358–361.

Romero, D. L., Busso, M., Tan, C.-K. & 8 other authors (1991). Nonnucleoside reverse transcriptase inhibitors that potently and specifically block human immunodeficiency type 1 replication. *Proc Natl Acad Sci U S A* **88**, 8806–8810.

Ruff, T. A. (1999). Immunisation strategies for viral diseases in developing countries. *Rev Med Virol* **9**, 121–130.

Schnipper, L. E. & Crumpacker, C. S. (1980). Resistance of herpes simplex virus to acycloguanosine: role of thymidine kinase and DNA polymerase loci. *Proc Natl Acad Sci U S A* **77**, 2270–2273.

Schuurman, R., Nijhuis, M., van Leeuwen, R. & 10 other authors (1995). Rapid changes in human immunodeficiency virus type 1 RNA load and appearance of drug-resistant virus population in persons treated with lamivudine (3TC). *J Infect Dis* **171**, 1411–1419.

Shirsaka, T., Kavlick, M. F., Ueno, T. & 8 other authors (1995). Emergence of human immunodeficiency virus type 1 variants with resistance to multiple dideoxynucleosides in patients receiving therapy with dideoxynucleosides. *Proc Natl Acad Sci U S A* **92**, 2398–2402.

Smith, K. O., Kennell, W. L., Poirier, R. H. & Lynd, F. T. (1980). *In vitro* and *in vivo* resistance of herpes simplex virus to 9-(2-hydroxyethoxymethyl)guanine (acycloguanosine). *Antimicrob Agents Chemother* **17**, 144–150.

Tenser, R. B. & Edris, W. A. (1987). Herpes simplex virus thymidine kinase expression in trigeminal ganglion infection. Correlation of enzyme activity with ganglion virus titre and evidence of *in vivo* complementation. *Virology* **112**, 328–341.

Tipples, G. A., Ma, M. M., Fischer, K. P., Bain, V. G., Knetman, N. M. & Tyrrell, D. L. (1996). Mutation of HBV RNA-dependent DNA polymerase confers resistance to lamivudine *in vivo*. *Hepatology* **24**, 1670–1677.

Tisdale, M., Kemp, S. D., Parry, N. R. & Larder, B. A. (1993a). Rapid *in vitro* selection of human immunodeficiency virus type 1 variants resistant to 3′-thiacytidine inhibitors due to a mutation in the YMDD region of reverse transcriptase. *Proc Natl Acad Sci U S A* **90**, 5653–5656.

Tisdale, M., Myers, R. E., Maschera, B., Parry, N. R., Oliver, N. M. & Blair, E. D. (1993b). Cross-resistance of human immunodeficiency virus type 1 variants individually selected for resistance to five different protease inhibitors. *Antimicrob Agents Chemother* **39**, 1704–1710.

Townsend, L. B., Devivar, R. V., Turk, S. R., Nassiri, M. R. & Drach, J. C. (1995). Design synthesis and antiviral activity of certain 2,5,6-tri-halo-1-(beta-D-ribofuranosyl)benzimidazoles. *J Med Chem* **38**, 4098–4105.

Underwood, M. R., Harvey, R. J., Stanat, S. C., Hemphill, M. L., Miller, T., Drach, J. C., Townsend, L. B. & Biron, K. K. (1998). Inhibition of human cytomegalovirus DNA maturation by a benzimidazole ribonucleoside is mediated through the UL89 gene product. *J Virol* **72**, 717–725.

Von Itzstein, M., Wu, W. K., Kok, G. B. & 15 other authors (1993). Rational design of sialidase based inhibitors of influenza virus infection. *Nature* **363**, 418–423.

Wade, J. C., McLaren, C. & Meyers, J. D. (1983). Frequency and significance of acyclovir-resistant herpes simplex virus isolated from marrow transplant patients receiving multiple courses of treatment with acyclovir. *J Infect Dis* **148**, 1077–1082.

Wade, J. C., Newton, B., Flournoy, N. & Meyers, J. D. (1984). Oral acyclovir for prevention of herpes simplex reactivation after marrow transplantation. *Ann Intern Med* **100**, 823–828.

Winters, M. A., Coolley, K. L., Girard, Y. A., Levee, D. J., Hamdan, H., Shafer, R. W., Katzenstein, D. A. & Merigan, T. C. (1998a). A 6 base-pair deletion in the reverse transcriptase gene of HIV-1 confers resistance to multiple nucleoside inhibitors. *J Clin Investig* **102**, 1769–1775.

Winters, M. A., Schapiro, J. M., Lawrence, L. & Merigan, T. C. (1998b). Human immunodeficiency virus type 1 protease genotypes and in vitro protease inhibitor susceptibilities of isolates from individuals who were switched to other protease inhibitors after long term saquinavir therapy. *J Virol* **72**, 5303–5306.
Wong, J. K., Hezareh, M., Gunthard, H. F., Havlir, D. V., Ignacio, C. C., Spina, C. A. & Richman, D. D. (1997). Recovery of replication-competent HIV-1 despite prolonged suppression of plasma viremia. *Science* **278**, 1291–1295.

INDEX

References to tables/figures are shown in italics